Proceedings of the 1st International Discussion Meeting on

SUPERIONIC CONDUCTOR PHYSICS

editors

Junichi Kawamura
Tohoku University, Japan

Shinzo Yoshikado
Doshisha University, Japan

Takashi Sakuma
Ibaraki University, Japan

Yoshitaka Michihiro
Tokushima University, Japan

Masaru Aniya
Kumamoto University, Japan

Yoshiaki Ito
Kyoto University, Japan

Kyoto, Japan 10 – 14 September 2003

Proceedings of the 1st International Discussion Meeting on

SUPERIONIC CONDUCTOR PHYSICS

World Scientific

NEW JERSEY • LONDON • SINGAPORE • BEIJING • SHANGHAI • HONG KONG • TAIPEI • CHENNAI

Published by

World Scientific Publishing Co. Pte. Ltd.

5 Toh Tuck Link, Singapore 596224

USA office: 27 Warren Street, Suite 401-402, Hackensack, NJ 07601

UK office: 57 Shelton Street, Covent Garden, London WC2H 9HE

British Library Cataloguing-in-Publication Data
A catalogue record for this book is available from the British Library.

SUPERIONIC CONDUCTOR PHYSICS
Proceedings of the 1st International Discussion Meeting on Superionic Conductor Physics (IDMSICP)

ISBN-13 978-981-270-565-5
ISBN-10 981-270-565-1

Printed in Singapore by World Scientific Printers (S) Pte Ltd

Organizing Committee

Junichi Kawamura (IMRAM, Tohoku University)
Shinzo Yoshikado (Doshisha University)
Takashi Sakuma (Ibaraki University)
Yoshiaki Ito (ICR, Kyoto University)
Yoshitaka Michihiro (Tokushima University)
Masaru Aniya (Kumamoto University)

International Committee

W. Dieterich (Universität Konstanz)
E. Sherman (Graz University)

Advisory Committee

S. Hoshino (Prof. Emeritus, University of Tokyo)
A. Ueda (Prof. Emeritus, Kyoto University)
T. Hattori (IMRAM, Tohoku University)
T. Ishii (Okayama University)

Chairpersons

Junichi Kawamura (IMRAM, Tohoku University)
Shinzo Yoshikado (Doshisha University)

Sponsors

The 1st International Discussion Meeting on Superionic Conductor Physics (IDMSICP) has been sponsored in part by:

Nitta Corporation

Rigaku International Corporation

Doshisha University

IDMSICP
1st International
Physics

1.S.Takeda
2.N.Inoue
3.H.Koyama
4.Zou
5.O.Kamishima
6.Y.Michiue
7.M.Onoda
8.J.Dygas
9.F.Fujishiro
10.T.Ishii
11.S.Mochizuki
12.S.Selvasekarapandian
13.S.Ono
14.T.Hattori
15.J.C.Dyre
16.T.Tomoyose
17.S.Sharma
18.S.Adams
19.T.Tojo
20.S.Yoshikado
21.W.Dieterich
22.J.Metoson
23.J.Kawamura
24.Y.Kowada
25.M.Aniya
26.T.Sakuma
27.T.Usuki
28.K.Kamada
29.M.Kobayashi
30.T.Michihiro
31.F.Shimojo
32.K.Takahashi
33.T.Kanashito
34.D.Sidebottom
35.A.Ueda
36.A.Imai
37.K.Ibuki
38.I.Katayama
39.K.Nakamura

Preface

A series of materials called superionic conductors, which exhibit high ionic conductivity, have been attracting attention in terms of their engineering application as solid electrolytes. In promoting such an application, the importance of basic research based on, for example, physics and chemistry, has been recognized, and has led to the establishment of an academic field called solid- state ionics in which science and engineering have been integrated. Superionic conductors are highly promising functional materials. As a stepping stone in the development of new superionic conductors that can be utilized as functional materials, efforts to reevaluate solid-interior diffusion and conduction phenomena of ions and molecules in a superionic conductor on the basis of basic physical properties, and to clarify mechanisms governing these phenomena from a microscopic standpoint are important. How are diffusing ions associated with material structures within a superionic conductor? What types of interaction are diffusing ions undergoing with the host ions surrounding them? How important is the correlation among diffusing ions in their motion? At present, we understand only some part of the structures and dynamic characteristics of superionic conductors, and accordingly, intensive research on these basic physical properties will be required in order to utilize superionic conductors as functional materials.

To date, the research on superionic conductors has focused on studies of engineering applications in which the transfer phenomenon and motion of ions within a solid are utilized. In addition, basic research seems to focus only on the behavior of ions. We consider that, in the future, such a superionic conduction phenomenon should be considered to be based on the dynamics of the ions composing a solid, and should be clarified on the basis of various types of information including electronic states. Among the basic physical properties of superionic conductors, electrical conductivity is a significant physical property. However, in addition to electrical conductivity, various material properties of superionic conductors, including magnetism, optical properties, structure, and thermal properties, should be reevaluated, and applied to roles envisioned by humans, which can become functionalities based on the behavior of ions in a solid. Through such processes, a new aspect of superionic conductors as functional materials should be sought in the future. Furthermore, without dwelling on materials that are conventionally classified as superionic conductors, members of the Discussion Meeting on Superionic Conductor Physics intend to pursue and search for possible new functionalities based on the behavior of ions in a solid.

The first Discussion Meeting on Superionic Conductor Physics was held in December 1987; since then, the meetings have been held every 3 or 4 years, with issues of interest being actively discussed. In April 2000, the 4th meeting entitled "Discussion Meeting on Ion Transport Properties of Ionic-Conductive Solid," was held in Okayama. The theme of this 4th meeting was that to further advance the research from the basic standpoint of physical properties, it would be necessary to provide a forum and sufficient time for the presentation of basic studies in this field. There was also the need to foster young researchers and to vitalize research. In May 2001, the 5th Discussion Meeting was held in Kyoto. The Ion Transport Society of Japan was inaugurated at this meeting. The concept behind the 5th meeting was the same as the previous concept; however, having established our society, albeit small, for the first time, we hoped to advance more active research activities with the impetus of this organization as a stepping stone. Thereafter, the 6th meeting was held in Kyoto in May 2002, the 7th meeting in Kumamoto in May 2003, and the 8th meeting in Ibaraki in May 2004.

To advance the research on the physical properties of ionic conductors from a more global viewpoint, we invited foreign researchers to participate in our Discussion Meeting and held the first International Research Meeting on Superionic Conductor Physics on September 10-14, 2003. It was held in Kyoto, where The Ion transport Society of Japan was first inaugurated. Thirty-eight domestic researchers (including five guest speakers), and eight foreign guest speakers from Austria, Denmark, Germany, India, New Zealand, Poland and United States, participated in the meeting. Lectures, questions and answer sessions and free discussions were actively carried out. The research topics ranged widely, including theory, modeling, simulation, crystalline structure, nuclear magnetic resonance, electric conduction, optical properties, and thermal measurement. To publicize the outcome of this international meeting to related researchers, we decided to publish these proceedings of the meeting. Almost all the participants submitted their papers, which were carefully reviewed and processed. The successful publication of these proceedings is due to the efforts of the organizers of the international meeting, whom we deeply thank.

We also express our appreciation to Nitta Corporation, Rigaku International Corporation, and Doshisha University for their significant support of this international meeting.

March 2005

Junichi Kawamura
Shinzo Yoshikado
Takashi Sakuma
Yoshitaka Michihiro
Masaru Aniya
Yoshiaki Ito

Contents

CHEMICAL BONDING OF MOVING CATIONS IN SUPERIONIC CONDUCTORS

Y. Kowada[1], M. Okamoto[1], I. Tanaka[2], H. Adachi[2], M. Tatsumisago[3], and T. Minami[3]

1 Hyogo University of Teacher Education, Yashirocho, Kato-gun, Hyogo 673-1494, Japan
2 Department of Materials Science and Engineering, Kyoto University, Kyoto 606-8501, Japan
3 Department of Applied Materials Science, Osaka Prefecture University, Sakai 599-8531, Japan
e-mail address: ykowada@sci.hyogo-u.ac.jp

Electronic state of monovalent ions in superionic conductors α-AgI and Li_3N was calculated by the DV-Xα cluster method. The movements of the cations were simulated by several model clusters with different position of the moving cation. The net charge of moving cations and the total bond overlap population between the moving cation and the other ions were used for discussion of chemical bonding of the moving cation. In both superionic conductors, the total bond overlap population of the moving cation along the conduction path less changed than those of the other paths. On the other hand, the change of the net charge of the moving cation was similar in any paths. These results suggest that the smaller change of the total bond overlap population of the moving cations should play an important role in the fast ion conduction in superionic conductors, rather than the change of the net charge of the moving cation.

Key words: DV-Xα method, molecular orbital calculation, chemical bonding, monovalent ion, ionic conductivity

1. Introduction

Superionic conductors are very attractive because of their application for a solid state secondary batteries, capacitors, sensors etc. There are several typical monovalent ions, which can move in solid state compounds, such as Ag^+, Li^+ etc. Especially AgI-based superionic conductors have very high ionic conductivity, compared with the other monovalent cation conductors, though the ionic radius and mass of Ag ion are larger than those of, for example, Li ion. This interesting behavior of Ag ion may be dependent on bonding interaction with surrounding atoms being present near conduction paths. Then it is very useful to calculate the electronic state of Ag ion in Ag ion conductors in order to understand such high conductivity. On the other hand, we have tried to study chemical bondings in the other typical solid state electrolyte, Li_3N, containing a different monovalent ion, Li^+. In both materials, analysis of bonding nature between moving monovalent ions and the surrounding ions should be very important to understand their high ionic conductivities, since mobile monovalent ions might have strong covalent interaction with the surrounding ions. In the present work, we have calculated the electronic state of several model clusters of superionic conductors and discuss about change of bonding state of mobile monovalent ions with surrounding atoms.

2. Calculation method

The DV-Xα cluster method [1] has been used to calculate the electronic state of the superionic conductors. This method is one of the linear combination of atomic orbital (LCAO) molecular orbital methods. In this method the exchange potential V_{XC} [2] is described by

$$V_{XC}(r_\uparrow) = -3\alpha\left[\frac{3}{4\pi}\rho_\uparrow(r)\right]^{\frac{1}{3}}$$

Where ρ is the electron density of the cluster and α is a constant. α is fixed to be 0.7 throughout the present work [3,4]. The Mulliken population analysis was used for the evaluation of the net charge of each atom and the bond overlap population between atoms.

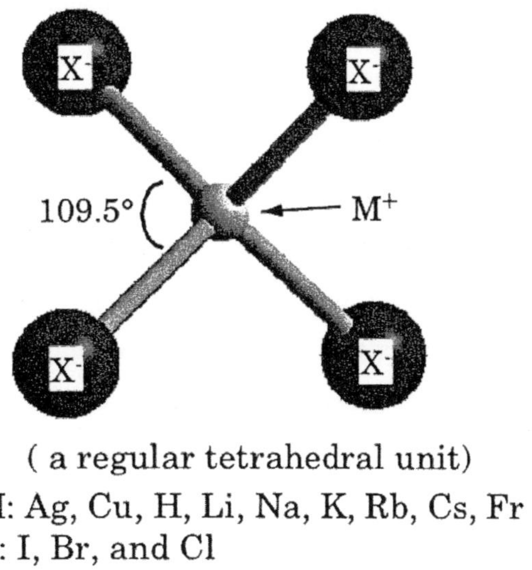

Fig. 1 Schematic diagram of MX_4 cluster.

3. Results and discussion

3.1. Fundamental bonding of monovalent cations

Generally chemical bonds have both ionic and covalent characters. The components of the bonding natures are varied by the combinations of the elements and the environment of the moving cation. Fig. 1 shows the model cluster for the calculation of the fundamental interaction of monovalent cations with halide anions, MX_4^{3-} (M= Ag, Cu, H, Li, Na, K, Rb, Cs, X= Cl, Br, I). This cluster is the simplest one to obtain the basic bonding nature of the moving cations in the halide. The distances of M-X were

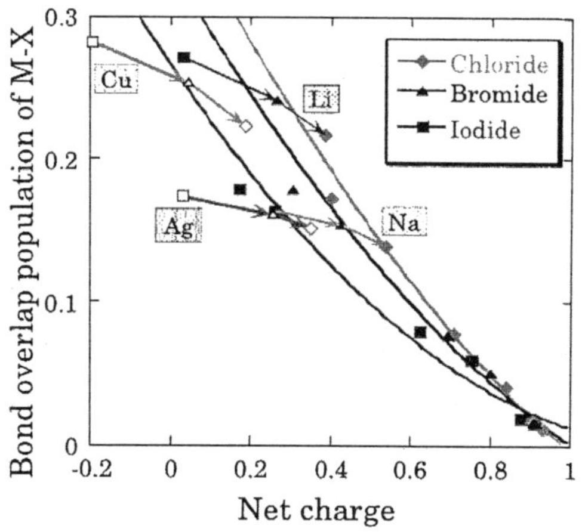

Fig. 2 Relationship between the net charge and the bond overlap population of M ions.

determined by the ionic radii reported by Shannon [5]. To discuss the bonding nature between the moving cation and surrounding anion, we have analyzed the bonding of the moving cation M in terms of ionicity and bond overlap population as a measure for the strength of the covalent M-X bond. We can expect the characteristic relationship between ionicity and covalency for the chemical bond of the moving cation in the superionic conducting materials. The results are shown in Fig. 2. In this figure the relationship between the bond overlap population and net charge of the M ions in the cluster with X = Cl, Br, and I ions are plotted. In the case of X = I ion, the bond overlap populations of the alkali ions are decreased with an increase in net charge of the M ion. On the other hand, the results of the Ag and Cu ions were located at smaller side of the net charge. The Ag ion has similar ionic radius to the Na ion in the four-coordinated environment and Cu ion has also almost the same radius as the Li ion. These two ions, however, have smaller charges and larger bond overlap populations than those of the alkali metal ions. This result suggests that the interaction between the Ag and I ions has less ionic and slightly more covalent than that between the alkali metal and I ions. There seems to be another line of the Ag and Cu ions. In the bromide clusters, the net charges of monovalent ions become larger than those in the iodide clusters. Furthermore, the net charges are larger in the chloride clusters. The net charge of the Ag and Cu ions, however, are smaller and the bond overlap population is slightly larger than those of the alkali metal ions in any halide clusters. This is interesting behavior of the Ag and Cu ions in the viewpoint of the chemical bonding of the monovalent ions in halides. The result suggests that the characteristics of the bonding state of the Ag and Cu ions could be estimated by using net charges and bond overlap populations.

3.2. α-AgI crystal

In the α-AgI crystal structure, there are three kinds of stable positions for Ag ions. The first is the tetrahedral site coordinated by four I ions. This position should be the most stable site for Ag ions from the neutron diffraction study.[6] The second is the octahedral site and the third is the trigonal site. Usually Ag ions are moving from a tetrahedral site to the other tetrahedral site

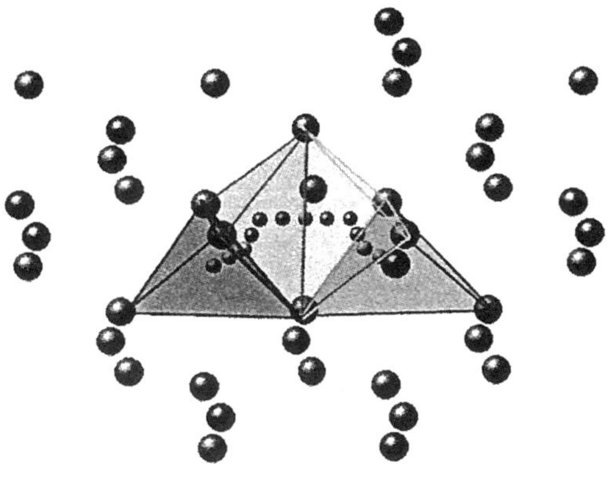

Fig. 3 Schematic diagram of the $M_{20}I_{20}$ model cluster.

through either an octahedral site or a trigonal site. Then there are several kinds of the conduction paths for Ag ions in the α-AgI crystal. In the present study, two kinds of the shortest paths among the conduction paths were adopted. Fig. 3 shows an example of model clusters used in the present work. The structure of this cluster was adopted from the α-AgI crystal. One M ion was moved from a tetrahedral site to the other tetrahedral site through a trigonal site, a tetrahedral site, an octahedral site, a tetrahedral site and a trigonal site. Na ion, which has almost the same ionic radius as Ag ion, was also adopted as the M ion in the cluster for comparison in the electronic state with the result of Ag ion. Fig. 4 shows the variation of the total bond overlap population of M ion in the $M_{20}I_{20}$ cluster with the position. Since the M ion should move in the interaction with all of the surrounding atoms, the sum of all the bond overlap populations, total bond overlap population, of the moving M ion with surrounding atoms is shown in the figure. In this figure, circles and rectangles display the results of the Ag and Na ions, respectively. The position of M ion is shown as a relative value, that is, the value of zero means the center of the initial tetrahedral site, 0.5 means the center of the octahedral site and 1 means the final tetrahedral site. In the case of the Ag ion, the total bond overlap population is 1.03 at the initial tetrahedral site and is slightly decreased with the movement. Though the population of the Na ion is also decreased with the movement and has a minimum around the center of the octahedral site, the difference between the minimum and the initial value for the Ag ion is much smaller than that for the Na ion. This decrease of the bond overlap population is corresponding to the decrease of covalent interaction between the M ion and surrounding ions. Then at the center of the octahedral site, both Ag and Na ions become more unstable than at the center of the tetrahedral site. The unstableness of the Ag ion, however, is much smaller than that of the Na ion. This result means that the Ag ion can easily move from the tetrahedral site to the other tetrahedral site compared with the Na ion. This smaller change of the covalent interaction of the moving cations should be one of the origins of the fast movement of cations in the superionic conductors.

On the other hand, the ionicity of the moving cation is also a very important property to discuss the ionic conductivities of superionic conductors. The ionicity of the cation is evaluated by the net charge of the cation.

Fig. 5 shows the net charges of the Ag and

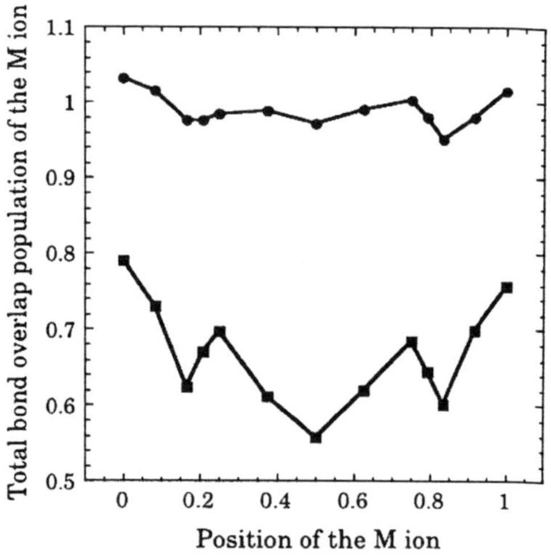

Fig. 4 The relationship between the position and the total bond overlap population of the M ion. ●: Ag ■: Na

Na ions with their movements. At the initial position, the net charge of the Na ion is 0.43 and is increased with the movement. There is the maximum at x = 0.5 and the net charge becomes 0.57. The difference between the maximum and the minimum values is 0.14.

The similar tendency is observed in the net charge of the Ag ion. At x = 0, the Ag ion has the charge of 0.04. The charge becomes larger with the movement of the Ag ion and takes a maximum values, 0.17 at x = 0.5. The difference between the maximum and minimum values of the net charge is 0.13. The change of the net charge of the Ag ion is very similar to that of the Na ion. This result suggests that the variation of the ionic interaction of the Ag ion with the movement along the path is almost the same as that of the Na ion.. So there are no characteristics in the change of the ionic interaction of the moving Ag ion.

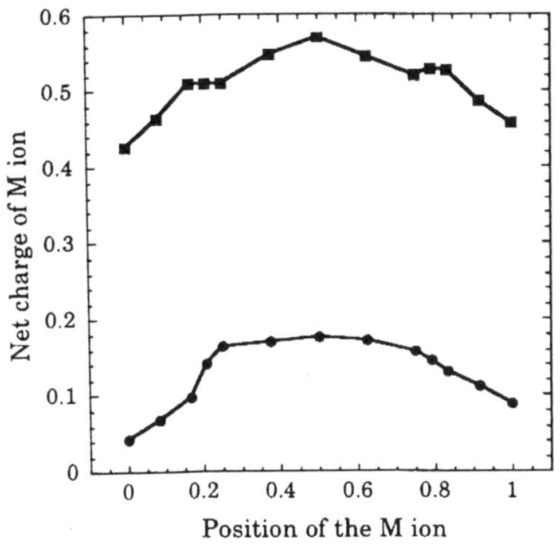

Fig. 5 The relationship between the position and the net charge of the moving M ion. ●: Ag ■: Na

3.3. Li_3N crystal

Fig. 6 shows model clusters for calculations of electronic state of Li_3N crystal. Large and small spheres show the position of the N and Li ions, respectively. In this crystal, there are two kinds of Li ions. The cluster (i) is a model for the moving Li ions on the Li_2N plane. Generally the Li ions on this plane can move easily as reported by several experimental studies.[7,8] Another Li ion exists on the Li plane between the Li_2N planes shown as cluster (ii). These Li ions cannot move to another. On the other hand, the Li_3N crystal has anisotropic ionic conductivity. The conductivity along the c axis is much smaller than that of the a and b axes. Then we have adapted two other model clusters (iii) and (iv). In the present study, we used these four model clusters corresponding to four kinds of movements of Li ions. We can discuss the relationship between the high Li ion conductivity and the chemical bonding of the moving Li ions in Li_3N crystals by these four clusters.

Fig. 7 shows the change of bonding state of the moving Li ions in four model clusters. In this figure abscissa shows the position of the moving Li ion in the relative value and ordinate shows total bond overlap population between the moving Li ion and surrounding ions.

In the case of the Li_2N plane, total bond overlap population of the moving Li ion changes little with the movement, since the bonding interaction between the moving Li ion and the nitride ion at side of the conduction path was kept constant during the movement of the Li ion. In contrary to this, total bond overlap populations of the moving Li ions in non-conducting paths have large change with the movement. These results suggest that the high ionic conductivity of Li_3N might be dependent on the small change of covalent bonding between the moving Li ion and surrounding ions. It is similar to the case of the α-AgI crystal.

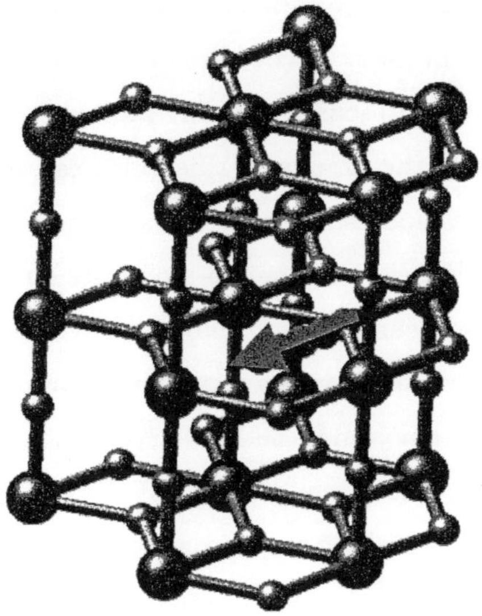

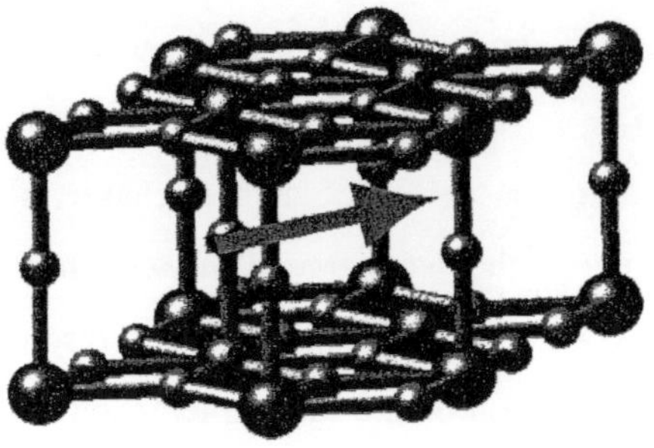

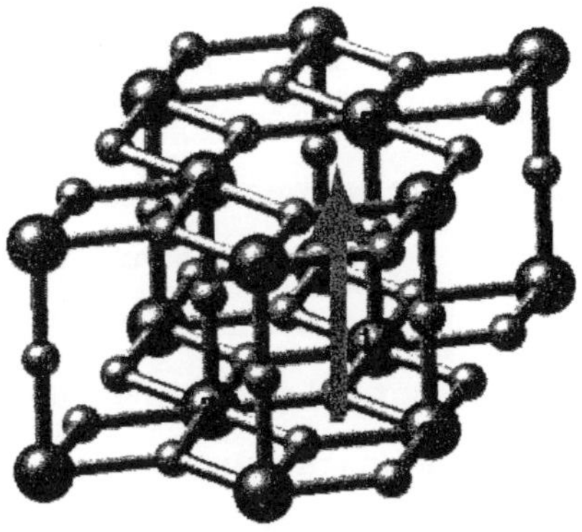

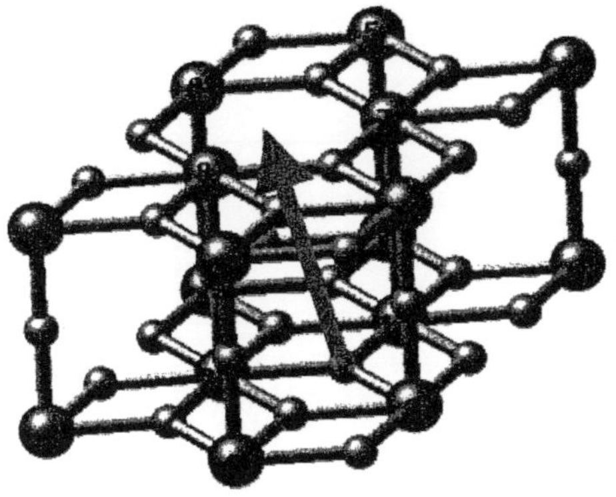

Fig. 6 Schematic diagram of the model clusters (i) Li_2N plane, (ii) Li plane, (iii) c axis 1, (iv) c axis 2.

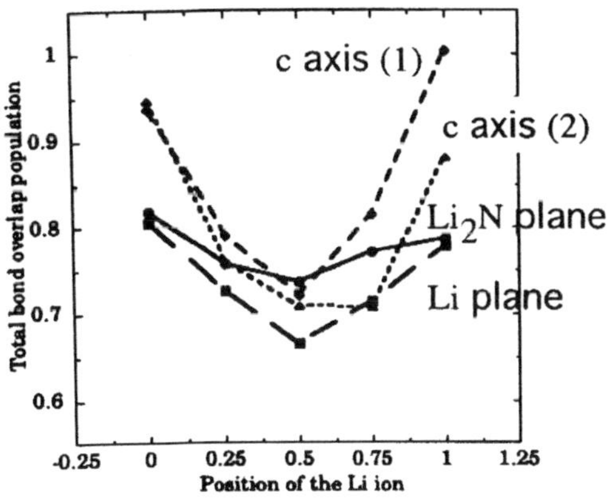

Fig. 7 The relationship between the position and the net charge of the moving Li ion.

4. Conclusion

Electronic states of fundamental bonding state of the monovalent ions in halides and two kinds of superionic conductors, α-AgI and Li_3N, were calculated by the DV-Xα cluster method.

In the simple MX_4 model cluster, the relationship between the net charge and bond overlap population of the monovalent ions in halides was discussed. The results for the alkali metal ions were located on the same line. It is suggested that the bonding interactions between alkali metal ions and the halide ions including both the covalent and the ionic interactions were similar to each other. On the other hand, the results of the Ag and Cu ions had weaker interaction than the alkali metal ions, which had similar ionic radii. This result suggested that the covalent interaction of the Ag and Cu ions should be important to discuss the fast movement of these cations.

In the α-AgI crystal, the variation of the total bond overlap population of the Ag ion was much smaller than that of the Na ion. On the other hand, the change of the net charge of the Ag ion with the movement was almost the same as that of the Na ion, while the absolute values of the net charge of the Na ion is larger than that of the Ag ion.

In the Li_3N crystal, the total bond overlap population of the moving Li ion on the Li_2N plane, which included the conduction path of the Li ion, was much smaller than those on the non-conduction paths. This result suggested that the covalent bonding of the mobile Li ion should be related to the fast movement of the Li ion in the Li_3N crystal.

The results obtained in the present work suggest that the smaller change of covalent bonding of the mobile cation should play an important role for the fast ion conduction in α-AgI and Li_3N crystals.

References

[1] H. Adachi, M. Tsukada, C. Satoko, J. Phys. Soc. Japan **45**, 875 (1978).

[2] J. C. Slater, Quantum Theory of Molecules and Solids, Vol. 4 (McGraw-Hill, New York, 1974).

[3] E. J. Baerends and P. Ros, Chem. Phys. **2**, 52 (1973).

[4] E. J. Baerends and P. Ros, Molec. Phys. **30**, 1735 (1975).

[5] J. E. Huheey, Inorganic Chemistry: Principles of Structure and Reactivity, 2nd Ed. (Harper and Row, New York, 1978) p. 71.

[6] R. J. Cava, F. Reidinger and B. J. Wuensch, Solid State Commun. **24**, 411 (1977).

[7] J. Wahl and U. Holland , Solid State Commun. **27**,

237 (1978).

[8] H. Schulz, K. H. Thiemann, Acta Crystallogr. A **35**, 309 (1979).

ELECTRONIC STATES OF SILVER AND COPPER HALIDES BASED ON DV-Xα METHOD

T. TOMOYOSE, H. WATANABE

Department of Physics, Ryukyu University, Okinawa 903-0213, Japan
E-mail: tomoyose@sci.u-ryukyu.ac.jp

M. KOBAYASHI

Department of Physics, Niigata University, Niigata 950-2181, Japan
E-mail: michi@phys.sc.niigata-u.ac.jp

The electronic states of AgX and CuX (X=Cl, Br, I) are studied by using the DV-Xα method. Several model clusters for AgX and CuX are used to obtain the density of states (DOS) and the bond overlap populations (BOP) changing the cluster-size. The cluster-size variations of DOS and BOP are discussed from the viewpoint of the chemical bond between neighboring ions. We found that the BOP of AgI and CuX are larger than those of AgCl and AgBr. This result is compared with the Phillips ionicity for $A^N B^{8-N}$ compounds.

1 Introduction

The study of electronic state for silver and copper halides has been attracted much interest from the viewpoint of ionic and covalent bonding in the superionic conductors[1,2]. The low temperature phases of AgX and CuX (X=Cl, Br, I) are distributed to ionic rock-salt and covalent zinc-blend structures so that their ionicities are clearly separated at the Phillips critical ionicity $F_i = 0.785$[3]. The ionic properties of AgX and CuX are characterized by the several peculiar properties, the high ionic conductivity and the low-frequency mode[4], and the quadrupolar deformability on phonon modes[5,6]. These peculiar properties of AgX and CuX are considered to be correlated to the d electron energy levels of Ag and Cu ions which are nearly closed to X-p electron energy level and strongly influences the valence band profiles in the electronic structure of AgX and CuX.

Kikuchi *et al.*[7,8] have studied the electronic states of silver and copper chalcogenides by using the first-principle density-functional method. Kobayashi *et al.*[9] and Ono *et al.*[10] have studied the energy bands of AgX and CuX by using the linear combination of atomic orbital (LCAO) method. Matsunaga *et al.*[11] have calculated the electronic states of AgCl and NaCl by using the DV-Xα method[12]. Kowada *et al.*[13] have studied the variation of electronic states of α-AgI and NaI with the moving cations assuming the model cluster Ag_6I_{10}. Ono and Kobayashi[14] also have studied the electronic states of AgX and NaX of rock-salt structure by using the DV-Xα method. However, to our knowledge, we have no the DV-Xα calculation for the electronic states of γ-AgI and γ-CuX (X=Cl, Br, I) of zinc-blend structure.

In order to obtain a unified insight for the electronic states of silver and copper halides, we have calculated the density of state (DOS) and the bond overlap population (BOP) of several model clusters A_mB_n where A denotes Ag and Cu, and B denotes Cl, Br, and I. In the present paper we have used the following model clusters: AB_4, $A_{13}B_4$, $A_{13}B_{16}$, $A_{16}B_{13}$, A_4B_{13}, A_4B for zinc-blend structures and AB_6, $A_{13}B_6$, $A_{13}B_{14}$, A_6B_{14}, A_6B for rock-salt structures, respectively.

2 DV-Xα Method

We here briefly review the DV-Xα Method. Following the Hartree-Fock-Slater approximation, many-electron system is reduced to the following one-electron Schrödinger equation:

$$H(\mathbf{r})\phi_k(\mathbf{r}) = \varepsilon_k \phi_k(\mathbf{r}), \tag{1}$$

where $\phi_k(\mathbf{r})$ is the wave function of k-th molecule, ε_k is the energy, and $H(\mathbf{r})$ is one-electron Hamiltonian containing the exchange and correlation potential $V_{xc}(\mathbf{r})$. In the Xα method, $V_{xc\uparrow}(\mathbf{r})$ is given by

$$V_{xc\uparrow}(\mathbf{r}) = -3\alpha \left\{ \frac{3}{4\pi}\rho_\uparrow(\mathbf{r}) \right\}^{1/3}, \tag{2}$$

where $\rho_\uparrow(\mathbf{r})$ is the local electron charge density and α is Slater's parameter which is fixed at 0.7.

We assume the LCAO approximation for $\phi_k(\mathbf{r})$,

$$\phi_k(\mathbf{r}) = \sum_j C_{jk}\chi_j(\mathbf{r}), \tag{3}$$

where $\chi_j(\mathbf{r})$ is the atomic orbital of j-th atom and the coefficient C_{jk} determines the probability amplitude of $\chi_j(\mathbf{r})$. From Eqs. (1) and (3) we get the secular equation,

$$\sum_j (H_{ij} - \varepsilon S_{ij})C_{jk} = 0, \tag{4}$$

$$H_{ij} = \int \chi_i^*(\mathbf{r})H(\mathbf{r})\chi_j(\mathbf{r})d\mathbf{r}, \tag{5}$$

$$S_{ij} = \int \chi_i^*(\mathbf{r})\chi_j(\mathbf{r})d\mathbf{r}, \tag{6}$$

where H_{ij} is the resonance integral and S_{ij} is the overlap integral. In the DV-Xα method, we replace these integrals by the numerical summations at discrete sample points $\mathbf{r}_k$ in space as follows:

$$H_{ij} = \sum_{k=1}^{N} \omega(\mathbf{r}_k)\chi_i^*(\mathbf{r}_k)H(\mathbf{r}_k)\chi_j(\mathbf{r}_k), \tag{7}$$

$$S_{ij} = \sum_{k=1}^{N} \omega(\mathbf{r}_k)\chi_i^*(\mathbf{r}_k)\chi_j(\mathbf{r}_k), \tag{8}$$

where N is the number of sample point and $\omega(\mathbf{r}_k)$ is the reciprocal of sample densities at $\mathbf{r}_k$.

By using the Mulliken's population analysis, the electron number n of molecule is presented as

$$\begin{aligned} n &= \int \rho(\mathbf{r})d\mathbf{r} = \sum_l f_l \int |\phi_l(\mathbf{r})|^2 d\mathbf{r}, \\ &= \sum_{ij} Q_{ij} = \sum_{ij} S_{ij} \sum_l f_l C_{il} C_{jl}, \end{aligned} \tag{9}$$

where f_l is the occupation number of l-th molecular orbital. Q_{ij} is the bond overlap population (BOP) between i and j-th orbital.

3 Model Clusters

We have assumed six model clusters of different atomic configuration for CuX and AgI of zinc-blend structure. Figure 1 shows the smallest model cluster AB_4 which is composed of the A ion at the center and four B ions at the tetrahedral corners. This tetrahedron has the symmetry of the point group T_d. Figure 2 shows the cluster $A_{13}B_4$ which contains twelve second-neighbor A ions. Figure 3 shows the more large cluster $A_{13}B_{16}$ which contains twelve third-neighbor ions. Furthermore we can consider three inverse configurational model clusters, A_4B, A_4B_{13}, $A_{16}B_{13}$ by exchanging A and B ions.

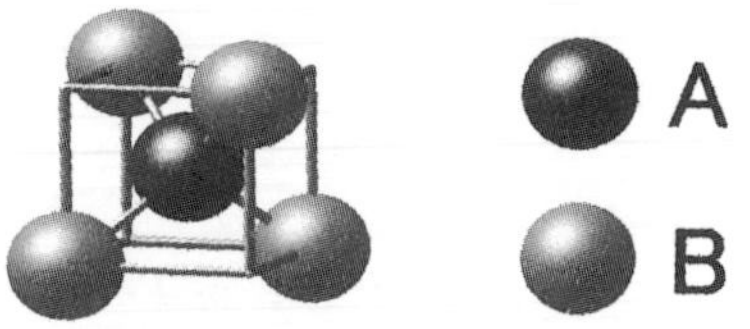

Figure 1. AB_4 model cluster.

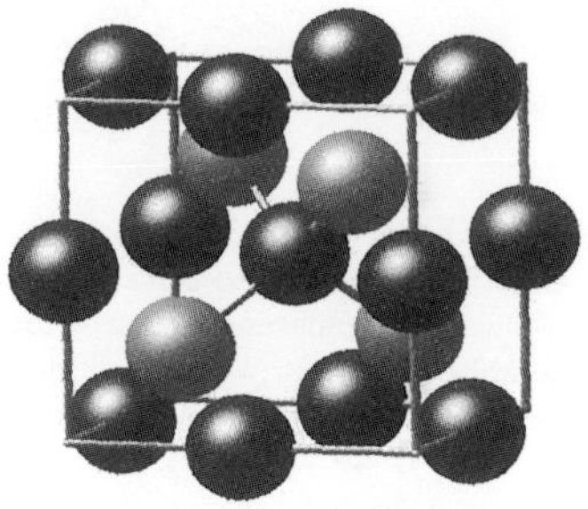

Figure 2. $A_{13}B_4$ model cluster.

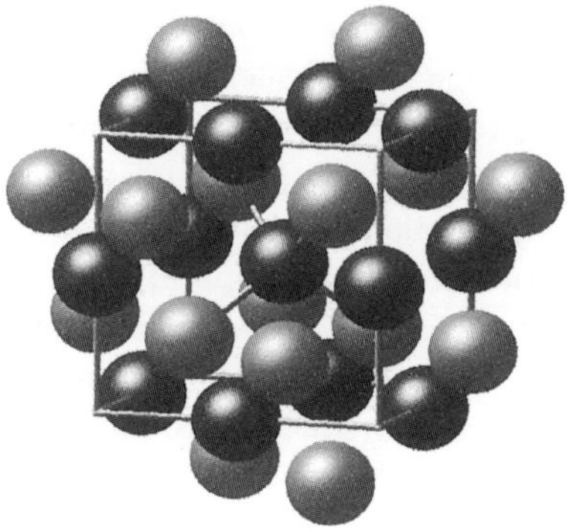

Figure 3. $A_{13}B_{16}$ model cluster.

For the rock-salt structure, we also consider six model clusters with different cluster size. The smallest cluster is the octahedral AB_6 shown in Fig. 4. This octahedron is composed of the A ion at the center and six B ions at the octahedral corners keeping the symmetry of the point group O_h. We can consider the large model clusters $A_{13}B_6$ and $A_{13}B_{14}$. The $A_{13}B_6$ cluster contains twelve second-neighbor ions as shown in Fig. 5. The $A_{13}B_{14}$ contains eight third-neighbor ions as shown in Fig. 6. These large clusters also satisfy the symmetry of the point group O_h. Furthermore we can consider three inverse configurational model clusters A_6B, A_6B_{13}, $A_{14}B_{13}$ by exchanging A and B ions.

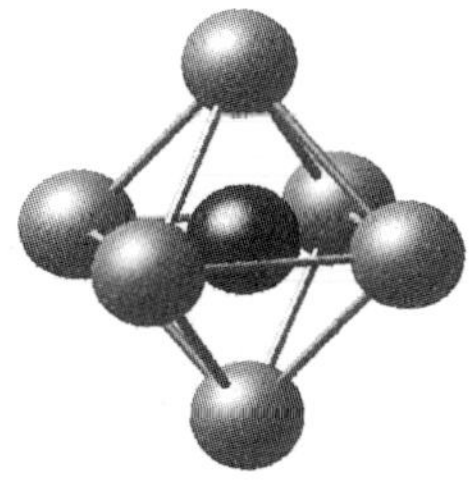

Figure 4. AB_6 model cluster.

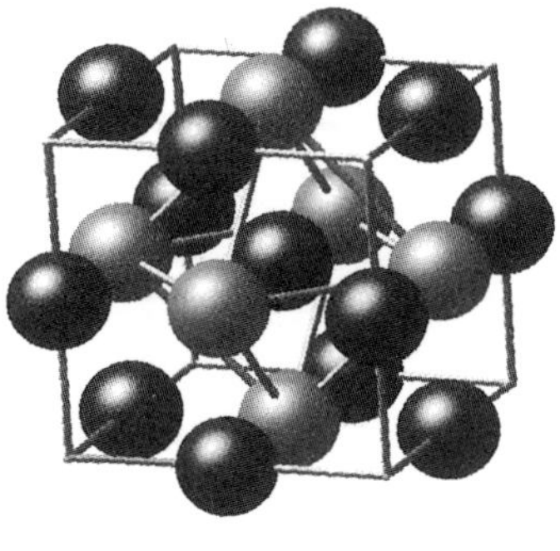

Figure 5. $A_{13}B_6$ model cluster.

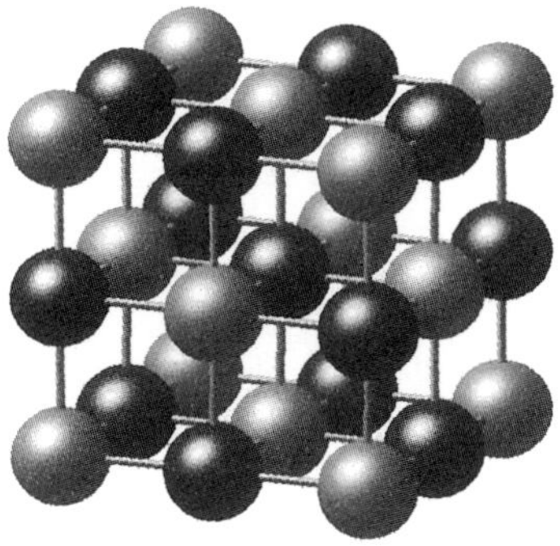

Figure 6. $A_{13}B_{14}$ model cluster.

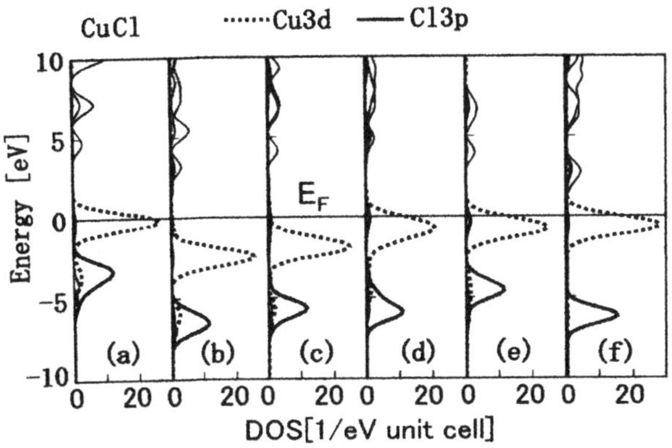

Figure 7. Density of states of CuCl.

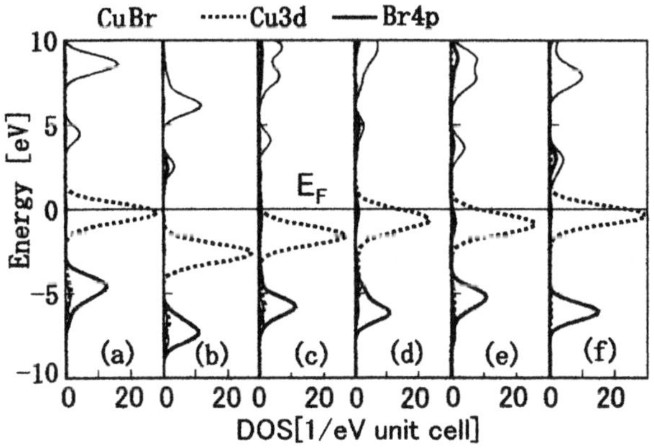

Figure 8. Density of states of CuBr.

4 Numerical Results

We have calculated the density of states (DOS) and the bond overlap populations (BOP) of AgX and CuX (X=Cl, Br, I) by making use of the model clusters described in the preceding section. In this calculation, we have taken account of the Madelung potentials of crystals by arranging image cells of cluster around the model clusters.

4.1 DOS

Figs. 7, 8, and 9 show the DOS of CuX for six model clusters, (a) AB_4, (b) $A_{13}B_4$, (c) $A_{13}B_{16}$, (d) $A_{16}B_{13}$, (e) A_4B_{13}, (f) A_4B. The dot lines denote the partial DOS of Cu-3*d* electrons and the solid lines are the partial DOS of X-*p* electrons. The line E_F denotes the Fermi energy level.

From Figs. 7, 8, and 9, we can evaluate the cluster-size variation of DOS of CuX. All model clusters show that the partial DOS of Cu-*d* and X-*p* electrons are clearly separated. For the cluster (b) $Cu_{13}X_4$, both *p* and *d* energy levels become low with the increase of the second neighbor Cu ions. This effect comes from the reason that the second neighbor Cu ions act as the positive crystal field on electrons of the inner ions inside the cluster. Furthermore, for model clusters (c) $Cu_{13}X_{16}$, the increase of third neighbor anions make the *p* and *d* energy levels rise up slightly. On the other hand, the inverse configurational clusters, Cu_4X, Cu_4X_{13} and $Cu_{16}X_{13}$ lead to the different results that the partial DOS of Cu-*d* electrons are nearly unchanged but the partial DOS of X-*p* electrons are changed by the increasing of the second neighbor X ions and third neighbor Cu ions.

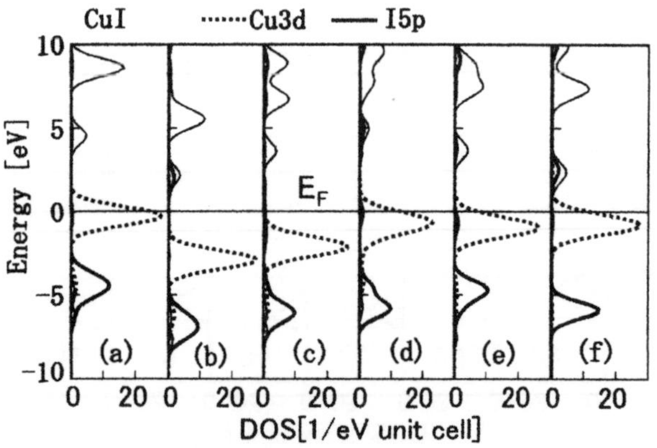

Figure 9. Density of states of CuI.

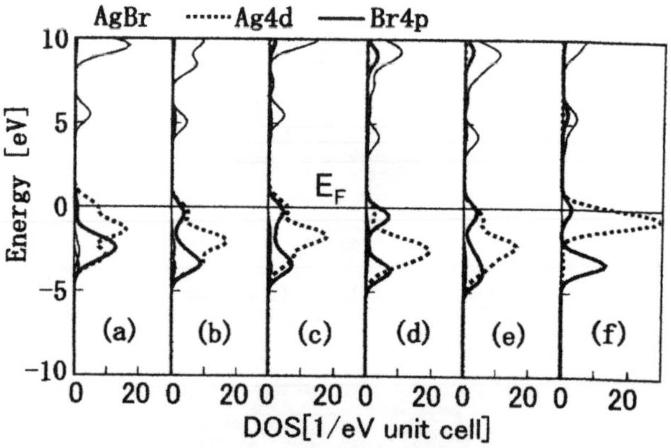

Figure 11. Density of states of AgBr.

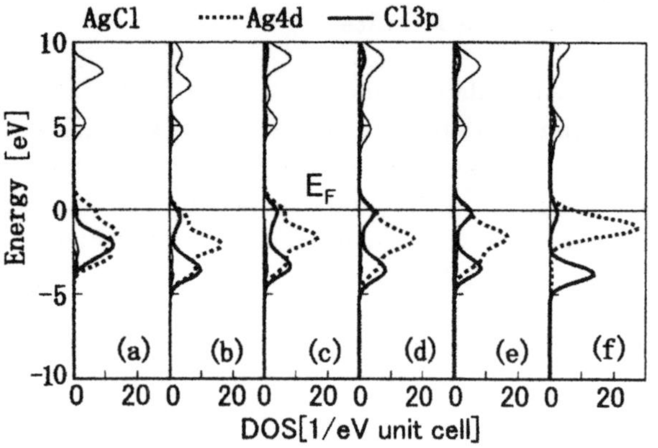

Figure 10. Density of states of AgCl.

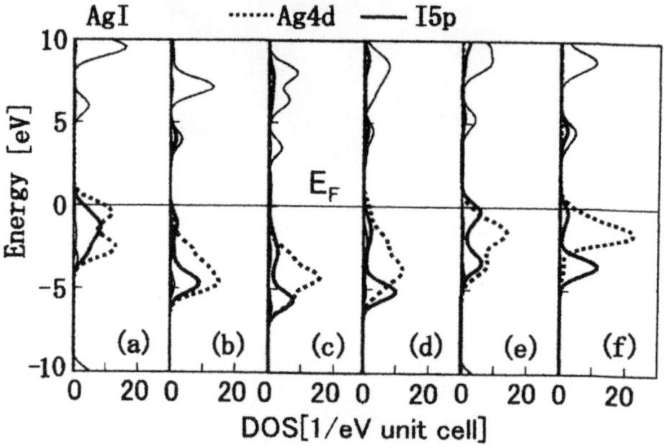

Figure 12. Density of states of AgI.

Figures.10, 11 and 12 show DOS of AgX for six model clusters. The dot and solid lines denote the partial DOS of Ag-4*d* and X-*p* electrons, respectively. For AgI, six model clusters are the same used for CuX. For AgBr and AgCl, we have used the following six model clusters: (a) AB_6, (b) $A_{13}B_6$, (c) $A_{13}B_{14}$, (d) $A_{14}B_{13}$, (e) A_6B_{13}, (f) A_6B.

From Figs. 10, 11 and 12, we find that the partial DOS of Ag-*d* and X-*p* electrons are broad for all model clusters. Since the *p* and *d* energy levels in AgX are nearly closed, these broad DOS are caused by the mixing between the nearly closed *p* and *d* states. For the model cluster (b), the second neighbor Ag act as the interactive potential on electrons of the inner X-anions and the Ag ion at the cubic center so that both Ag-*d* and X-*p* energy levels become low. Furthermore the increase of third neighbor anions make the *p* and *d* electron energy levels rise up slightly. On the other hand, the inverse configurational model cluster (f) lead to the clear separation between the partial DOS of Ag-*d* and X-*p* because of the strong repulsive crystal fields by much anions. Further, the large clusters (e) and (d) lead to the broadening of these partial DOS with the increase of the second and third neighbor ions.

4.2 BOP

We have calculated the bond overlap populations (BOP) of silver and copper halides by making use of six model clusters (a), (b), (c), (d), (e), (f) which are the same used in the Sec. 4.1. The calculated BOP of AgX and CuX are respectively listed in Table 1 and Table 2 with six model clusters. The BOP(A-B), BOP(A-A), and BOP(B-B) correspond to the cation-anion BOP, cation-cation BOP, and anion-anion BOP, respectively. Both Tables 1 and 2 show for all model clusters that BOP(A-A) and BOP(B-B) are smaller than BOP(A-B). This is reasonable since the BOP(A-B) is responsible to the overlap integral between the nearest neighbor ions. From Tables 1 and 2, we can evaluate the variation of the BOP(A-B) of AgX and CuX when the cluster-size changes.

Table 1. BOP of AgX.

		AgCl	AgBr	AgI
(a)	BOP(A-B)	0.205	0.188	0.310
	BOP(B-B)	0.003	−0.003	−0.001
(b)	BOP(A-B)	0.183	0.174	0.339
	BOP(B-B)	−0.001	−0.003	0.000
(c)	BOP(A-B)	0.176	0.171	0.316
	BOP(B-B)	−0.001	−0.003	0.000
(d)	BOP(A-B)	0.170	0.173	0.299
	BOP(A-A)	0.006	0.006	0.000
(e)	BOP(A-B)	0.172	0.164	0.320
	BOP(A-A)	0.006	0.005	0.000
(f)	BOP(A-B)	0.181	0.160	0.307
	BOP(A-A)	0.004	0.003	0.000

Table 2. BOP of CuX.

		CuCl	CuBr	CuI
(a)	BOP(A-B)	0.371	0.297	0.276
	BOP(B-B)	0.002	−0.001	−0.002
(b)	BOP(A-B)	0.370	0.338	0.362
	BOP(B-B)	0.000	0.000	−0.001
(c)	BOP(A-B)	0.348	0.317	0.321
	BOP(B-B)	0.000	0.000	0.000
(d)	BOP(A-B)	0.345	0.294	0.304
	BOP(A-A)	0.000	0.000	0.000
(e)	BOP(A-B)	0.345	0.329	0.330
	BOP(A-A)	0.001	0.001	0.000
(f)	BOP(A-B)	0.363	0.315	0.338
	BOP(A-A)	0.001	0.000	0.000

Figure 13 shows the cation-anion BOP of AgX and CuX for six model clusters, (a), (b) , (c), (d), (e), (f). The line $F_c = 0.215$ denotes the critical covalency estimated from the Phillips's critical ionicity $F_i = 0.785$ for $A^N B^{8-N}$ compounds[3].

From Fig. 13 we find that the order of BOP of AgX and CuX is of CuCl > CuI > CuBr > AgI > AgCl > AgBr for (f) model cluster and of CuCl > CuI > AgI > CuBr> AgCl > AgBr for (a) model cluster. The order of BOP is slightly different because of the variation of the atomic configuration and cluster size. However, we find that the BOP of AgI and CuX are always higher than those of AgBr and AgCl for all model clusters. Since the BOP measures the covalency of these compounds, this result of BOP is natural from the reason that the covalency of zinc-blend silver and copper halides are stronger than those of rock-salt one.

It is interesting that the BOP of AgBr and AgCl are lower than the critical covalency $F_c = 0.215$ and the BOP of CuX and AgI are higher than F_c for all model clusters. This result is consistent with Phillips's picture for the criticial ionicity $F_i - 0.785$ of $A^N B^{8-N}$ compounds[3].

5 Conclusion

We have studied the electronic states of silver and copper halides based on the DV-Xα method. Six model clusters are used to investigate the cluster-size variation of DOS and BOP of AgX and CuX. We have found that the partial DOS of Cu-d and X-p are clearly separated for all model clusters but that the partial DOS of Ag-d and X-p are nearly close and broad for all model clusters.

We have found that the BOP of AgBr and AgCl are lower than the critical covalency $F_c = 0.215$ estimated from the Phillips's critical ionicity for all model clusters. On the other hand the BOP of CuX and AgI are higher than F_c for all model clusters. This result is consistent with Phillips's picture for the ionicity of $A^N B^{8-N}$ compounds.

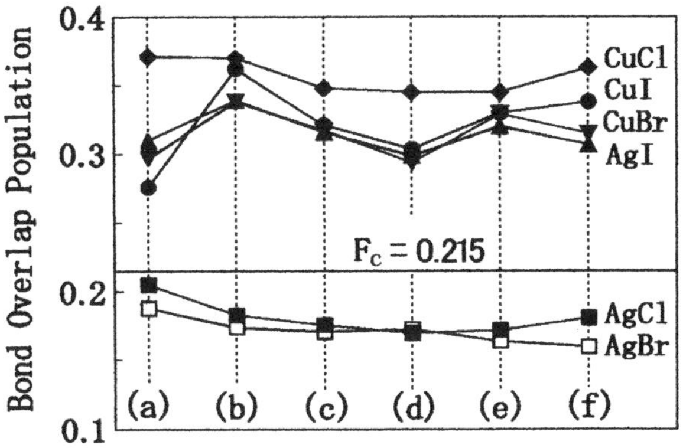

Figure 13. Variation of BOP(A-B) for six model clusters.

Acknowledgements

The numerical calculations in this study were executed on the basis of SCAT program due to Professor H. Adatch, Professor Y. Kowada, and their coworkers.

References

1. M. Aniya, *Solid State Ionics* **139**, 125 (1992).
2. T. Tomoyose, *J. Phys. Soc. Jpn* **64**, 1616 (1995).
3. J. C. Phillips, *Bonds and Bands in Semiconductors*, (Academic Press, New York, 1973).
4. K. Wakamura, *Phys. Rev.* B **59**, 3560 (1999).
5. K. Fischer, H. Bilz, R. Haberkorn and W. Weber, *Phys. Status Solidi* B **54**, 285 (1972).
6. W. G. Kleppmann and W. Weber, *Phys. Rev.* B **20**, 1669 (1979).
7. H. Kikuchi, H. Iyetomi, A. Hasegawa, *J. Phys. Condens. Matter* **10**, 11439 (1998).
8. H. Kikuchi, H. Iyetomi, A. Hasegawa, *J. Phys. Condens. Matter* **9**, 6031 (1997).
9. M. Kobayashi, S. Ono, H. Iyetomi S. Kashida and T. Tomoyose, *Solid State Ionics* **154-155** 209 (2002).
10. S. Ono, M. Kobayashi, H. Iyetomi and T. Tomoyose, *Solid State Ionics* **139** 249 (2001).
11. K. Matsunaga, I. Tanaka and H. Adachi, *J. Phys. Soc. Jpn.* **65** 3582 (1996).
12. H. Adachi, M. Tsukada, and C. Satoko, *J. Phys. Soc. Jpn*, **45** 875 (1978).
13. Y. Kowada, Y. Yamada, M. Tatsumisago, T. Minami, *Solid State Ionics* **136-137** 393 (2000).
14. S. Ono and M. Kobayashi, submitted.

PHYSICAL STUDIES OF ELECTRONIC STRUCTURE AND IONIC DYNAMICS IN SUPERIONIC CONDUCTORS

M. KOBAYASHI[a], S. ONO[b], T. TOMOYOSE[c], K. NOMURA[b], Y. YOKOYAMA[b], F. SAITO[b] and H. OGAWA[b]
[a]*Dept. of Phys., Niigata Univ., Niigata 950-2181, Japan,*
[b]*Graduate School of Sci. and Tech., Niigata Univ., Niigata 950-2181, Japan,*
[c]*Dept. of Phys. and Earth Science, Ryukyu Univ., Okinawa 903-0129, Japan,*

The electronic states of silver halides and sodium halides are calculated by the DV-Xα cluster method to get more microscopic evidence for the p-d hybridization in noble metal halides. It is found that both components of anti-bonding and bonding exist in the diagram of overlap population for AgX (X=halogen) and these two components are made up of the $4d$ band of Ag^+ and the p band of halogen ion, which form the p-d hybridization. Next a computer simulation by a molecular dynamics method has been performed to a model material which is composed of accumulating two different fluoride conductors: $\cdots BaF_2$-CaF_2-BaF_2-$CaF_2 \cdots$. It is obtained that the diffusion coefficient and ionic conductivity of F ions in the layered fluoride conductors increases with decreasing periods, which coincide with experiments. Lastly the dynamical correlation between diffusing cations and tetrahedra of the anion bcc sublattice in an α-AgI is studied using a molecular-dynamics method. When a silver ion is approaching the boundary surface of an empty tetrahedron(TH), the boundary plane area of the TH increases to faciliate the uptake of a silver ion. THs repeat contraction and expansion with motion of mobile silver ions. The ionic conductivity in superionic conductors with two kinds of mobile ions is investigated by general formulation of multicomponent lattice gas model in one dimension. The Hamiltonian includes a weak hopping term for its interacting-ion system. It is shown that the mixing of correlational two kinds of mobile ions leads to a decrease of the ionic conductivity, compared with the case of excluding the inter-ionic interaction.

Keywords: superionic conductor, p-d hybridization, DV-Xα method, molecular dynamics, layered fluoride conductors, nano-ionics, tetrahedron analysis, lattice gas model, two kinds of mobile ions

PACS numbers: 71.20.-b, 71.15.Fv, 71.15.Nc, 66.10.Ed, 66.30.-h, 66.30.Dn

1 Introduction

Superionic conductors (SSI) are crystalline materials which exibit extremely high values of ionic conductivity comparable to those of liquid electrolytes at relatively low temperatures. Physical and chemical properties of SIC have been investigated by many scientists [1–8]. SIC are attractive materials because of their high ionic conductivities in the solid state phase. SIC have both lattice ions and mobile ions. The latter show high ionic conductivity and might be thought as sub-lattice melting. Then those materials are applicable to high reliable batteries because we expect no liquid leak from battery compared with liquid solution.

In the 1st international discussion meeting on superionic conductor physics, which was held in Kyoto, we reported our recent works on SIC. We started from electronic structure studies of SIC, such as electronic band structure, electronic density of state, p-d hybridization, bond overlap population, diagram of overlap population *etc.* Next we introduced transport property studies on SIC, such as ionic conduction in superlattice, the dynamical correlation btween diffusing cations and the tetrahdra of the anion bcc sublattice. In the following we summarized our presentation in the meeting.

2 Electronic Structure

AgI is known as the most typical material among SIC. Many scientists have been studying to elucidate the origin of high ionic conductivity of AgI. This material shows superionic phase transition at 147°C and Ag ion can move fast through the lattice of I ions. Other similar family of superionic conductors are copper halides of CuCl, CuBr and CuI. These four superionic conductors have the coordition number four. On the other hand, other noble metal halides, such as AgCl, AgBr and AgF, do not show superionisity. They have the coordination number six. They are known as high ionic conductors. Considering from the viewpoint of coordination number, the low coordination number seems to be favorable for the ion transport through the lattice ions [9]. Even these materials have much larger values of ionic conductivities than those of alkali halides [10].

We have been interested in the difference of ionic

conductivity between silver halides and alkali halides and also have been interested in the difference of ionic conductivity among similar noble metal halide families. The qualitative differences between noble metal halides and alkali halides are mainly attributable to the difference of electron configuration between noble metal ions and alkali metal ions. They must have different kinds of electronic structures. A silver ion has the filled outer shell of $4d$ electrons and a copper ion has the filled outer shell of $3d$ electrons. They have both filled outer shell of d electrons. But alkali ions have no filled outer shell of d electrons. They have the outer shells of p electrons. These d electrons often play important roles to elucidate the mechanism of superionic conductivity and high ionic conductivity in noble metal halides [11].

We can see the difference in an electronic structure between noble metal halides and alkali halides using first principles band calculations. Experimental results also show the difference of electronic structure between noble metal halides and alkali halides. These results always show the existence of d electrons of a noble metal ion in the upper valence band and are well admixed with p electrons of a halogen ion [12]-[18]. The importance of this p-d hybridization has been also reported by many researchers. It has been recently proved that a local lattice instability occurs with a double well (DW) formation [19]-[21]. The coupling of electronic excitations (p-d dipoles) with the transverse optical phonon (at $k = 0$) results in the local DW formation, which, in turn, seems to be the prerequisite of the high ionic mobility. Thus, it seems to be possible to conclude that the appearance of p-d hybridization is an essential prerequisite to the formation of a local DW for some part of crystal constituent ions. If we are able to promote such a mixing, we would induce the formation of local dipoles corresponding to the DW with a consequent reduction in the activation energy for ion hopping. The promotion of p-d hybridization at $k = 0$ means the breaking down of selection rules to allow the electronic states to mix [19]. Recently, Kikuchi, Iyetomi and Hasegawa have studied the electronic properties of a series of tellurides with the antifluorite structure, Ag_2Te, Cu_2Te, M_2Te(M=Li, Na, K), to unveil the outstanding diffusivity of Ag ion [20]. They have carried out 1st principles density-functional calculations for these systems using the linearized augmented plane-wave (LAPW) method, where the exchange-correlation effects of electrons are treated in the local density approximation. They have elucidated that Ag_2Te and Cu_2Te have remarkably different degree of the p-d hybridization and the d-states of Ag ion are much more weakly coupled with the p-states of Te ion [20, 21]. If this p-d hybridization was characteristic among noble metal halides, we could deduce that the weakness of this p-d hybridization decides the values of ionic conductivity among noble metal halides. Because the values of ionic conductivity are owing to some kind of weakness of chemical bonding between mobile ions and lattice ions.

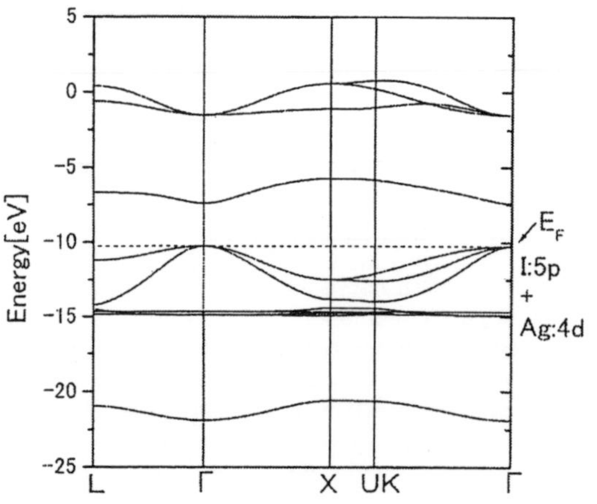

FIG. 1: Band structure of γ−AgI. E_F denotes the Fermi energy. The band indices of interest are shown (from [29]).

Studies of electronic states of cluster materials using the DV-Xα method have been carried out in material science [22]. DV-Xα stands for the discrete variational-Xα. This method is one of the useful technique for solving the Hartree-Fock-Slater equation. We can get the electronic structure of material exactly and easily by preparing the appropriate size cluster. Moreover, if we would like to know the electronic structures of ionic crystal, we can add the effect of madelung potential and good results are expected to the local electronic structure of bonding. This method has been widely applied to the electronic state calculations in the field of metals, ceramics, glasses, and so on. This method has been also applied to calculate the electronic structures of silver halides [23]-[28].

We performed the band calculations of noble metal halides to reveal the high ionic conductivity of Ag^+ and Cu^+ in noble metal halides using the linear combination of atomic orbitals theory [29]. Fig.1 shows the results of band calculations of γ-AgI. The figure is almost the same as that by Goldman *et al.*[30]. The upper region of valence band of AgX is mainly occupied by p-band of X^-, while that of CuX is mainly occupied by d-band of Cu^+. It was found that the d states of Ag^+ are much more weakly coupled with the p states of halogen ions, while those of Cu^+ are much more strongly coupled with the p bands. The strength of p-d hybridization was discussed to connect with the activation energy for the ionic conduction. It might be shown that the high ionic conductivity of AgX(X=halogen) primarily stems from the combination of the deformability of d shell and the weakness of p-d hybridization.

With the electronic knowledge, we have calculated also the phonon dispersion relation of γ-Ag_3SI, assuming an ideal perovskite-type structure [31]. The method of DAF model and the extra dynamical matrix of the virtual d-s excitations such as quadrupolar deformability force have been taken into consideration in calculations. The optical phonon modes have been affected strongly by introducing the extra dynamical matrix. The lowerings of accoustic modes also have been found. We could find the lowering of the phonon frequencies in γ-Ag_3SI, as in silver halides. When we investigate the effect of the quadrupolar deformability quantitatively, we will have to consider the extra interactions such as I-I, I-S, Ag-Ag, Ag-I, and Ag-S to connect with the ionic conductivity.

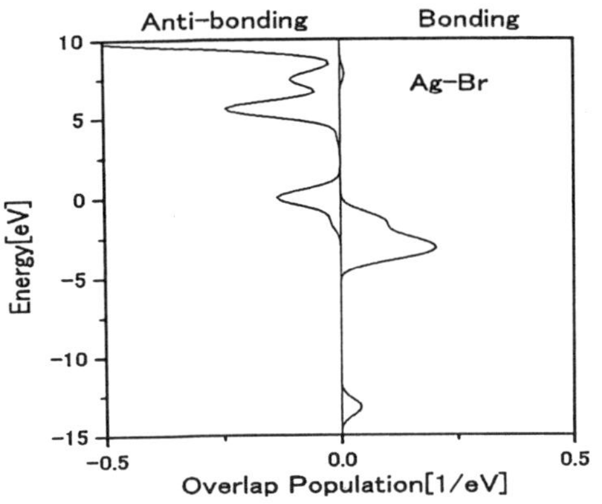

FIG. 2: Diagrams of overlap populations for AgBr of $(A_{13}B_{14})^{1-}$ cluster.

The DV-Xα cluster method is used to calculate the electronic state of $(A_{13}B_{14})^{1-}$ cluster, in which A and B ions are distributed to compose 2×2×2 unitcells of the NaCl crystal structure [28]. We use the Mulliken's population analysis in order to get the orbital population and the overlap population. The influence of the Madelung potential is included in our calculations. The DV-Xα method is powerful and useful to calculate the electronic energy of materials from the viewpoint of the 1st principle calculation. Fig.2 shows the diagrams of overlap populations [24] for AgBr in $(A_{13}B_{14})^{1-}$ cluster.

3 Transpot Properties

3.1 Superionic Superlattice

Sata *et al.* [32] have carried out an epoch-making experiment on the mesoscopic fast ion conduction in nanometre-scale planar heterostructures composed of CaF_2 and BaF_2. The conductivity increases progressively with decreasing period in the period range 500 to 16nm. Here the period means the thickness of CaF_2 plus BaF_2 unit layer.

To get the microscopic information about transport properties of superionic superlattice AgI-Ag_2S system, Kobayashi *et al.* [33] tried a computer simulation to the layered system by the molecular dynamics (MD) method. The calculated ionic conductivity of the superlattice system has been larger than that of either case of AgI-system or Ag_2S system. Their calculation suggested the possibility of the existance of the new material which has a larger ionic conductivity.

To investigate the experiment by Sata *et al.* [32] microscopically, we have devised a model material which is composed of accumulating two different fluoride conductors: $\cdots BaF_2$-CaF_2-BaF_2-$CaF_2\cdots$. We have performed a computer simulation to the layered fluoride conductors by the MD method at constant volume to get the microscopic information about transport properties of the material [34]. We have made use of the effective pair potentials used by Kaneko and Ueda [35]. Those are given by

$$V_{ij}(r) = A_{ij}\left(\frac{\sigma_i+\sigma_j}{r}\right)^7 + \frac{z_i z_j e^2}{r}, \tag{1}$$

where i, j describe the type of ions, A_{ij} the repulsive strength, σ_i, σ_j the particle radii, z_i, z_j the effective valence, and e the elementary charge. The value of A_{ij} is taken as 0.28eV. The values of parameters are shown in Table I.

TABLE I: Charge and radius [34].

	z_i	σ_i(Å)
Ba^{2+}	+2	1.37
Ca^{2+}	+2	0.99
F^-	-1	1.36

When the system has reached an equilibrium state, CaF_2 region is compressed and BaF_2 region is stretched along the c-axis (z-asis). This phenomenon may be due to taking the average value for a lattice constant of a layered superlattice system. It might be a trivial result as the lattice constant of BaF_2 is higher than that for CaF_2 both in nature and in the chosen force-field. The values of lattice constants of the z-direction change as shown in Table II. These changes may be caused to remove the pressure difference.

TABLE II: Lattice constants in Å [34].

a_z = 6.41 (+5%)	for BaF_2 part
a_z = 5.70 (-6.6%)	for CaF_2 part

The mean-square displacement (MSD) is defined by

$$\langle |\boldsymbol{r}^2|(t)\rangle = \frac{1}{N}\sum_{i=1}^{N}\langle |\boldsymbol{r}_i(t) - \boldsymbol{r}_i(0)|^2\rangle \qquad (2)$$

where $\langle\cdots\rangle$ is to be understood as an average over time. The linear regions in the functions $\langle |\boldsymbol{r}^2|(t)\rangle$ may be related to the diffusion coefficients D by the well-known equation

$$\langle |\boldsymbol{r}^2|(t)\rangle = 6Dt + C, \qquad (3)$$

where C is a constant.

The ionic conductivity σ is given by the Kubo formula using the current-current correlation function by

$$\sigma(\omega) = \frac{1}{3Vk_BT}\int_0^\infty dt\ e^{i\omega t}\ \langle \boldsymbol{J}_i(t)\cdot \boldsymbol{J}_i(0)\rangle, \qquad (4)$$

where V is the volume.

The ionic conductivity of F^- in the layered fluoride conductors increases with decreasing period, which is shown in Table III. Correctly for a fixed total size and composition of the system the conductivity increases with decreasing thickness of the MF_2 layers.

TABLE III: Ionic conductivity of F^- in the layered fluorite conductors at $T = 1400$K in the unit of $10^{-2}(\Omega\text{cm})^{-1}$ [34].

conductivity	20 period	10 period	4 period
σ	1.65	3.74	3.87

The period dependence on those transport coefficients is in accord with the experiment [32] roughly.

We can see dark region as a whole, which are marked by ∗1, ∗2 and ∗3 in Fig.3. Those regions represent a random distributin of F ions which are mobile ions.

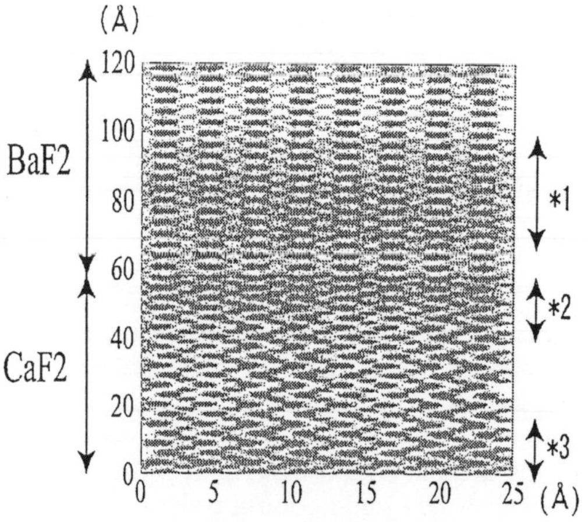

FIG. 3: Integrated distribution of ions in a limited time for a 20 period system at T=1400K. A vertical and horizontal lines are the c-axis (z-axis) and x-axis in unit of Å, respectively (from [34]).

3.2 Tetrahedral Analysis

We studied the particle correlation functions in α−AgI by making use of molecular-dynamics (MD) calculation [36]. The quantities $< |\vec{\Delta r_i}|^2 > + < |\vec{\Delta r_j}|^2 >$ were calculated for I and Ag ions. Here $\vec{\Delta r_i}$ is the displacement of the ith ion and $< \cdots >$ denotes a statistical average. These quantities might give important contribution to investigate the structure analysis. We showed that there is a close connection between the particle correlation function and the radial distribution function. We defined the correlation functions λ_{ij} as

$$\lambda_{ij} =< \frac{\vec{\Delta r_i}\cdot\vec{\Delta r_j}}{|\vec{\Delta r_i}||\vec{\Delta r_j}|} >. \qquad (5)$$

We showed that the values of particle correlation functions λ_{II} and λ_{IAg} at the nearest neighbor site are below 0.3. The values of λ_{AgAg} at the nearest neighbor site and λ_{II} at the next nearest neighbor site are about $1/2 \sim 1/3$ of it of λ_{II} (and λ_{IAg}) at the nearest neighbor site. Structure factors $S_I(\boldsymbol{k})$ and $S_{Ag}(\boldsymbol{k})$ have peaks at (2,0,0), (1,1,0), (4,0,0) which reflect the bcc lattice.

Ueda *et al.*[37, 38] studied the static and dynamic correlations among cations and anions using a MD method. They examined the correlation between diffusing cations and tetrahedra of anion bcc sublattice. They calculated the potential energy experienced by cations which move from one tetrahedron(TH) to the neighbor TH. They also emphasized that the main driving force responsible for cation diffusion is the Coulomb force.

3.3 Wandering Ion

Next we investigated the dynamical correlation between the diffusing cations and the TH of the anion bcc sublattice in the typical SIC α-AgI, using the MD simulation [39]. The dynamical correlation was investigated in detail. All space in α-AgI is filled with many tetrahedra which are composed of four I ions. One TH allows only one Ag ion to enter into it. When a Ag ion approaches at a distance d to a TH, the height, area and volume of the TH will change with a function of a distance d. We calculated these quantities with nano-scale level in detail using enormous data obtained by computer simulation. Using the correlation function λ_{ij} defined by eq.(5), the ref.[36] investigated the correlation between ions in a nano-scale level. On the other hand the refs.[37, 38] discussed about a relation

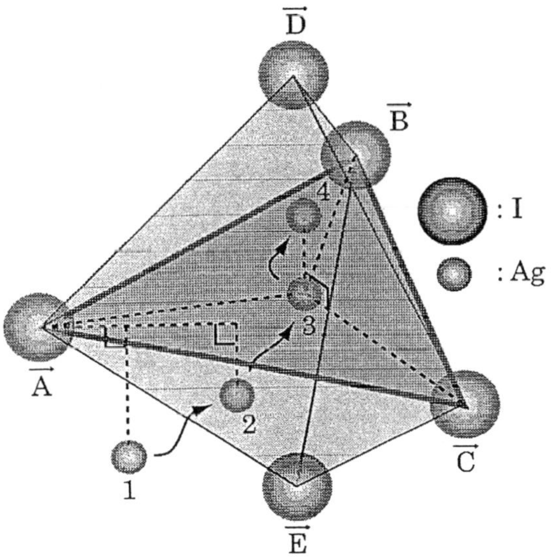

FIG. 4: The plane(ABC) of TH(ABCD) is held in common with TH(ABCE). This shows the movement of the marked Ag ion. Each number shows in time order the position of the mobile ion. (from [39]).

between potential barrier and the migration of mobile ions. Original points of the work were to carry out research into geometric quantities, such as volume, area, and height, of a TH made of anions, which were calculated in detail as a function of distance d by making use of the MD simulation.

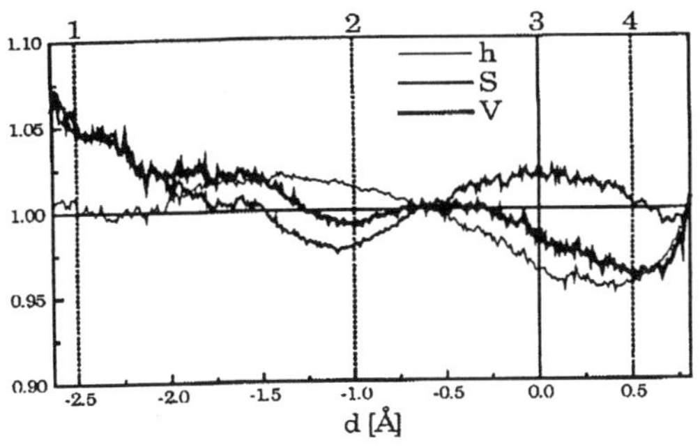

FIG. 5: d dependence on h, S and V. h is the height of TH(ABCD), S the area of plane(ABC), V the volume of TH(ABCD). Each site number, which shows a position of the marked mobile Ag ion (from [39]).

The MD method was applied to study the dynamical correlation between diffusing cations and tetrahedra of the anion bcc sublattice in a superionic conductor α-AgI [39].

Figure 4 shows a schematic view of a diffusing Ag ion through a TH which is made up of I ions. Each number corresponds to that in Fig.5. Geometric quantities such as a volume and an area of a TH are calculated as a function of a distance between a mobile Ag ion and a TH. As is shown in Fig.5, when a silver ion is approaching to the boundary surface of an empty tetrahedron , the boundary plane area of the TH increases to faciliate the uptake of a silver ion. Tetrahedrons composed by four iodine ions repeat contraction and expansion correlating with motion of mobile silver ions.

3.4 Lattice Gas Model

In order to understand the various physical properties of SIC, several theoretical models have been proposed. Mahan introduced the lattice gas theory of the ionic conductivity [40]. The lattice gas model in SIC has been studied also by the master equation approach on the transition rate [41, 42]. The lattice gas model is mathematically equivalent to the Ising model with the ion on or off a site corresponding to spin up or down, respectively [43]. The lattice gas model is based on the crystal structural characteristics of SIC. We can represent a solid as a network of lattice sites. A mobile ion can stay at each site, where at most one ion can occupy a given lattice site and only ions on nearest neighbor lattice sites interact with each other.

Following Mahan's approach, the lattice gas model in SIC has been investigated by several theoretical works. Tanaka *et al.* [44, 45] developed a theory of the ionic conductivuty for the two-dimensional SIC on the honey comb lattice and applied it to the $AgCrS_2$. They calculated the dc conductivity by the Kubo formula, in which the Fourier transform of the retarded Green's function was evaluated by means of the second-order perturbation applied to the analytic continuation of the thermal Green's function. They expressed the dc conductivity in terms of some 50 even-spin correlation functions. Recently Yonashiro *et al.* studied the dependence of the ionic conductivity on the mobile ion concentration [46]. Their result is that the ionic conductivity has symmetry with respect to concentration $\rho = 1/2$ and maximum value is at $\rho = 0.3$ and $\rho = 0.7$.

Ogawa *et al.* [47] have extended the lattice gas model to that with N_s species following Wetting *et al.* [48]. They have constructed a simple multicomponent lattice gas model in one dimension in which each site can either be empty or occupied by one particle at most of any one of D species. Particles interact with a nearest neighbor interaction which depends on the species involved.

We applied our result to calculate the concentration dependence of the conductivity in a binary system in one dimension [47].

4 Conclusion

In the meeting we could report and discuss about our recent works on electronic structure studies and transport property studies of superionic conductors. The details of our works have been written in published papers. Please refer them, if readers are interested in our works. We would like to thank attendants and referees for many useful discussions and comments.

Acknowledgments

We would like to acknowledge Professor S. Yoshikado for his organization of the meeting. One of the authors (M.K.) was supported by a Grant-in-Aid for Scientific Research from the Ministry of Education and Culture of Japan. He was also funded in part by the Uchida Energy Science Promotion Foundation.

References

[1] G.D. Mahan and W.L. Roth, eds., "*Superionic Conductors*" (Plenum, New York, 1976).
[2] J.B. Boyce and B.A. Huberman, Phys. Rep. **51** (1979) 189.
[3] W. Dieterich, P. Fulde and I. Peschel, Adv. Phys. **29** (1980) 527.
[4] S. Chandra, ed., "*Superionic Solids*" (North-Holland, Amsterdam, 1981).
[5] H. Bottger and V.V. Bryksin, "*Hopping Conduction in Solids*" (Academic, Berlin, 1985).
[6] T. Kudo and K. Fueki, "*Solid State Ionics*" (Kodansha, Tokyo, 1986) [in Japanese].
[7] T. Takahashi, ed., "*High Conductivity Solid Ionic Conductors: Recent Trends and Applications*" (World Scientific, Singapore, 1989).
[8] M. Kobayashi, Solid State Ionics **39** (1990) 121.
[9] M. Aniya, Solid State Ionics **50** (1992) 125.
[10] P. Brüesch, "*Phonons Theory and Experiments III*" (Springer,)
[11] G. D. Mahan, Solid State Commun. **33** (1980) 797.
[12] A. Goldmann, J. Tejeda, N. J. Shevchik, and M. Cardona, Phys. Rev. B **10** (1974) 438.
[13] P. V. Smith, J. Phys. Chem. Solids **37** (1976) 589.
[14] J. Shy-Yih Wang, M. Schluter, and M. L. Cohen, Phys. Stat. Sol. (b) **77** (1976) 295.
[15] A. B. Gordienko, Yu. N. Zhuravlev, and A. S. Poplavnoi, Phys. Stat. Sol.(b) **168** (1991) 149.
[16] G. S. Nunes, P. B. Allen, and Jose Luis Martins, Solid State Commun. **105** (1998) 377.
[17] B. N. Onwuagba, Solid State Commun. **97** (1996) 267.
[18] S. Ves, D. Glotzel, and M. Cardona, Phys. Rev. B **24** (1981) 3073.
[19] A. Rakitin and M. Kobayashi, Phys. Rev. B **53** (1996) 3088.
[20] H. Kikuchi, H. Iyetomi and A. Hasegawa, J. Phys.: Condensed Matter **10** (1998) 11439.
[21] H. Kikuchi, H. Iyetomi and A. Hasegawa, J. Phys.: Condensed Matter **9** (1997) 6031.
[22] H. Adachi, M. Tsukada, and C. Satoko, J. Phys. Soc. Jpn. **45** (1978) 875.
[23] Y. Kowada, H. Adachi, M. Tatsumisago and T. Minami, J. Non-Cryst. Solids **232-234** (1998) 497.
[24] K. Matsunaga, I. Tanaka, and H. Adachi, J. Phys. Soc. Jpn. **65** (1996) 3582.
[25] K. Matsunaga, I. Tanaka, and H. Adachi, J. Phys. Soc. Jpn. **67** (1998) 2027.
[26] K. Matsunaga, N. Narita, I. Tanaka, and H. Adachi, J. Phys. Soc. Jpn. **65** (1996) 2564.
[27] Y. Kowada, Y. Yamada, M. Tatsumisago, T. Minami and H. Adachi, Solid State Ionics **136-137** (2000) 393-397.
[28] S. Ono and M. Kobayashi, submitted.
[29] S. Ono, M. Kobayashi, H. Iyetomi and T. Tomoyose, Solid State Ionics **139** (2001) 249.
[30] A. Goldman, J. Tejeda, N.J. Shevchik and M. Cardona, Phys. Rev. B **10** (1974) 4388.
[31] S. Ono, H. Ogawa, M. Kobayashi, and T. Tomoyose, Proceedings of 8th Asian Conf. on Solid State Ionics (Malaysia, 2002), *Solid State Ionics: Trends in the New Millennium* p.801.
[32] N. Sata, K. Eberman, K. Eberl and J. Maier, Nature **408** (2000) 946.
[33] M. Kobayashi, F. Shimojo F. Tachibana and H. Okazaki, J. Phys. Soc. Jpn. **60** (1991) 245.
[34] K. Nomura and M. Kobayashi, Proceedings of 9*th* Euro Conf. on Ionics (Rhodes, 2002), Ionics **9** (2003) 64.
[35] Y. Kaneko and A. Ueda, J. Phys. Soc. Jpn. **57** (1988) 3064.
[36] K. Nakamura, K. Ihata, Y. Yokoyama, K. Nomura and M. Kobayashi: Ionics **7** (2001) 178.
[37] M. Hokazono, A. Ueda and Y. Hiwatari: Solid State Ionics **13** (1984) 151.
[38] Y. Kaneko and A. Ueda: J. Phys. Soc. Jpn. **55** (1986) 3924.
[39] Y. Yokoyama, and M. Kobayashi, Solid State Ionics **159** (2003) 79.
[40] G. D. Mahan, Phys. Rev. B **14** (1976) 780.
[41] L. Pietronero and S. Strässler, Z. Phys. B**36** (1980) 263.
[42] T. Ishii, Prog. Theo. Phys. **73** (1985) 1084.
[43] C. Domb and M. S. Green, eds., *Phase Transition and Critical Phenomena,* (Academic, London and New York, 1976).
[44] T. Tanaka, M. A. Sawtarie, J. H. Barry, N. L. Sharma, and C. H. Munera, Phys. Rev. B**34** (1986) 3773.
[45] N. L. Sharma and T. Tanaka, Phys. Rev. B**28** (1983) 2146.
[46] K. Yonashiro and M. Iha, J. Phys. Soc. Jpn. **70** (2001) 2958.
[47] H. Ogawa, F. Saito and M. Kobayashi, submitted.
[48] T. Wetting and A. D. Jackson, Phys. Rev. D **49** (1994) 157.

LDA STUDY ON POLARIZABILITIES OF IONS IN COPPER HALIDE CRYSTALS WITH THE ZINC-BLENDE STRUCTURE

YOSHITAKA MICHIHIRO, MD. MAHBUBAR RAHMAN, KOICHI NAKAMURA, and TATSUO KANASHIRO

Department of Physics, Faculty of Engineering, Tokushima University, Tokushima 770-8506, Japan
E-mail: yositaka@pm.tokushima-u.ac.jp

By the local density approximation (LDA), the electron states are calculated for the closed-shell ions in copper halide crystals with the zinc-blende structure. The crystalline environment is expressed by the spherical solid model. The modified Sternheimer equation is employed to calculate the perturbed electron states. The dipole polarizability and the quadrupole polarizability of ions are calculated. The calculated value of the dipole polarizability is in reasonable agreement with the experimental value in each crystal. The size effect is seen in the values of the dipole polarizability. The values of the dipole polarizability of the copper ion show only a slight change in different crystals. In contrast, the values of the anion show a significant change. A similar trend is also seen in the values of the quadrupole polarizability. A discussion is given about the effect of the crystalline environment on the electron states.

1 Introduction

The polarizabilities of ions are important parameters of ionic crystals. The dipole polarizability can be related to the high frequency dielectric constant through the Clausius-Mossotti relation. When the binary ionic crystal AX with the high frequency dielectric constant $\epsilon_\infty(\mathrm{AX})$ is composed of the cation A^+ with the dipole polarizability $\alpha_1(\mathrm{A}^+,\mathrm{AX})$ and the anion X^- with the dipole polarizability $\alpha_1(\mathrm{X}^-,\mathrm{AX})$, then

$$\alpha_1(\mathrm{A}^+,\mathrm{AX}) + \alpha_1(\mathrm{X}^-,\mathrm{AX}) = \frac{3v_\mathrm{u}}{4\pi}\frac{\epsilon_\infty(\mathrm{AX})-1}{\epsilon_\infty(\mathrm{AX})+2}, \quad (1)$$

where v_u is the volume of the formula unit. However, this relation yields the summation of the dipole polarizability of ions. The dipole polarizability of the individual ion cannot be measured directly. Some theory is needed to decompose the summation.

Most electron states calculations for solids are based on the density functional theory.[1,2] The local density approximation method is used in the majority of the density functional theory calculations for solids. When the external perturbation is applied, the electrons partially screen the field. This screening effect can be recast as the modified Sternheimer equation with the self-consistent field potential.[3,4] The use of the self-consistent field potential causes the reduction of calculated values of the dipole polarizability of free cations and greatly improves the accuracy.[4–6] The spherical solid model[7] was used to represent the crystalline environment in cubic ionic crystals. The values of the dipole polarizability of ions in the rock-salt structure alkali halide crystals were calculated[8] by using the spherical solid model and the local density approximation with the screening effect. These calculations found the size effect of the dipole polarizability in the rock-salt structure alkali halides: (i) the values of the dipole polarizability of a cation are nearly the same in different crystals; (ii) the values of the dipole polarizability of an anion vary from crystal to crystal depending mainly on the lattice constant. The experimental results support these theoretical predictions. The calculated values[8,9] of the dipole polarizability are in close agreement with the experimental values. The assessment of the accuracy of the perturbed electron density is provided by the calculation of the dipole shielding factor.[10] It was found that the quadrupole polarizability and the quadrupole shielding factor also show the size effect.[8,10] Later, the calculation was applied to the cesium-chloride structure crystals and the fluorite structure crystals. The size effect is seen again in these crystals. These calculations were extended to evaluate the van der Waals coefficient,[11] the hyperpolarizability,[12] and so on.

Here, the calculation is applied to copper halide crystals with the zinc-blende structure. The effect of the crystalline environment on the electron states is discussed.

2 Numerical Method

The electron states are calculated by the standard local density approximation method.[1,2] The calculation is composed of three steps. A brief outline of the procedure is presented for completeness.

In the first step the unperturbed electron states are calculated. The unperturbed eigenfunction $\psi_j^{(0)}(\boldsymbol{r})$ and eigenvalue $\lambda_j^{(0)}$ obey the Kohn-Sham equation,

$$\left(H^{(0)} - \lambda_j^{(0)}\right)\psi_j^{(0)}(\boldsymbol{r}) = 0. \tag{2}$$

The unperturbed electron density $n^{(0)}(\boldsymbol{r})$ is given by summing up all occupied states j,

$$n^{(0)}(\boldsymbol{r}) = \sum_j \left(\psi_j^{(0)}(\boldsymbol{r})\right)^* \psi_j^{(0)}(\boldsymbol{r}). \tag{3}$$

The unperturbed electron potential $V^{(0)}(\boldsymbol{r})$ is,

$$V^{(0)}(\boldsymbol{r}) = -\frac{Ze^2}{|\boldsymbol{r}|} + e^2 \int \mathrm{d}\boldsymbol{R} \frac{n^{(0)}(\boldsymbol{R})}{|\boldsymbol{R}-\boldsymbol{r}|} + V_{\mathrm{xc}}[n^{(0)}(\boldsymbol{r})] + V_{\mathrm{c}}^{(0)}(\boldsymbol{r}). \tag{4}$$

Here the terms are the Coulomb interaction with the nucleus, the Hartree potential, the exchange-correlation potential $V_{\mathrm{xc}}[n^{(0)}]$, and the spherically averaged crystalline potential $V_{\mathrm{c}}^{(0)}(\boldsymbol{r})$. The exchange-correlation potential $V_{\mathrm{xc}}[n^{(0)}]$ is taken to be a local function of the electron density $n^{(0)}(\boldsymbol{r})$. Following the previous calculations,[4–11] we adopt the form suggested by Gunnarsson and Lundqvist.[13] The crystalline potential $V_{\mathrm{c}}(\boldsymbol{r})$ is simulated by the summation of the electron-ion interaction over the surrounding ions. The electron-ion interaction is expressed by the pseudopotential. For halogen ions and alkali ions, the parameters of the pseudopotential of Heine-Abarenkov[14] are given in Ref. 7. In the calculation, these parameterizations of the pseudopotential are employed. For the copper ion, to avoid complexity, the empty-core pseudopotential of Ashcroft[15] with the core radius 0.96 Å by Pauling is applied. In view of the cubic arrangement of the closed-shell ions, the spherical solid model (SSM)[7] is used. The fundamental assumptions of the SSM are the ones that the electrons are localized on a single ion and that the local environment of each ion is spherically symmetric. Thus the crystalline potential $V_{\mathrm{c}}(\boldsymbol{r})$ is approximated by its spherical average $V_{\mathrm{c}}^{(0)}(\boldsymbol{r})$. The charge distributions of ions are spherically symmetric in the SSM.

In the second step the perturbation due to the infinitesimal external field is treated. By the infinitesimal perturbation, each eigenfunction $\psi_j^{(0)}(\boldsymbol{r})$ is perturbed. It causes the first order change $\psi_j^{(1)}(\boldsymbol{r})$. The first order change in the electron density $n^{(1)}(\boldsymbol{r})$ is given by,

$$n^{(1)}(\boldsymbol{r}) = 2\mathrm{Re}\sum_j \left(\psi_j^{(0)}(\boldsymbol{r})\right)^* \psi_j^{(1)}(\boldsymbol{r}). \tag{5}$$

The first order change in the electron density $n^{(1)}(\boldsymbol{r})$ causes the first order change in $V^{(0)}(\boldsymbol{r})$ in Eq. (4). This is the screening effect of the external field. The first order self-consistent field (SCF) potential $V_{\mathrm{SCF}}^{(1)}(\boldsymbol{r})$ acting on the electron is given by combining the external potential $V_{\mathrm{ext}}^{(1)}(\boldsymbol{r})$ and the first order change in $V^{(0)}(\boldsymbol{r})$,

$$V_{\mathrm{SCF}}^{(1)}(\boldsymbol{r}) = V_{\mathrm{ext}}^{(1)}(\boldsymbol{r}) + e^2 \int \mathrm{d}\boldsymbol{R} \frac{n^{(1)}(\boldsymbol{R})}{|\boldsymbol{R}-\boldsymbol{r}|} + n^{(1)}(\boldsymbol{r}) \frac{\delta V_{\mathrm{xc}}[n^{(0)}]}{\delta n^{(0)}}. \tag{6}$$

The equation for the first order change in the eigenfunction $\psi_j^{(1)}(\boldsymbol{r})$ and the eigenvalue $\lambda_j^{(1)}$ is derived from the perturbation theory,

$$\left(H^{(0)} - \lambda_j^{(0)}\right)\psi_j^{(1)}(\boldsymbol{r}) = -\left(V_{\mathrm{SCF}}^{(1)} - \lambda_j^{(1)}\right)\psi_j^{(0)}(\boldsymbol{r}). \tag{7}$$

This equation is defined as the modified Sternheimer equation.[3,4]

In the third step the polarizabilities are evaluated. The response of the ion to the external perturbation is expressed by the correlation. The correlation $\langle K, L|M, N\rangle$ links the weight function $F_{K,L}(\boldsymbol{r})$ of the form $P_K(\cos\theta)r^L$ to the external field $F_{M,N}(\boldsymbol{r})$ of the form $P_M(\cos\theta)r^N$ with the Legendre polynomials $P_M(\cos\theta)$ and $P_K(\cos\theta)$. It is given by the integral of the product of the function $F_{K,L}(\boldsymbol{r})$ and the first order change in the electron density $n_{M,N}^{(1)}(\boldsymbol{r})$ caused by the field $F_{M,N}(\boldsymbol{r})$,

$$\langle K, L|M, N\rangle = \int \mathrm{d}\boldsymbol{r} F_{K,L}(\boldsymbol{r}) n_{M,N}^{(1)}(\boldsymbol{r}). \tag{8}$$

The dipole polarizability is expressed by the correlation between the external field of the form $P_1(\cos\theta)r$ and the weight function of the form $P_1(\cos\theta)r$. Thus the dipole polarizability α_1 is given by,

$$\alpha_1 = \left(\frac{\hbar^2}{me^2}\right)^3 \langle 1, 1|1, 1\rangle. \tag{9}$$

In a similar way, the quadrupole polarizability α_2 is given by,

$$\alpha_2 = \left(\frac{\hbar^2}{me^2}\right)^5 \langle 2, 2|2, 2\rangle. \tag{10}$$

3 Results and Discussion

Table 1 shows the calculated values of the dipole polarizability α_1 of ions in copper halide crystals with the zinc-blende (ZB) structure. Also shown are the calculated values in alkali halide crystals with the rock-salt (RS) structure. Here the values in alkali halide crystals are taken from Refs. 8 and 9. The calculated values[4,5] of free cations are shown for comparison. The first column lists the crystal AX (A = Cu, Li, Na, K, and X = Cl, Br, I). The second column and the third column are the calculated values $\alpha_1^+ = \alpha_1(A^+, AX)$ for the cation A^+ and $\alpha_1^- = \alpha_1(X^-, AX)$ for the anion X^-, respectively. The next column gives the summation $\alpha_1^{tot}(AX) = \alpha_1(A^+, AX) + \alpha_1(X^-, AX)$. The last column lists the experimental value[16–18] α_1^{exp} derived from the Clausius-Mossotti relation (1).

In each crystal, the calculated value α_1^{tot} is in reasonable agreement with the experimental value α_1^{exp}. The error is less than 20%. As seen, the values of the dipole polarizability $\alpha_1(Cu^+, CuX)$ of the copper ion in CuX crystals are very similar. The values are insensitive to the partner anion X^-. Moreover, the values are close to the value $\alpha_1(Cu^+, \text{free})$ of the free copper ion.

The value of the dipole polarizability $\alpha_1(Cl^-, CuCl)$ of the chlorine ion in CuCl crystal is similar to the value $\alpha_1(Cl^-, NaCl)$ in NaCl crystal. Both of the values are smaller than the value $\alpha_1(Cl^-, KCl)$ in KCl crystal, and larger than the value $\alpha_1(Cl^-, LiCl)$ in LiCl crystal. Similarly the values of the dipole polarizability of the bromine ion and the iodine ion vary from crystal to crystal. The values of the dipole polarizability of the anion are sensitive to the crystalline environment; the partner cation and the crystal structure.

Table 1. Dipole polarizability in Å^3 of cations and anions in copper halide crystals AX with the zinc-blende structure. The values in alkali halide crystals are taken from Refs. 8 and 9. α_1 is calculated by Eq. (9). $\alpha_1^+ = \alpha_1(A^+, AX)$. $\alpha_1^- = \alpha_1(X^-, AX)$. $\alpha_1^{tot} = \alpha_1^+ + \alpha_1^-$. α_1^{exp} is the experimental value.[16–18]

Crystal AX		α_1^+	α_1^-	α_1^{tot}	α_1^{exp}
Cu^+	(free)	1.14	/		0.98
CuCl	(ZB)	1.14	2.97	4.11	4.36
CuBr	(ZB)	1.14	3.87	5.01	5.60
CuI	(ZB)	1.10	5.38	6.48	7.44
Li^+	(free)	0.03	/		0.03
LiCl	(RS)	0.03	2.86	2.89	2.94
LiBr	(RS)	0.03	3.97	4.00	4.09
LiI	(RS)	0.03	5.67	5.70	5.90
Na^+	(free)	0.16	/		0.16
NaCl	(RS)	0.15	3.06	3.21	3.24
NaBr	(RS)	0.15	3.98	4.13	4.38
NaI	(RS)	0.15	6.37	6.52	6.35
K^+	(free)	0.84	/		0.85
KCl	(RS)	0.80	3.39	4.19	4.15
KBr	(RS)	0.79	4.36	5.15	5.29
KI	(RS)	0.78	6.68	7.46	7.44

Table 2. Quadrupole polarizability in Å^5 of the copper ion and anions in copper halide crystals with the zinc-blende structure. The value α_{2i} of the free copper ion is that calculated previously.[19,20] The value α_{2i} of the free chlorine ion is that summarized by Dalgarno.[21] α_2 is calculated by Eq. (10). The values in alkali chloride and alkali bromide crystals are taken from Ref. 8.

Ion	α_{2i}	Crystal AX		α_2
Cu^+	0.83	CuCl	(ZB)	2.46
	~2.47	CuBr	(ZB)	2.48
		CuI	(ZB)	2.31
Cl^-	13.1	CuCl	(ZB)	5.37
	~13.8	LiCl	(RS)	4.32
		NaCl	(RS)	5.33
		KCl	(RS)	6.34
Br^-		CuBr	(ZB)	7.90
		LiBr	(RS)	6.45
		NaBr	(RS)	7.79
		KBr	(RS)	9.01
I^-		CuI	(ZB)	13.60
		LiI	(RS)	11.20
		NaI	(RS)	13.44
		KI	(RS)	15.65

Table 2 shows the calculated results of the quadrupole polarizability of ions in copper halide crystals with the ZB structure. Again, the calculated values of ions in alkali chloride,[8] alkali bromide,[8] and alkali iodide crystals with the RS structure are also shown for comparison. Here α_{2i} is the value of the quadrupole polarizability of the free ion calculated previously.[19–21] Our value of the quadrupole polarizability $\alpha_2(\mathrm{Cu}^+, \mathrm{free}) = 2.46$ Å^5 of the free copper ion is the same order of magnitude to the value[19] $\alpha_{2i} = 0.83$ Å^5 and the value[20] $\alpha_{2i} = 2.47$ Å^5. The values of the quadrupole polarizability $\alpha_2(\mathrm{Cu}^+, \mathrm{CuX})$ of the copper ion in CuX crystals are almost independent of the partner anion X^-.

Our calculated values α_2 of the chlorine ion in crystals are smaller than the value α_{2i} of the free chlorine ion calculated by many authors using many techniques, such as the Hartree-Fock method, and summarized by Dalgarno.[21] Moreover, the calculated value $\alpha_2(\mathrm{Cl}^-, \mathrm{CuCl})$ of the chlorine ion in CuCl crystal is similar to the value $\alpha_2(\mathrm{Cl}^-, \mathrm{NaCl})$ in NaCl crystal. Again, both of the values are smaller than the value $\alpha_2(\mathrm{Cl}^-, \mathrm{KCl})$ in KCl crystal, and larger than the value $\alpha_2(\mathrm{Cl}^-, \mathrm{LiCl})$ in LiCl crystal. A similar trend is also seen in the bromine ion and the iodine ion.

Table 3. Under the point charge approximation, the dipole polarizability $\alpha_{1\mathrm{M}}$ in Å^3 and the quadrupole polarizability $\alpha_{2\mathrm{M}}$ in Å^5 of cations and anions in copper halide crystals with the zinc-blende structure and alkali halide crystals with the rock-salt structure.

Crystal AX		$\alpha_{1\mathrm{M}}^+$	$\alpha_{1\mathrm{M}}^-$	$\alpha_{2\mathrm{M}}^+$	$\alpha_{2\mathrm{M}}^-$
CuCl	(ZB)	1.14	3.75	2.58	7.38
LiCl	(RS)	0.03	3.92	0.01	7.92
NaCl	(RS)	0.16	4.37	0.08	9.57
KCl	(RS)	0.84	4.87	0.75	11.72
CuBr	(ZB)	1.13	4.94	2.48	11.02
LiBr	(RS)	0.03	5.26	0.01	12.21
NaBr	(RS)	0.16	5.76	0.08	14.32
KBr	(RS)	0.84	6.33	0.75	17.05
CuI	(ZB)	1.11	6.77	2.38	18.77
LiI	(RS)	0.03	7.44	0.01	21.82
NaI	(RS)	0.16	8.05	0.08	24.97
KI	(RS)	0.84	8.70	0.75	28.75

The size effect is evident for copper halide crystals. The values of the dipole polarizability and the quadrupole polarizability of the copper ion in copper halide crystals are almost independent of the crystalline environment. On the other hand, the values of the dipole polarizability and the quadrupole polarizability of the anion are dependent on the crystalline environment. Since the quadrupole polarizability is more sensitive to the electron density tail than the dipole polarizability, it is more strongly affected by the crystalline environment.

The Hartree interaction and the exchange interaction tend to expel the electrons from the regions with high electron density at the neighboring ions. The electron states are calculated under the approximation that all surrounding ions are replaced by point charges to address the effect of the overlap repulsive interaction. There is no overlap of the electron cloud of the central ion and those of the neighboring ions in the point charge approximation (PCA).

Table 3 shows the calculated results of the dipole polarizability $\alpha_{1\mathrm{M}}$ and the quadrupole polarizability $\alpha_{2\mathrm{M}}$ of ions in copper halide and alkali halide crystals with the PCA. As seen, the value of the dipole polarizability $\alpha_{1\mathrm{M}}(\mathrm{Cu}^+, \mathrm{CuCl})$ of the copper ion in CuCl crystal with the PCA is equal to the value $\alpha_1(\mathrm{Cu}^+, \mathrm{CuCl})$ without the PCA. This is consistent with the result that the values of the dipole polarizability $\alpha_1(\mathrm{Cu}^+, \mathrm{CuX})$ of the copper ion in CuX crystals are close to the value of the free copper ion. This feature is also seen for the values of the quadrupole polarizability. The electrons on the cation are tightly bound, and they are scarcely influenced by the crystalline environment. In contrast, the value of the dipole polarizability $\alpha_{1\mathrm{M}}(\mathrm{Cl}^-, \mathrm{CuCl})$ of the chlorine ion in CuCl crystal with the PCA is larger than the value $\alpha_1(\mathrm{Cl}^-, \mathrm{CuCl})$ without the PCA. Furthermore, the value $\alpha_{1\mathrm{M}}(\mathrm{Cl}^-, \mathrm{CuCl})$ in CuCl crystal is smaller than the value $\alpha_{1\mathrm{M}}(\mathrm{Cl}^-, \mathrm{KCl})$ in KCl crystal. Thus, the PCA predicts that the value $\alpha_{1\mathrm{M}}^{\mathrm{tot}}(\mathrm{CuCl}) = \alpha_{1\mathrm{M}}(\mathrm{Cu}^+, \mathrm{CuCl}) + \alpha_{1\mathrm{M}}(\mathrm{Cl}^-, \mathrm{CuCl})$ is smaller than the value $\alpha_{1\mathrm{M}}^{\mathrm{tot}}(\mathrm{KCl}) = \alpha_{1\mathrm{M}}(\mathrm{K}^+, \mathrm{KCl}) + \alpha_{1\mathrm{M}}(\mathrm{Cl}^-, \mathrm{KCl})$. This prediction contradicts with the experimental result that $\alpha_1^{\mathrm{exp}}(\mathrm{CuCl})$ is larger than $\alpha_1^{\mathrm{exp}}(\mathrm{KCl})$. On the other hand, without the PCA, the value $\alpha_1(\mathrm{Cl}^-, \mathrm{CuCl})$ is comparable to the value $\alpha_1(\mathrm{Cl}^-, \mathrm{KCl})$. A similar situation holds for the bromine ion and the iodine ion. This trend is also

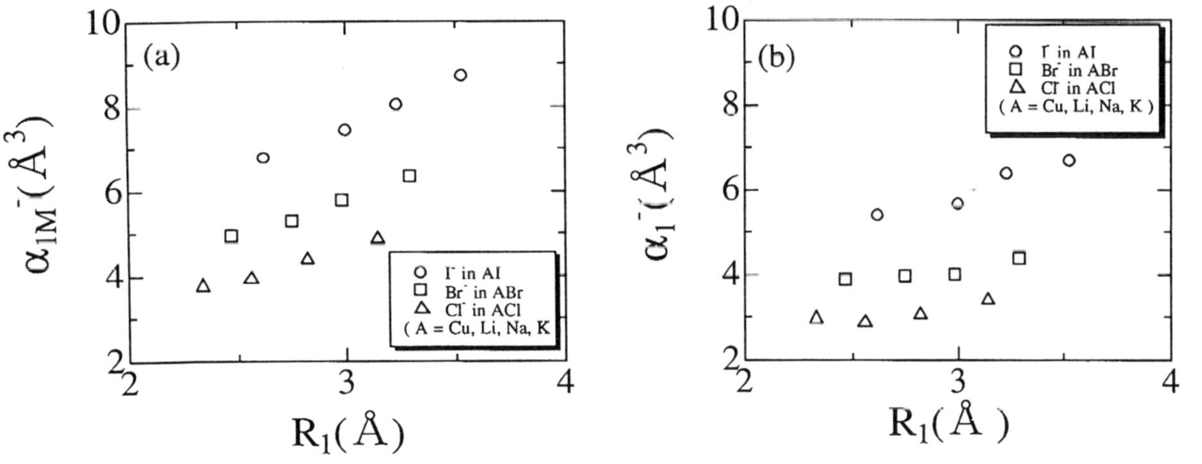

Figure 1. The dipole polarizabilities of anions in binary crystals. The horizontal axis is the nearest neighbor distance R_1. (a) in the lattice of point charges and (b) in the lattice of pseudopotentials.

found for the values of the quadrupole polarizability. These results show that the electron states of the anion are sensitive to the overlap repulsive interaction.

The SSM with the PCA showed[7] that the electrons on the anion are effectively confined in the spherically symmetric crystalline potential well, whose depth is given by the Madelung potential U_M and whose radius is determined by the first neighbor distance R_1. The Madelung potential $U_M(\mathrm{Cl}^-, \mathrm{CuCl}) = -10.064$ eV in CuCl crystal is deeper than that $U_M(\mathrm{Cl}^-, \mathrm{LiCl}) = -9.656$ eV in LiCl crystal. The nearest neighbor distance $R_1(\mathrm{CuCl}) = 2.336$ Å in CuCl crystal is shorter than that $R_1(\mathrm{LiCl}) = 2.560$ Å in LiCl crystal. Therefore, the PCA predicts that the electrons on the chlorine ion in CuCl crystal are more tightly bound than those in LiCl crystal.

In Fig. 1, the dipole polarizability of the anion X^- in the AX crystal is plotted against the nearest neighbor distance $R_1(\mathrm{AX})$: (a) in the lattice of point charges and (b) in the lattice of pseudopotentials. Since the value of the Madelung constant $M_{\mathrm{RS}} = 1.7476$ in the RS structure is similar to that $M_{\mathrm{ZB}} = 1.6381$ in the ZB structure, the electron states in the SSM with the PCA are sensitive to the nearest neighbor distance but insensitive to the coordination number. Figure 1(a) shows that the dipole polarizability $\alpha_{1M}^- = \alpha_{1M}(X^-, \mathrm{AX})$ of the anion X^- in the AX crystal with the PCA almost scales with the nearest neighbor distance $R_1(\mathrm{AX})$.[7,22]

The SSM showed[7] that with replacing the lattice of point charges by the lattice of pseudopotentials, the range of the confinement crystalline potential is modified. The overlapping between nearby ions makes the potential well of the electrons on the anion narrower. The anion behaves as if it is confined in an effective potential well with a radius smaller than R_1. As seen in Fig. 1(b), the value $\alpha_1(\mathrm{Cl}^-, \mathrm{CuCl})$ without the PCA is less reduced from the value $\alpha_{1M}(\mathrm{Cl}^-, \mathrm{CuCl})$ with the PCA. The ratio $\alpha_1(\mathrm{Cl}^-, \mathrm{CuCl})/\alpha_{1M}(\mathrm{Cl}^-, \mathrm{CuCl})$ of the values in CuCl crystal with the ZB structure is larger than those in alkali chloride crystals with the RS structure. This applies also to the bromine ion and the iodine ion.

In the region between the neighboring ions, the charge distributions overlap. However the electrons on the cation are tightly bound. They are almost unaffected by the overlap repulsive interaction. When an anion is surrounded by a cage of cations with tightly bound electron distributions, its diffuse electron distribution will be more compressed. The overlap repulsive interaction makes the electrons on the anion more tightly bound to the nucleus. In the ZB and RS structure crystals, the anions are surrounded by four and six nearest neighbor cations, respectively. In CuX crystals with the ZB structure, the electrons on the anion are more loosely bound. They exhibit larger dipole polarizability. This result is in accordance with the widely accepted view that the halogen ions are highly polarizable in copper halide crystals with the ZB structure.

References

1. P. Hohenberg and W. Kohn, *Phys. Rev.* **136**, B864 (1964).
2. W. Kohn and L. J. Sham, *Phys. Rev.* **140**, A1133 (1965).
3. R. M. Sternheimer, *Phys. Rev.* **96**, 951 (1954).
4. G. D. Mahan, *Phys. Rev.* A **22**, 1780 (1980).
5. M. J. Stott and E. Zaremba, *Phys. Rev.* A **21**, 12 (1980) [Errata A **22**, 2293 (1980)].
6. A. Zangwill and P. Sovan, *Phys. Rev.* A **21**, 1561 (1980).
7. G. D. Mahan, *Solid State Ionics* **1**, 29 (1980).
8. G. D. Mahan, *Phys. Rev.* B **34**, 4235 (1986) [Errata B **38**, 7841 (1988)].
9. S. Pettersson and K. R. Subbaswamy, *Phys. Rev.* B **42**, 5883 (1990).
10. Y. Michihiro and G. D. Mahan, *Phys. Rev.* B **56**, 12151 (1997).
11. G. D. Mahan, *J. Chem. Phys.* **76**, 763 (1982).
12. G. Senatore and K. R. Subbaswamy, *Phys. Rev.* A **34**, 3619 (1986).
13. O. Gunnarsson and B. I. Lundqvist, *Phys. Rev.* B **13**, 4274 (1976).
14. V. Heine and I. V. Abarenkov, *Philos. Mag.* **9**, 451 (1964); **12**, 529 (1965).
15. N. W. Ashcroft, *Phys. Lett.* **23**, 48 (1966).
16. R. P. Lowndes and D. H. Martin, *Proc. R. Soc. London, Ser.* B **308**, 473 (1969).
17. K.-H. Hellwege and A. M. Hellwege, *Landolt-Börnstein Zahlenwerte und Funktionen aus Physik, Chemie, Astromie, Geophysik und Technik*, II *Eigenshaften der Materie in Ihren Aggregatzuständen* (Springer-Verlage, Berlin, 1962), Vol. 8 Optische Konstanten [in German].
18. N. Wilson and R. M. Curtis, *J. Phys. Chem.* **74**, 187 (1970).
19. P. C. Schmidt, A. Weiss, and T. P. Das, *Phys. Rev.* B **19**, 5525 (1979).
20. M. J. Stott, E. Zaremba, and D. Zobin, *Can. J. Phys.* **60**, 210 (1982).
21. A. Dalgarno, *Adv. Phys.* **11**, 281 (1962).
22. P. W. Fowler and P. Tole, *Rev. Solid State Sci.* **5**, 149 (1991).

CRYSTAL STRUCTURE OF THE SUPERIONIC PHASE OF CuAgSe

TOMOTAKA SHIMOYAMA, MASAJI ARAI, TAKASHI SAKUMA
Department of Physics, Faculty of Science, Ibaraki University, Mito 310-8512, Japan
Fax: +81-29-228-8357, E-mail: sakuma@mx.ibaraki.ac.jp

The phase transition and the crystal structure of CuAgSe were investigated by differential thermal analysis and X-ray diffraction measurements above room temperature. Heat-flux differential scanning calorimetry (DSC) measurement of CuAgSe was performed from 300 K to 520 K at a scanning rate of 10 K/min. The phase transition temperature to the superionic phase and the enthalpy of transition are 473 K and 8.01 kJ/mol, respectively. The X-ray diffraction pattern of CuAgSe was measured with the powder sample at 523 K. Several Bragg lines and a large diffuse scattering were observed. The wide and strong peak of the diffuse scattering appears at $2\theta\sim35°$. The Bragg lines can be denoted by face-centered cubic with the lattice constant $a=6.112\pm0.001$ Å. Various structural models including ordered and disordered arrangement of copper and silver atoms were examined for the quantitative interpretation of the measured X-ray diffraction profile by the Rietveld analysis. In consequence, the space group of Fm3m with disordered arrangement of copper and silver ions was found to be most plausible. All copper and silver ions are statistically distributed on the 8(c) and 32(f) sites in the unit cell. The disordered arrangement of cations shows a wide and strong peak in the diffuse scattering. The shortest interatomic distance between cation sites is 1.556 Å. The path connecting two nearest neighboring sites of 8(c) and 32(f) could be suggested as an easiest channel of cation movement relevant to the fast ion diffusion in α-CuAgSe.

1. Introduction

There has been much interest in the structure and dynamic properties of superionic conductors, which show unusual high ionic conductivity in the high temperature phase. The high temperature phase of AgI, Ag_2Se and Ag_3SI indicating a bcc type anionic structure has been widely investigated. However, the quantitative information on the crystal structure of high temperature phases of superionic conductors such as Ag_2Te and CuI having an fcc type structure has not yet been well elucidated.

The high temperature phase (superionic phase) of CuAgSe shows a mixed conductor with high ionic conductivity of ~1 S/cm [1]. The structure of the high temperature phase of CuAgSe was first investigated by Frueh et al. using the X-ray photographic method in 1956. The X-ray diffraction pattern showed four Debye lines [2]. The relative intensities and the *d*-spacing of these lines were discussed. However, the detailed atomic structure of the high temperature phase has not been determined.

In the present study, specific heat and X-ray diffraction measurements of CuAgSe have been carried out. The phase transition temperature and the enthalpy of transition from the room temperature phase to the high temperature phase were obtained. The structure of the superionic phase of CuAgSe was determined by Rietveld analysis.

2. Experimental

Stoichiometric quantities of copper, silver and selenium were weighted and enclosed in a pyrex tube in vacuum and kept at 773 K for 1 d. After solidification the product was annealed at 623 K for 2 d and 373 K for 1 d. The compound obtained was crushed to a fine powder. The sample was enclosed in a pyrex tube again and was kept at 773 K for 1d. It was then gradually cooled to room temperature.

Heat-flux differential scanning calorimetry (DSC) measurement of CuAgSe was performed from 300 K to 520 K at a scanning rate of 10 K/min using a Shimadzu DSC-50 thermal analyzer. The measurement was made in a flowing N_2 atmosphere. A powder sample of 60.2 mg was encapsulated in an aluminum pan. The same calorimetry measurements of Cu_2Se and Ag_2Se were also performed.

The X-ray diffraction pattern of CuAgSe was measured with the powder sample at 523 K. X-ray diffraction data were collected using Cu-Kα radiation monochromatized with a PG 002 reflection for 10 sec per step at 0.05° intervals over the 2θ range of 10° to 90° in a step-scan mode.

3. Results and Discussion

Fig. 1 shows the DSC traces for CuAgSe in the temperature range between 420 K and 520 K. An endothermic peak T_{c1} is observed at 480 K on heating and an exothermic peak T_{c2} is observed at 467 K on cooling. The transition temperature is 473 K. This value coincides with that of the electrical conductivity measurement [1]. The enthalpy of the phase transition in CuAgSe was estimated to be 8.01 kJ/mol from a computation of the peak area. The enthalpy of transition in CuAgSe is greater than that in Cu_2Se at 395 K (3.39 kJ/mol) and in Ag_2Se at 400 K (6.19 kJ/mol).

The X-ray diffraction pattern of the high temperature phase of CuAgSe measured by Cu-Kα radiation at 523 K is given by the mark (+) in Fig. 2. Several Bragg lines and a large diffuse scattering contribution were observed readily interpreted as a face-centered cubic structure with the lattice constant $a=6.112\pm0.001$ Å. Various structural models including an ordered and disordered arrangement of copper and silver atoms were examined for the quantitative interpretation of the measured X-ray diffraction profile by the Rietveld analysis [3]. As a consequence, the space group of Fm3m with a disordered arrangement of copper and silver ions was found to be most plausible. The number of atoms and the atomic positions for the analysis are as follows;

(000, 0 1/2 1/2, 1/2 0 1/2, 1/2 1/2 0) +

4 Se in 4(a) 000,

n (Cu+Ag)/2 in 8(c) 1/4 1/4 1/4; 3/4 3/4 3/4,

(8-n) (Cu+Ag)/2 in 32(f) xxx; $x\bar{x}\bar{x}$; $\bar{x}x\bar{x}$; $\bar{x}\bar{x}x$;

$\bar{x}\bar{x}\bar{x}$; $\bar{x}xx$; $x\bar{x}x$; $xx\bar{x}$.

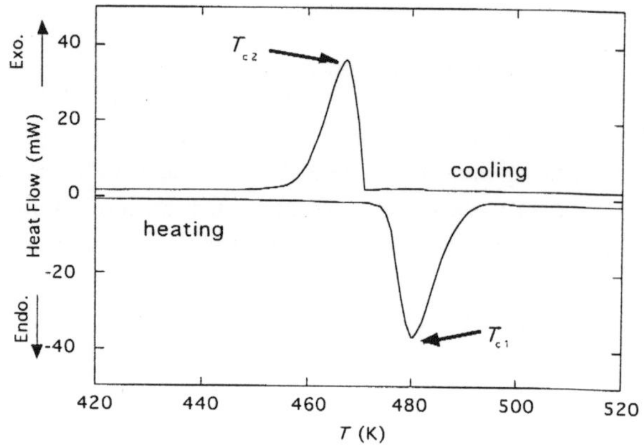

Figure 1 Differential scanning calorimetry (DSC) heating and cooling traces for CuAgSe.

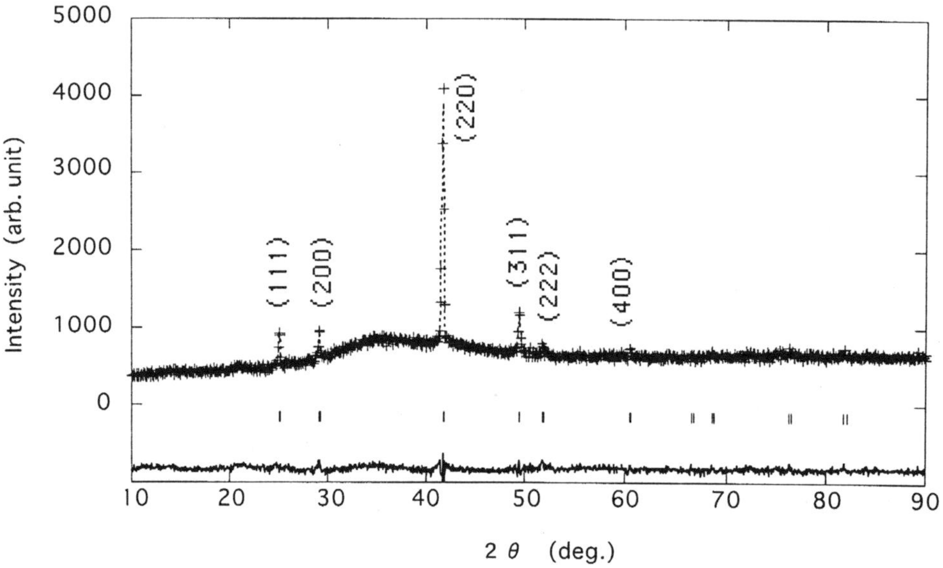

Figure 2. Rietveld refinements pattern for CuAgSe at 523 K by Cu-Kα X-ray diffraction measurement. The broken lines are calculated intensities, the crosses (+) are observed intensities and the lower short vertical lines mark the positions of possible Bragg diffraction of CuAgSe. Solid lines are the difference between calculated and observed intensities.

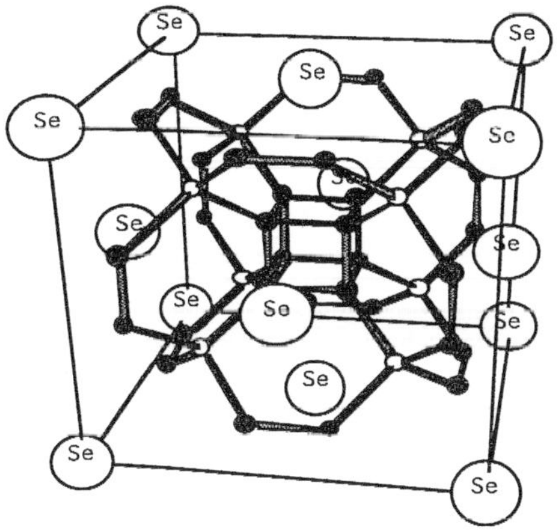

Figure 3. Atomic arrangements in the unit cell of CuAgSe at 523 K.

The atomic positions in the unit cell are illustrated in Fig. 3. The least squares refinement was carried out for five structural parameters, that is, the number of cations distributed on the 8(c) sites (n), the position of 32(f) sites for cations (x) and three thermal parameters for cations and anions (B_{Se}, $B_{cat(c)}$, $B_{cat(f)}$). The values obtained from the Rietveld analysis are as follows; n=5.4, x=0.397, B_{Se}=3.0 Å^2, $B_{cat\ 8(c)}$=1.8 Å^2 and $B_{cat\ 32(f)}$=6.6 Å^2. Thecalculated intensities with these parameters are shown by broken lines in Fig. 2. The reliability factor (R_I) of the present calculation is 5.38 %. The interatomic distances of the CuAgSe calculated from the atomic positions and the lattice constant are given in Table 1.

The present structural analysis indicates that all copper and silver ions are statistically distributed on the 8(c) and 32(f) sites in the unit cell. The shortest interatomic distance between these sites is 1.56 Å. The distance between cation and cation sites plays an important role in atomic diffusion in α–AgI, α–Ag_3SI and α–Cu_2Se [4,5]. The path connecting the two nearest neighboring sites of 8(c) and 32(f) could be suggested as the easiest channel of cation movement relevant to fast ion diffusion in α-CuAgSe.

The high temperature phase of α-Ag_2Se has a bcc lattice and silver atoms are randomly distributed over many atomic sites [6]. On the other hand the high temperature phase Cu_2Se has an fcc lattice of the space group denoted by Fm3m and all copper atoms are distributed over two 32(f) sites [5]. The structure obtained for the high temperature phase of CuAgSe is similar to that of Cu_2Se. However, one of the 32(f) sites

Table 1 Interatomic distances r (Å) and the number of nearest neighboring sites Z in CuAgSe at 523 K.

	r	Z
8(c)-32(f)	1.556	4
-32(f)	2.504	12
-Se	2.647	4
-8(c)	3.056	6
-32(f)	3.181	12
-32(f)	3.737	4
32(f)–32(f)	1.259	3
-8(c)	1.556	1
-32(f)	1.781	3
-32(f)	2.181	1
-8(c)	2.504	3
-32(f)	2.542	3
-Se	2.585	3
-32(f)	2.836	3
-8(c)	3.181	3
-32(f)	3.668	12
-8(c)	3.737	1
-Se	3.792	3
-32(f)	3.876	12
Se-32(f)	2.585	24
-8(c)	2.647	8
-32(f)	3.792	24

for cations in Cu_2Se was replaced by an 8(c) site in CuAgSe. If we used the two 32(f) sites for cations in CuAgSe, we could not explain the intensities of (200) and (222) Bragg lines.

The observed X-ray diffuse scattering intensities at 523 K are shown in Fig. 4 obtained by changing the scales in Fig. 2. In addition to the strong peak at 2θ~35° a very-weak second peak of the oscillatory diffuse scattering could be observed at 2θ~75°. The relative intensity of these peaks in CuAgSe obtained with Cu-Kα radiation is similar to that in α-AgI [7]. The first strong peak in α-AgI corresponded to the static disordered arrangement of silver atoms and a correlation of the thermal vibration among atoms. The main contribution of the weak peak in α-AgI corresponded to a correlation of thermal vibration among atoms.

Two contributions, a thermal vibration and a disordered arrangement of Cu and Ag atoms, compose the diffuse scattering in the high temperature phase of CuAgSe. As the intensity from the atomic correlation

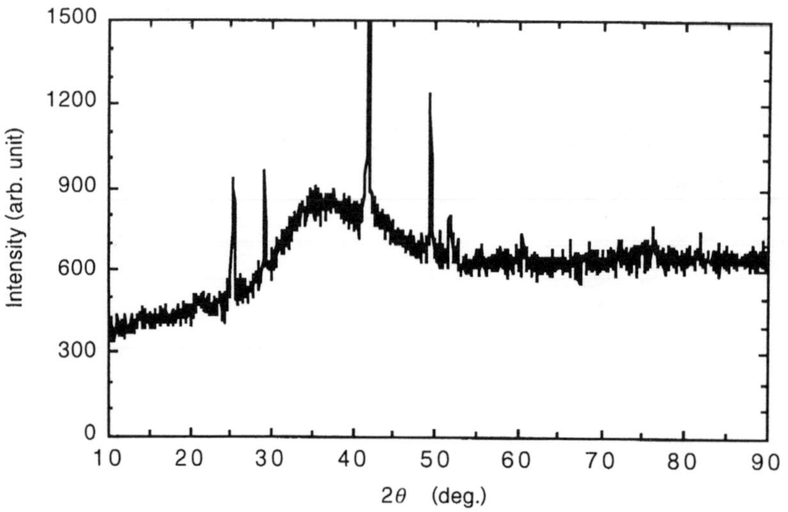

Figure 4. Observed intensities of CuAgSe at 523 K by changing the scales in Fig. 2.

of the thermal vibration is proportional to $\sin(Qr)$, where the distance r between cation and Se is ~2.65 Å in Table 1, the maximum values of the diffuse scattering intensity occur at Q (=$4\pi \sin\theta/\lambda$) ~3.0 and 5.4 $Å^{-1}$ and minimum values at Q~1.8 and 4.2 $Å^{-1}$. Therefore we could expect the peaks in the diffuse scattering at 2θ~42° and 80° from the contribution of thermal vibration. From the disordered arrangements of cations a strong peak appears at $\sin(Qr)$ ~1, where the distance r between 8(c) and 8(c) sites is 3.06 Å in Table 1. The wide and strong peak at Q~2.0 (2θ~36°) would correspond to the contribution from the disordered arrangements of cations. In addition to the cation-vacant disordered arrangements in α−AgI, the diffuse scattering from Cu-Ag disordered arrangements also contributes in CuAgSe. For the quantitative calculation of diffuse scattering in α-AgI the theoretical treatment of Ag and vacancies similar to a binary alloy was needed. In the case of CuAgSe a more complicated theoretical treatment utilizing Cu, Ag and vacancies similar to a ternary alloy would be necessary.

References

1. S. Miyatani, J. Phys. Soc. Jpn., **34**, 423 (1973).
2. A. J. Frueh Jr., G. K. Czamanske and Ch. Knight, Zeit. Kristall., **108**, 389 (1957).
3. F. Izumi in *"The Rietveld Method"*, edited by R. A. Young, (Oxford University Press, Oxford, 1993) Chap. 13.
4. T. Sakuma, J. Phys. Soc. Jpn., **62**, 1080 (1993).
5. T. Sakuma, K. Sugiyama, E. Matsubara and Y. Waseda, Materials Trans., JIM, **30**, 365 (1989).
6. T. Sakuma, K. Iida, K. Honma and H. Okazaki, J. Phys. Soc. Jpn., **43**, 538 (1977).
7. T. Sakuma, J. Phys. Soc. Jpn., **61**, 4041 (1992).

APPLICATION OF ION-CODUCTING MICROELECTRODES FOR CATION DOPING INTO OXIDE MATERIALS

SHUICHI YAMASHITA, KAI KAMADA, and YASUMICHI MATSUMOTO

Department of Applied Chemistry and Biochemistry, Faculty of Engineering, Kumamoto University, 2-39-1 Kurokami, Kumamoto 860-8555, Japan

Ion-conducting microelectrodes were applied for solid-state electrochemical cation doping, where M-β"-Al_2O_3 was used mainly as a cation source. The fundamental doping system consists of anode / M-β"-Al_2O_3 microelectrode / doping target / Na-β"-Al_2O_3 / cathode. M^{n+} in the M-β"-Al_2O_3 can be doped into the doping target under applied electric field, and then the cation (Q^{m+}) from the doping target migrates to the cathode side Na-β"-Al_2O_3 for maintenance of electrical neutrality in the target. That is, M^{n+} is exchanged for Q^{m+} in the doping target during the electrolysis. Therefore, this doping system corresponds to the "electrosubstitution" mechanism. In the case where the β"-Al_2O_3 microelectrode was fixed onto the surface of the doping target, we have achieved pinpoint doping on a 10^2 μm scale because of the small contact radius (10 μm) of the solid-solid interface, and the hemispherical dopant distribution centered on the microcontact between β"-Al_2O_3 and the target was obtained. The distribution state of the dopant may be dominated by the potential distribution around the point contact electrode. The distribution diameter of the dopant on the anodic surface depended on the current density and/or the doping time, which reflect the doping amount. Furthermore, scanning the β"-Al_2O_3 microelectrode resulted in the fine-patterned metal distribution in the doping target. In other words, electrochemical design of metal distribution can be carried out using an ion-conducting microelectrode.

1. Introduction

Ion conducting microelectrodes have been widely used in the research field of *solid state ionics*. Microelectrodes play a significant role in the measurements of local conductivities in ceramic materials [1], transference numbers in mixed conductors [2,3], and in the determination of redox potentials in metal oxides [4,5]. The use of a microelectrode allows better control of electrochemical parameters and contact areas at the interface compared to a conventional planar electrode, which often shows time-dependent changes of the contact area because of its inhomogeneous interface [6,7].

Recently, we have developed a solid state electrochemical doping method using a solid electrolyte, which is called the solid oxide electrochemical doping (SOED) method [8]. The fundamental doping system in this method consists of an anode / M-β"-Al_2O_3 (M = metal cation) / doping target / M'-β"-Al_2O_3 or YSZ / cathode sandwich electrolysis system. The SOED method has several advantages, such as selective doping into the grain boundary of ceramics [9], controlling the amount of the dopant by the electric charge [10], etc., compared with the usual method for doping. Moreover, materials graded in the dopants can easily be made with the SOED methods when the metal cation and oxide ion are simultaneously injected into the materials [11]. These techniques can therefore be important for the development of electroceramics used in microelectronics.

The SOED method will be very useful for pinpoint doping into the desired position of samples if the contact area between β"-Al_2O_3 and the doping target remains small during the electrolysis. In this contribution, pinpoint cation doping into a selected area of the doping target has been carried out using a M-β"-Al_2O_3 microelectrode to develop a new field of applications of the SOED method. To our knowledge, ion-conducting microelectrodes have never been applied to a doping study. This paper summarizes the some experimental results of pinpoint doping and expanded technique.

2. Pinpoint Doping using β"-Al_2O_3 microelectrode [12, 13]

Fig. 1 illustrates the basic cation migration mechanism in the present system for pinpoint doping. Ag is electrochemically oxidized to Ag^+ at an Ag (anode) / M-β"-Al_2O_3 interface, and then M^{n+} is injected into the target material. Q^{m+} in the target is released to the cathode side Na-β"-Al_2O_3 for maintenance of electrical neutrality in the target material. Na deposited at the Na-β"-Al_2O_3 / Ag (cathode) interface then immediately reacts with O_2 and CO_2 in the air to form Na_2CO_3. Consequently, M^{n+} is substituted for Q^{m+} under an electric field. Therefore, this doping scheme corresponds to an "electrosubstitution" mechanism. On the other

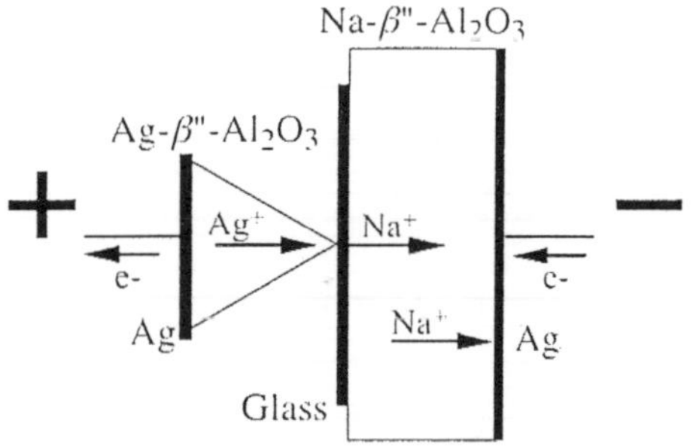

Figure 1. Model of pinpoint Ag doping using an Ag-β"-Al_2O_3 microelectrode.

hand, oxide ion conducting YSZ is also used at the cathodic side instead of Na-β"-Al_2O_3. In this case, M^{n+} and O^{2-} are simultaneously injected into the doping target under the electric field. Therefore, this doping is called the "electro-bi-injection" mechanism.

The pinpoint doping strongly depended on the conductive properties of the doping target and the valence of the dopant cation. Therefore, cation doping into the alkali silicate glass occurs only using the former system (electrosubstitution) because the glass shows pure cationic conduction. In contrast, a cation can be doped into superconducting $Bi_2Sr_2CaCu_2O_y$ ceramics using only the latter system (electro-bi-injection) because the electrosubstitution of a cation is difficult in electron-conducting ceramics. In this case, the migration of the metal cations and the oxide ions primarily proceeds through defects (pore surfaces, grain boundaries, etc.). The monovalent cation can be easily injected into the doping target, while the divalent cation cannot be doped.

Figure 2 shows the elemental distribution maps of the anodic surface (a) and the cross section (b) of the Ag-doped glass after applying 10 μA for 1 h at 673 K. Judging from these maps, pinpoint doping using the fixed β"-Al_2O_3 microelectrode caused a hemispherical dopant distribution with its center at the M-β"-Al_2O_3 / doping target microcontact. In general, metal distribution in the target was domin9ated by the applied electric charge and the potential distribution around the microcontact between M-β"-Al_2O_3 and the target. The equipotential surface generated in the target will assume a hemispherical configuration centered around the microcontact. Thus, M^{n+} is observed to diffuse radially from the M-β"-Al_2O_3 / doping target microcontact during doping [14].

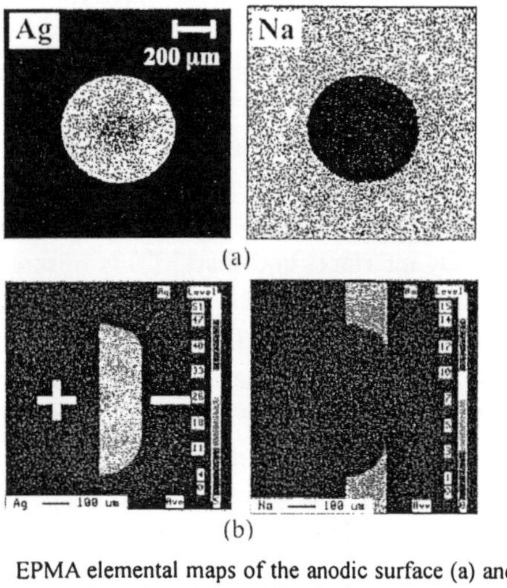

Figure 2. EPMA elemental maps of the anodic surface (a) and the cross section (b) of Ag-doped glass. The doping was carried out at 10 μA for 1 h at 673 K.

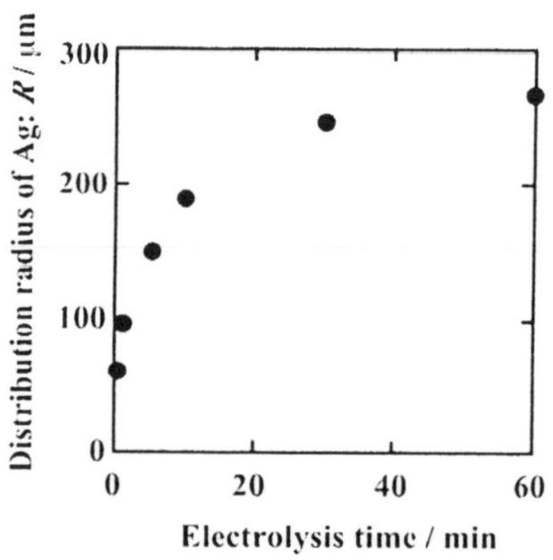

Figure 3. Time dependence of distribution radius of Ag in glass at a constant current of 10 μA and 673 K.

Figure 3 shows the doping time dependence of the silver distribution radius of Ag-doped glass at a constant current of 10 μA at 673 K, estimated based on the EPMA elemental distribution maps of the anodic surface of the doped glass. The radius of the hemisphere was strongly dependent on the applied electric charge, which reflects the electrolysis time under galvanostatic conditions, rather than the doping temperature.

3. Quantitative assessment of pinpoint doping [15]

That the radius of dopant distribution can be controlled by the electric charge suggests the doping amount may also be controllable. Therefore, we evaluated the current efficiencies of silver doping by the Ag-β"-Al_2O_3 microelectrode. Figure 4 shows the typical change observed in applied voltage (a) and the amount of Ag doping into sodium borate glass (b) as a function of doping time at 1 μA and 673 K. The solid line indicates the theoretical amount of Ag doping calculated based on Faraday's law. The applied voltage decreased sharply during the initial stage of electrolysis (< 30 min) and remained roughly constant after 30 min. Others have reported that the voltage response in the ion-conducting microelectrode technique depends on ionic conductivity only in the small region near the microcontact [16-18]. Therefore, the voltage change shows that the ionic

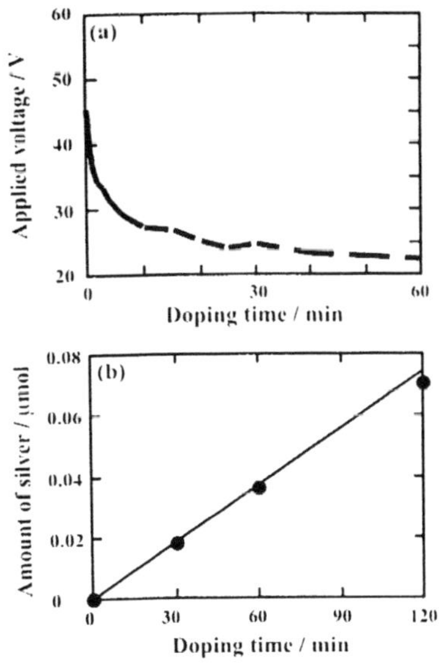

Figure 4. Doping time dependence of the applied voltage (a) and the amount of Ag dopant (b) during Ag doping into alkali glass at a constant current of 1 μA at 673 K.

conductivity of the glass around the microcontact increased with the substitution of Ag^+ for Na^+ in the borate glass during the initial stage of electrolysis. This is mainly due to the higher ionic conductivity of Ag^+ in the amorphous glass network as compared to Na^+ [19]. According to Fig. 4(b), the Ag amount increases linearly with doping time indicating that electrolysis time is an important factor for the control of dopant amount under galvanostatic conditions. The current efficiencies for Ag doping, calculated from the electric charge and the amount of Ag using Faraday's law, were above 90% over wide range of constant current. These results mean that the amount of Ag doping can be easily controlled on a µmol scale by the electric charge during electrolysis. The high Faraday efficiencies in Ag-doping are due to the pure alkali cation conduction in the alkali borate glass because the alkali metal cations are easily substituted by Ag^+ during electrolysis.

4. Electrochemical design of metal distribution [19, 20]

Our method would conceivably enable micropatterning of dopants in glass if the β"-Al_2O_3 microelectrode were to be moved along the glass surface during doping. The technique could be used to prepare optical devices, such as photowaveguides and microlenses, since pinpoint or patterned doping induces selective refractive index profiles in glass. Therefore, we have devised an electrochemical method for achieving metal cation distribution in the surface and interior of alkali silicate glass by manipulating an Ag-β"-Al_2O_3 microelectrode under an electric field.

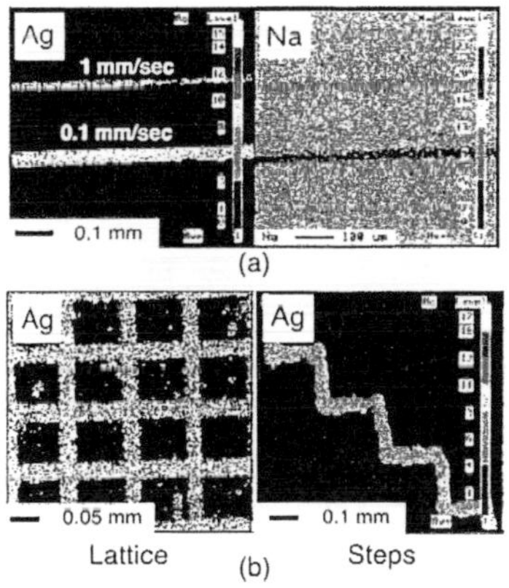

Figure 5. (a) EPMA elemental distribution maps of the anodic surface of Ag-doped glass with scanning the Ag-β"-Al_2O_3 microelectrode in a single direction at 403 K (scan rate: 1 mm/s (upper line) and 0.1 mm/s (lower line)). (b) More complex Ag pattern sketched in the glass surface.

Electrochemical design was performed at a constant current doping similar to pinpoint doping, where the β"-Al_2O_3 microelectrode was scanned using a PC-operated, automated XYZ microstage. EPMA elemental distribution maps of the anodic surface of the Ag-doped glass with scanning of the Ag-β"-Al_2O_3 microelectrode in a single direction at 403 K are shown in Fig. 5(a). The scan rates of the microelectrode were 1 mm/s (upper line) and 0.1 mm/s (lower line). These results indicate that Ag was dispersed in a line along the path of the microcontact, and that the line width was easily controlled by the electric charge per unit area, i.e., scan rate and/or current. Since the Ag-β"-Al_2O_3 microelectrode is attached to the automated XYZ stage, we can draw various silver patterns in the glass surface, not only simple structures (points and lines) but also more complicated forms (Fig. 5(b)). The lowest patterned Ag line width achieved to date is about 10 µm. However, the use of a microelectrode with a smaller contact radius and/or lower constant current (applied

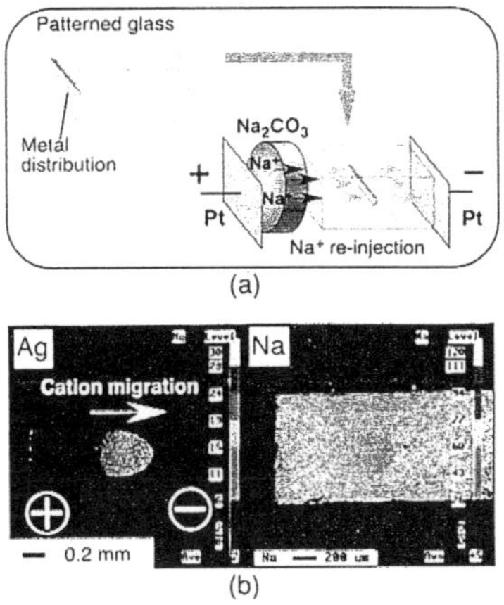

Figure 6. (a) Schema of a method for encapsulating the Ag-distribution inside the glass. (b) Ag mapping after Na^+ re-injection at 1 mA and 673 K for 15 h, following pinpoint Ag doping at 1 μA and 573 K for 60 min.

voltage) may enable the fabrication of finer and/or more complex patterns at the nanometer scale.

Expanding the present technique, we can control the glass composition not only in the surface region, but also in the bulk glass. In other words, the electrochemical design of glass composition is also possible in three dimensions. More specifically, following dopant patterning in the glass surface, Na^+ generated by Na^+-conducting planar anode is re-injected into the doped glass in a direction that is parallel to the anodic surface as schematized in Fig. 6(a). EPMA elemental maps of the cross section of the glass containing the Ag distribution are shown in Fig. 6(b). In this case, first, Ag^+ was doped using a fixed Ag-β"-Al_2O_3 microelectrode under 1 μA, for 60 min at 573 K. This was followed by Na^+ doping at 1 mA for 15 h at 673 K. Thanks to this two-step doping process, the Ag distribution is encapsulated within the glass sample by the Na^+ doping, during which it retains its hemispherical shape. It was found that the Ag concentration increased toward the direction of ion migration. Again, this is due to the larger ionic conductivity of Ag^+ in the glass compared with that of Na^+. As expected, the penetration depth of the metal distribution depended on the electric charge of Na^+ doping in the second step. It was also confirmed that the line patterned Ag dispersion could exist only within the glass. Qiu et al. have substantiated the possibility of selectively inducing a change of valence state of metal ions on a micrometer scale inside a glass sample by using a focused pulsed laser [21]. By contrast, our electrochemical technique can cause a space-selective change of chemical composition inside the glass.

5. Conclusions

It should be noted that the pinpoint doping into the selected area of the doping target on a 10^2 μm scale was accomplished using an ion-conducting microelectrode. The efficiency of pinpoint doping was determined by the electrical properties of the doping target and/or the valence of the dopant. Thus, submicron ~ nanoscale doping may be possible if the contact radius of microelectrode and/or the electric charge are small. For Ag doping into alkali glass, the measured doping amount almost exactly matched the theoretical value calculated from Farady's law. Moreover, we have investigated the possibility of the electrochemical design of metal distribution in alkali silicate glass using a β"-Al_2O_3 microelectrode. Various patterns of metal distribution were constructed near the surface or in the bulk. The size and shape of these patterns could be easily controlled by adjusting electrolysis conditions.

References

1. J. Fleig, S. Rodewald and J. Maier, *Solid State Ionics* **136-137,** 905 (2000).
2. H.-D. Wiemhöfer, *Ber. Bunsenges. Phys. Chem.* **97,** 461 (1993).
3. W. Zipprich, S. Waschilewski, F. Rocholl and H.-D. Wiemhöfer, *Solid State Ionics* **101-103,** 1015 (1997).
4. G. Fafilek, *Solid State Ionics* **113-115,** 623 (1998).
5. G. Fafilek and S. Harasek, *Solid State Ionics* **119,** 91 (1999).
6. H. Rickert and H.-D. Wiemhöfer, *Ber. Bunsenges. Phys. Chem.* **87,** 236 (1983).
7. S. Lübke and H.-D. Wiemhöfer, *Solid State Ionics* **117,** 229 (1999).
8. Y. Matsumoto, *Solid State Ionics* **100,** 165 (1997).
9. Y. Matsumoto, M. Koinuma, H. Yamamoto and T. Nishimori, *Solid State Ionics* **95,** 309 (1997)s.
10. Y. Matsumoto, T. Nishimori, H. Yamamoto, K. Nishimura, K. Kamada and A. Ogata, *Solid State Ionics* **107,** 41 (1998).
11. K. Kamada and Y. Matsumoto, *J. Solid State Chem.* **146,** 406 (1999).
12. K. Kamada, S. Udo and Y. Matsumoto, *Electrochem. Solid-State Lett.* **5,** J1 (2002).
13. K. Kamada, S. Udo, S. Yamashita and Y. Matsumoto, *Solid State Ionics* **146,** 387 (2002).
14. J. Fleig and J. Maier, *Electrochim. Acta* **41,** 1003 (1996).
15. K. Kamada, S. Udo, S. Yamashita, Y. Tsutsumi and Y. Matsumoto, *Solid State Ionics* **160,** 389 (2003).
16. H.-D. Wiemhöfer, *Ber. Bunsenges. Phys. Chem.* **97,** 461 (1993).
17. W. Zipprich, S. Waschilewski, F. Rocholl and H.-D. Wiemhöfer, *Solid State Ionics* **101-103,** 1015 (1997).
18. J. Fleig, S. Rodewald and J. Maier, *Solid State Ionics* **136-137,** 905 (2000).
19. K. Kamada, S. Yamashita and Y. Matsumoto, *J. Mater. Chem.* **13,** 1265 (2003).
20. K. Kamada, S. Yamashita and Y. Matsumoto, *J. Electrochem.* **151,** J33 (2004).
21 J. Qiu, C. Zhu, T. Nakaya, J. Si, K. Kojima, F. Ogura and K. Hirao, *Appl. Phys. Lett.* **79,** 3567 (2001).

RELIABILITY AND LIMITATIONS OF DIELECTRIC CONTINUUM MODEL FOR IONIC MOBILITY IN SUBCRITICAL LIQUID METHANOL AND WATER

KAZUYASU IBUKI, TAKA-AKI HOSHINA, and MASAKATSU UENO

Department of Molecular Science and Technology, Faculty of Engineering, Doshisha University, Kyo-Tanabe, Kyoto, 610-0321, Japan

E-Mail: kibuki@mail.doshisha.ac.jp

The reliability of the Hubbard–Onsager (HO) dielectric friction theory for ionic mobility in liquid solutions based on the sphere-in-continuum model has been examined by comparing experimental results for the potassium ion with theoretical ones in subcritical methanol and sub- and supercritical water. In methanol, the density dependence of the ionic mobility is well explained by the HO theory at densities higher than $1.8\rho_c$, where ρ_c is the critical density of the solvents, though qualitative difference between the theory and the experiments is observed at densities below $1.8\rho_c$. In water, on the other hand, the HO theory cannot explain the ionic mobility at high liquid densities, while the theory agrees well with the experiments at lower densities down to $1.4\rho_c$. At densities below $1.4\rho_c$, however, the theory predicts the density dependence of the ionic mobility in the opposite direction to the experimental results.

Keywords: Ionic mobility, Dielectric friction, Viscous friction, Continuum model, Subcritical fluids.

1 Introduction

The ionic transport processes in medium-density fluids are of great interest from theoretical and experimental points of view. At high liquid densities, the mobilities of simple ions has been measured extensively in aqueous and non-aqueous solutions, and the experimental results are well explained in terms of the sphere-in-continuum model based on the macroscopic hydrodynamics. lIn the low density limit, the molecular transport properties can be interpreted by the binary-collision theory. The reliability of the continuum and the binary-collision theories for ions has not been known in the intermediate density region. In this work, we discuss the reliability and limitations of the continuum theory for the ionic mobility in the medium-density region by comparing experimental results with the Hubbard–Onsager (HO) dielectric friction theory.[1,2]

The medium-density region can be studied by using sub- and supercritical fluids. Here we study the mobilities of the potassium ion, as a typical example, in subcritical liquid methanol and water along the liquid-vapor coexistence curves up to the critical temperatures,[3,4] and in supercritical water at 400 and 600°C.[4] The critical parameters for methanol[5] and water[6] are listed in Table 1.

Table 1. Critical parameters for methanol and water.

		methanol	water
Temperature	T_c/K	239.5	647
Density	ρ_c/ g cm^{-3}	0.276	0.315
Pressure	P_c/MPa	8.10	22.1

2 Hubbard–Onsager Theory

In an ionic solution, the polarization field around a fixed ion has a spherical symmetry. When the ion moves, the polarization becomes asymmetric because the dielectric relaxation takes some duration. The frictional force exerted on the ion by the interaction between the ionic charge and the asymmetric polarization is called the dielectric friction.

Hubbard and Onsager[1,2] (HO) calculated the friction on a moving ion taking account of the coupling between the dielectric and viscous frictions within the level of the sphere-in-continuum model. In the HO theory, the ion is a rigid non-polarizable sphere with a radius R and a charge e at its center. The solvent is a uniform continuum with a viscosity η, a static and high-frequency dielectric constants ε_S and ε_∞, and a dielectric relaxation time τ_d; a Debye-type dielectric relaxation is assumed. The slow steady flow of solvent around the ion is calculated by solving the following electrohydrodynamic equation in which the viscous and electric stress tensors are taken into account simultaneously:

$$\eta\nabla^2\mathbf{v} = \nabla p - \frac{1}{2}\left[\mathbf{E}_\circ \times (\nabla \times \mathbf{P}^*) + \mathbf{E}_\circ(\nabla \cdot \mathbf{P}^*)\right] \quad (1)$$

$$\frac{\mathbf{P}^*}{\tau_D} = (\mathbf{v}\cdot\nabla)\mathbf{P}_D + \frac{1}{2}\mathbf{P}_D \times (\nabla \times \mathbf{v}) \quad (2)$$

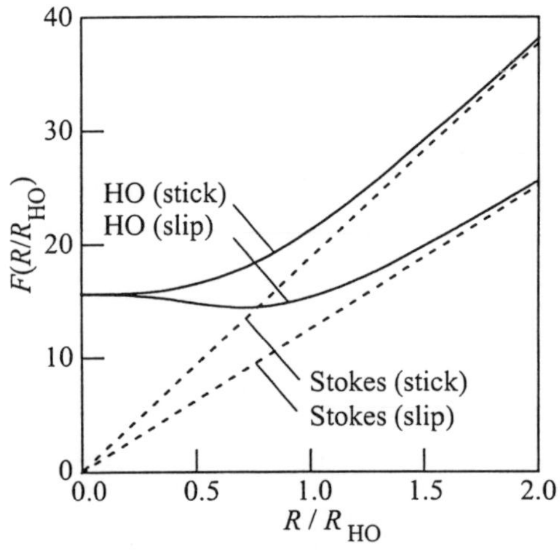

Figure 1. Ion-size dependence of the dimensionless friction coefficient $F(R/R_{HO})$ by the HO and the Stokes theory

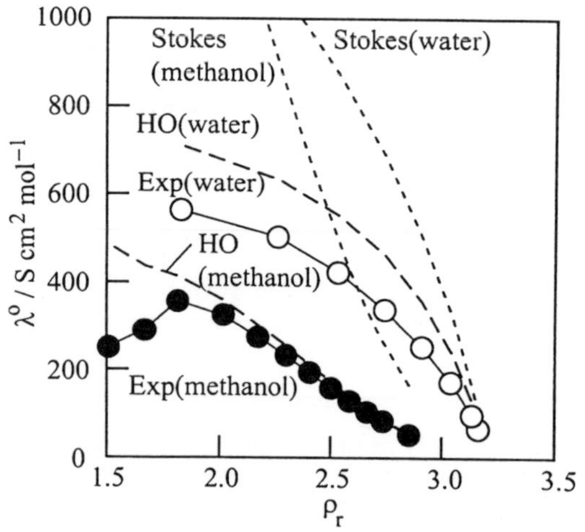

Figure 2. Density dependence of limiting molar conductivity for the K^+ ion in methanol and water along the liquid–vapor coexistence curve.

Here $\mathbf{v}$ is the velocity field, p the pressure field, $\mathbf{E}_o$ the equilibrium electric field at $\mathbf{v} = 0$, $\mathbf{P}^*$ the polarization deficiency, and $\mathbf{P}_D$ the orientational (Debye) polarization. $\mathbf{P}^*$ is the driving force of the dielectric relaxation and is caused by the solvent motion in the present problem.

The frictional force exerted on the ion is calculated from the velocity field of the solvent. The friction coefficient ζ_{HO} is given by

$$\zeta_{HO} = \eta R_{HO} F(R/R_{HO}) \tag{3}$$

$$R_{HO} = \left[\frac{e^2(\varepsilon_S - \varepsilon_\infty)\tau_d}{64\pi\eta\varepsilon_o\varepsilon_S^2}\right]^{1/4} \tag{4}$$

Here R_{HO} is the solvent parameter called the HO radius and $F(R/R_{HO})$ is the dimensionless friction coefficient shown in Fig. 1. $F(R/R_{HO})$ depends on the boundary condition on the ionic surface; the solvent moves on the ionic surface frictionlessly in the slip condition, while the solvent does not move on the surface in the stick condition. For monatomic ions in solutions, the slip boundary condition has usually been employed as in the present work.

In the absence of the electric field by the ionic charge, Eq. 1 is reduced to the linearlized Navier–Stokes equation for a steady flow, and the HO friction coefficient reduces to the well-known Stokes friction coefficient ζ_S. For the slip condition, ζ_S is given by

$$\zeta_S = 4\pi\eta R \tag{5}$$

As shown in Fig. 1, the HO friction coefficient is always larger than the Stokes one due to the dielectric friction effect, and the former asymptotically approaches the latter in the limits of large ionic radius or small HO radius. In the limit of small ionic radius, the HO theory predicts a finite friction.

The friction coefficient ζ is related to the limiting molar conductivity λ° of a monovalent ion as follows:

$$\zeta = \frac{|eF|}{\lambda^\circ} \tag{6}$$

where F is the Faraday constant.

3 Results and Discussion

3.1 Subcritical Methanol

First we examine the validity of the HO theory for the K^+ ion in subcritical liquid methanol near the liquid–vapor coexistence curve. The temperature range studied here is $0.58 \le T/T_c \le 1.00$ (25 – 240°C), and the corresponding density range is $2.85 \ge \rho/\rho_c \ge 1.51$ (0.786 – 0.415 g cm^{-3}). The limiting molar conductivities λ° for the K^+ ion are taken from Hoshina *et al.*[3] The viscosity,[7] the dielectric constant,[8] and the dielectric relaxation properties[9] are taken from the literature.

In Fig. 2, λ° for the K^+ ion in methanol are plotted against the reduced density $\rho_r(= \rho/\rho_c)$ together with the predictions of the HO and Stokes theories. The viscosity and the HO radius used in

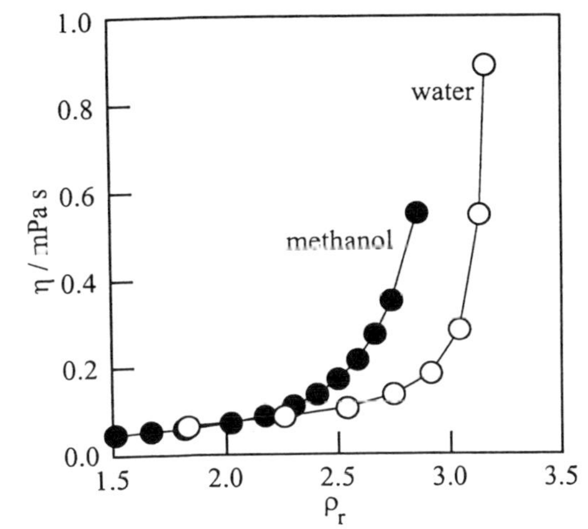

Figure 3. Density dependence of viscosity of methanol and water along the liquid–vapor coexistence curve.

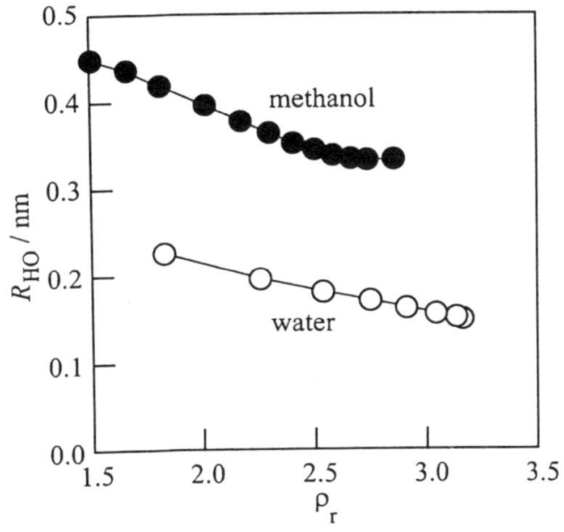

Figure 4. Density dependence of Hubbard–Onsager radius of methanol and water along the liquid–vapor coexistence curve.

the theoretical calculations are shown in Figs. 3 and 4, respectively. The ionic radius is assumed to be given by the crystallographic radius ($R = 0.133$ nm for the K^+ ion[10]).

At $\rho_r > 1.8$, the experimental molar conductivity increases with decreasing density, and the HO theory well reproduces the density dependence. Although the Stokes theory predicts the density dependence in the same direction, it significantly overestimates the conductivity. At $\rho_r < 1.8$, on the other hand, the density dependence of $\lambda°$ predicted by the

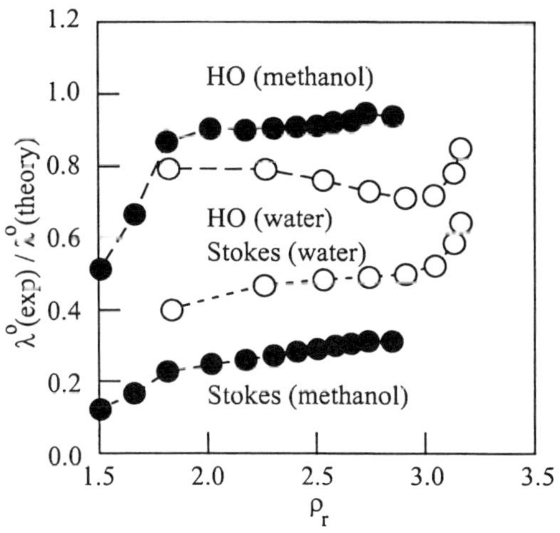

Figure 5. Density dependence of $\lambda°(\text{exp})/\lambda°(\text{theory})$ for the K^+ ion in methanol and water along the liquid–vapor coexistence curve.

HO theory is in the opposite direction to that obtained experimentally.

In the conditions studied here, the large density dependence of $\lambda°$ at $\rho_r > 1.8$ is dominated by that of the viscosity, and the steep density dependence of the viscosity somewhat conceals the effect of the dielectric friction. In order to discuss the significance of the dielectric friction effect on the ionic motion more clearly, the density dependence of $\lambda°(\text{exp})/\lambda°(\text{theory})$ is shown in Fig. 5. As can be seen from Eq. 6, this quantity is the ratio of the theoretical friction to the experimental one.

$$\frac{\lambda°(\text{exp})}{\lambda°(\text{theory})} = \frac{\zeta(\text{theory})}{\zeta(\text{exp})} \tag{7}$$

At $\rho_r > 1.8$, $\lambda°(\text{exp})/\lambda°(\text{HO})$ is close to unity and $\lambda°(\text{exp})/\lambda°(\text{Stokes})$ is much smaller. The dielectric friction effect is very large in subcritical methanol; the HO radius of methanol is much larger than the ionic radius as shown in Fig. 4. The sphere-in-continuum model is considered to be valid at high liquid densities, and the observed application limit of $\rho_r = 1.8$ in methanol may be lower than usually expected. The success of the HO theory indicates that the strong long-ranged polarization effect is well described by the continuum model even in medium-density fluids. In spite of the larger HO radius at lower densities, however, the limitation of the HO theory at $\rho_r < 1.7$ is obvious from Fig. 5.

3.2 Subcritical Water

Next we examine the validity of the HO theory for the mobility of the K^+ ion in subcritical liquid water along the liquid–vapor coexistence curve. The temperature range studied here is $0.46 \leq T/T_c \leq 0.96$ (25 – 350°C), and the corresponding density range is $3.17 \geq \rho/\rho_c \geq 1.82$ (0.997 – 0.575 g cm^{-3}). The experimental limiting ionic conductivities $\lambda°$ for the K^+ ions in water are taken from Oelkers and Helgeson.[11] The viscosity and the static dielectric constant are taken from the international steam tables.[6] The dielectric relaxation properties by Okada *et al.*[12] are used after slight modifications.[4]

In Fig. 2, $\lambda°$ for the K^+ ion in water are plotted against the reduced density together with the predictions of the HO and Stokes theories. The viscosity and the HO radius used in the theoretical calculations are shown in Figs. 3 and 4, respectively.

At first sight of Fig. 2, the HO theory seems to explain qualitatively the experimental result of the density dependence of $\lambda°$ in subcritical water. If we examine $\lambda°(\text{exp})/\lambda°(\text{theory})$ shown in Fig. 5, however, the limitation of the HO theory can be observed. At high liquid densities, $\lambda°(\text{exp})/\lambda°(\text{HO})$ rapidly decreases with decreasing density. Liquid water is known for interesting properties arising from the hydrogen-bonded network structure. Since the HO radius of water is about a half of that of methanol and is comparable to the radius of the K^+ ion, the dielectric friction effect on the K^+ ion is weak in liquid water. In fact, $\lambda°(\text{exp})/\lambda°(\text{Stokes})$ is larger and $\lambda°(\text{exp})/\lambda°(\text{HO})$ is smaller in water than in methanol. In such a case, the hydrogen-bonded network of water is considered to be broken by the presence of a potassium ion. The discrepancy between the HO theory and the experiment in high-density water is ascribed to the effect of the hydrogen-bonded liquid structure which is not taken into account in the sphere-in-continuum theory.

At $\rho_r < 3.0$ (corresponding to the temperature above 100°C), on the other hand, the hydrogen-bonded structure becomes weaker with decreasing density, and the nature of water approaches that of ordinary fluids. At the same time, the HO radius of water increases with decreasing density, and the dielectric friction effect becomes more significant. This is the reason why the HO theory is more successful at lower densities than at high liquid densities in subcritical water.

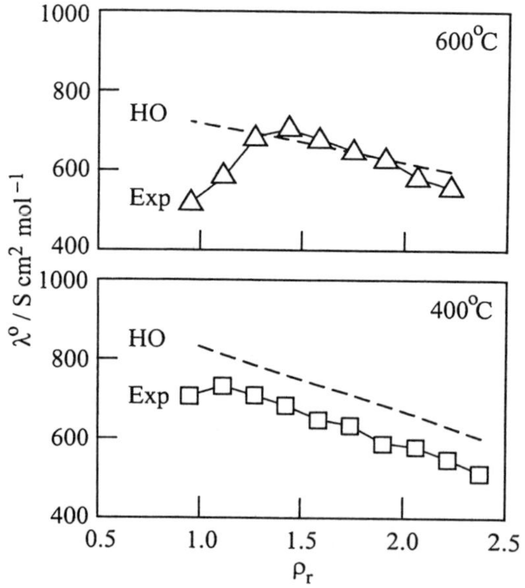

Figure 6. Density dependence of limiting molar conductivity for the K^+ ion in water at constant temperature.

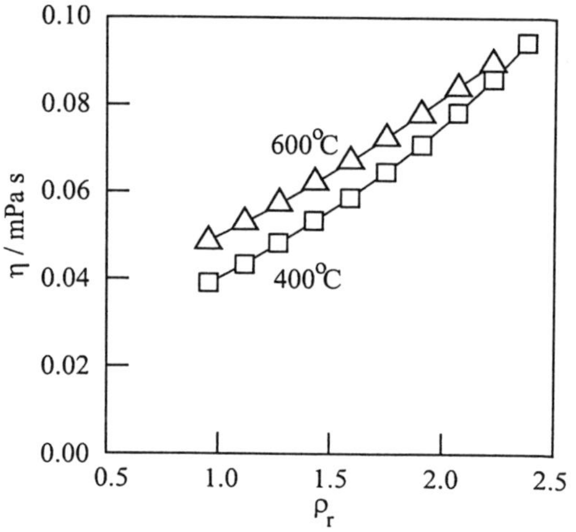

Figure 7. Density dependence of viscosity of water at constant temperature.

3.3 Supercritical Water

In order to clarify the application limit of the HO theory at low densities in water, next we examine the mobility of the K^+ ion in supercritical water at 400 and 600°C ($T/T_c = 1.04$ and 1.35, respectively). The density range studied here is $2.38 \geq \rho/\rho_c \geq 0.95$ (0.75 – 0.30 g cm^{-3}). The experimental limiting mo-

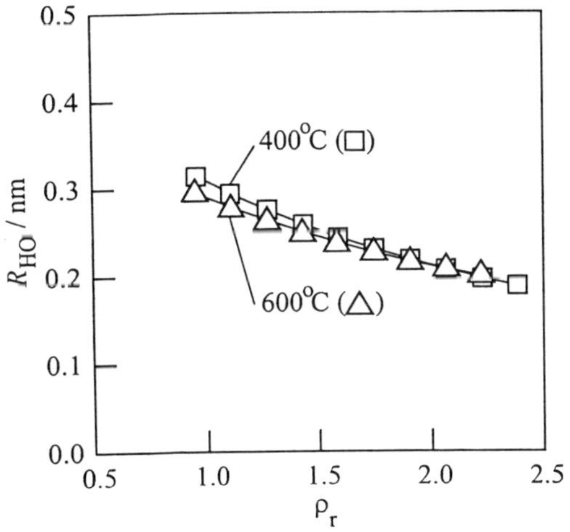

Figure 8. Density dependence of Hubbard–Onsager radius of water at constant temperature.

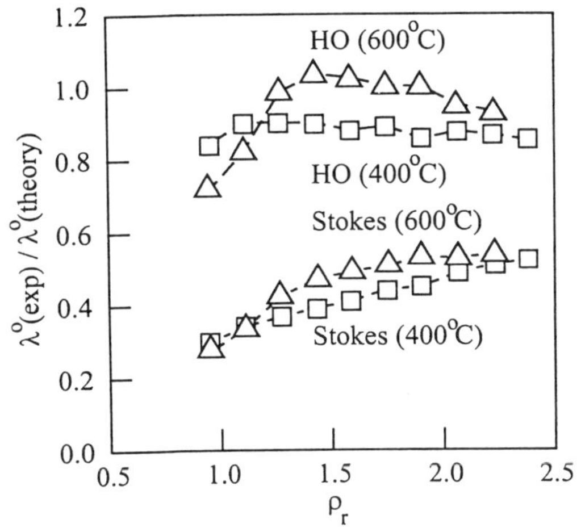

Figure 9. Density dependence of λ°(exp)/λ°(theory) for the K^+ ion in water at constant temperature.

lar conductivities $\Lambda_\circ$ for KCl are taken from Ibuki *et al.*[13] in which the experimental data for molar conductivities at finite concentrations by Ho *et al.*[14] are analyzed. The ionic values of conductivities are estimated by an assumption that $\lambda^\circ(K^+) = \lambda^\circ(Cl^-)$. The data sources for the properties of water are the same as those for subcritical water.

In Fig. 6, λ° for the K^+ ion in water are plotted against ρ_r together with the predictions of the HO theory. The Stokes theory predicts much larger λ° values because of the low viscosities. The viscosity and the HO radius used in the theoretical calculations are shown in Figs. 7 and 8, respectively.

It can be seen from Fig. 6 that the density dependence of λ° obtained by the HO theory is qualitatively different from the experimental results at $\rho/\rho_c < 1.4$ and 600°C, though the experiment and the HO theory show similar density dependence of λ° at 400°C. We can conclude that the application limit of the HO theory lies about $\rho/\rho_c = 1.4$ in supercritical water.

The density dependences of λ°(exp)/λ°(theory) are shown in Fig. 9. From Figs. 5 and 9, we can see that λ°(exp)/λ°(HO) becomes closer to unity with increasing temperature at $\rho/\rho_c > 1.4$. The continuum theory is more reliable at higher temperature if the dielectric friction effect is properly taken into account. This is what is usually expected because the effect of the hydrogen-bonded network is much less important at higher temperatures. At $\rho/\rho_c < 1.4$, on the other hand, the limitation of the HO theory is also obvious from the density dependence of λ°(exp)/λ°(HO).

4 Summary

In the present paper, we have examined the reliability of the Hubbard–Onsager dielectric friction theory in subcritical methanol and sub- and supercritical water by comparing the limiting molar conductivities for the potassium ion obtained by the theory and experiment. The HO theory is reliable in explaining the density dependence of the conductivity at densities above $1.8\rho_c$ and $1.4\rho_c$ in methanol and water, respectively. The success of the HO theory at such low densities indicates that the dielectric friction effect is significant in low-density fluids with a low dielectric constant, and that the non-equilibrium polarization field is well described by the HO theory even at low densities. At densities below the application limits, however, the HO theory predicts the density dependence of λ° in the direction opposite to the experimental results. In water, moreover, the HO theory does not explain the observed behavior of λ°, because the effect of the hydrogen-bonded network is important in liquid water.

Acknowledgments

This work was supported by the grant to the research project at Doshisha University entitled "Intelligent

Information Science and Its Applications to Problem Solving in Engineering Fields" from the Ministry of Education, Culture, Sports, Science and Technology (MEXT).

References

1. J. Hubbard and L. Onsager, *J. Chem. Phys.* **67**, 4850 (1977).
2. J. Hubbard, *J. Chem. Phys.* **68**, 1649 (1978).
3. T. Hoshina, N. Tsuchihashi, K. Ibuki, and M. Ueno, *J. Chem. Phys.* **120**, 4355 (2004).
4. K. Ibuki, M. Ueno, and M. Nakahara, *J. Mol. Liq.* **98/99**, 129 (2002).
5. K. M. de Reuck and R. J. B. Craven, *Methanol, International Thermodynamic Table of the Fluid State*, Vol. 12 (Blackwell, Oxford, 1993).
6. *JSME Steam Tables* (The Japan Society of Mechanical Engineers, Tokyo, 1999), based on the IAPWS (International Association for the Properties of Water and Steam) Industrial formulation 1997 for the Thermodynamic Properties of Water and Steam.
7. N. B. Vargaftik, *Tables on the Thermophysical Properties of Gases and Liquids*, (Hemisphere, Washington, 1975).
8. W. Dannhauser and L. W. Bahe, *J. Chem. Phys.* **40**, 3058 (1964).
9. Y. Hiejima, Y. Kajihara, H. Kohno, and M. Yao, *J. Phys.: Condens. Matter* **13**, 10307 (2001).
10. L. Pauling, *The Nature of the Chemical Bond*, 3rd Ed. (Cornell University, Ithaca, 1959).
11. E. H. Oelkers and H. C. Helgeson, *J. Solution. Chem.* **18**, 601 (1989).
12. K. Okada, Y. Imashuku, and M. Yao, *J. Chem. Phys.* **107**, 9302 (1997).
13. K. Ibuki, M. Ueno, and M. Nakahara, *J. Phys. Chem. B* **104**, 1539 (2001).
14. P. C. Ho and D. A. Palmer, *Geochim. Cosmochim. Acta* **61**, 3027 (1997).

PROTON CONDUCTIVITY OF SUPERIONIC CONDUCTOR $Tl_3H(SO_4)_2$

YASUMITSU MATSUO, KEISUKE TAKAHASHI AND SEIICHIRO IKEHATA
1-3 Kagurazaka, Shinjyukuku, Tokyo, Japan, 162-8601 Fax: +81-3-3260-4772
E-mail: ymatsuo@rs.kagu.tus.ac.jp

Measurements of thermoelectric power, electrical conductivity and ^{1}H-NMR absorption line have been carried out in order to investigate the role of proton to electrical conductivity on $Tl_3H(SO_4)_2$. From the measurement of the thermoelectric power $S = (\Delta V / \Delta T)$ it is found that the sign of S is a positive in a superionic phase. This result indicates that the majority carrier for electrical conductivity in the superionic phase is protonic. Moreover from the analyses of the second moment of ^{1}H-NMR absorption line we found that the activation energies of the hopping motion of protons in the superionic phase agree well with those observed in the electrical conductivity. These results indicate that the electrical conductivity of the $Tl_3H(SO_4)_2$ crystal observed in the superionic phase results from mobile proton accompanied by the breaking of the hydrogen bonds.

1. Introduction

Recently, zero-dimensional hydrogen-bonded $M_3H(XO_4)_2$ (M: K, Rb, Cs, NH_4; X: S, Se) type compounds have been studied with much interest with regard to a superionic conduction in conjunction with the development of fuel cell[1-11]. The $M_3H(XO_4)_2$ type compounds undergo a superionic phase transition from the lower temperature ferroelastic phase to the higher temperature paraelastic phase. The thermal rotational displacement of the XO_4 tetrahedra is closely related to the superionic phase transition. It was expected for the $M_3H(XO_4)_2$ type compounds that the superionic phase transition would be observed at the higher temperature (such as 400 K).

However, recently, it was found from dielectric and thermal studies that the $Tl_3H(SO_4)_2$ crystal, which is one of the $M_3H(XO_4)_2$ type compounds, undergoes the superionic phase transition at 239 K[12]. That is, this crystal becomes a superionic conductor at room temperature. The crystal structure at room temperature for this crystal is shown in Fig. 1[13]. This crystal belongs to the trigonal system with a space group of $R\bar{3}m$ at room temperature. The most interesting feature of this structure is the existence of three equivalent sites of the oxygen closely related with the hydrogen bond. From this fact, it is speculated in the $Tl_3H(SO_4)_2$ crystal that even at room temperature the breaking and rearrangement of the hydrogen bond are iterated accompanied by the rotational displacement of SO_4 tetrahedra. These results for the $Tl_3H(SO_4)_2$ crystal are interesting, because this crystal shows the superionic conductivity even at room temperature and therefore provides the further knowledge for the mechanism of the superionic phase transition in the $M_3H(XO_4)_2$ type compounds.

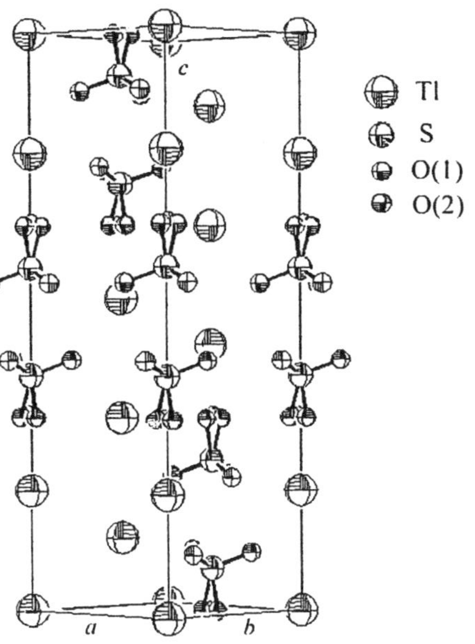

Figure 1. Crystal structure in $Tl_3H(SO_4)_2$ at room temperature. The coordinates of H atom do not determine. The O(2) atom, which is closely related with the hydrogen bond, occupies three equivalent sites with probability 1/3.

In the present paper, we report the experimental results of the thermal, electrical and ^{1}H-NMR measurements and discuss the role of the proton in the superionic phase transition in which electrical conductivity increases drastically.

2. Experiments

The $Tl_3H(SO_4)_2$ crystals were grown by slow evaporation method from aqueous solution of Tl_2SO_4 and H_2SO_4 with a molar ratio of $Tl_2SO_4 : H_2SO_4 = 3:2$ at 313 K. The single crystals are of hexagonal thin plate with the predominant (001) faces and are transparent.

Measurements of the electrical conductivity along the hexagonal *c* axis were carried out below room temperature at the frequency of 10 kHz using a LCR meter (4284A : Hewlett Packard. Co. LtD.). The NMR absorption lines were observed by the Q-meter detective method at the resonance frequency of 10.6 MHz with powder specimens.

3. Results and Discussion

The temperature dependence of the electrical conductivity is shown in Fig. 2 as a plot of log σT against inverse temperature. It is evident that the anomalies of σT are observed at $T_{\text{I-II}} = 267$ K, $T_{\text{II-III}} = 239$ K and $T_{\text{III-IV}} = 196$ K. At the superionic phase transition temperature of $T_{\text{II-III}}$, σ increases drastically and becomes about 10^{-5} S/cm at room temperature. Moreover, we note that log σT is proportional to $1/T$ with increasing temperature in the superionic phases. From this result, we can calculate the activation energies for the superionic phases I and II using the following equation,

$$\sigma = A'/T \exp(-E_a / k_B T), \qquad (1)$$

where A' is the pre-exponential factor and k_B is Bolzmann's constant. It was found that the activation energies for the phases I and II obtained are 0.42 eV and 0.46 eV respectively.

In Fig.3, the thermoelectromotive force ΔV is shown as a function of temperature difference ΔT in the superionic phase (331 K). It is evident that the absolute value of the thermoelectromotive force $|\Delta V|$ increases linearly with increasing $|\Delta T|$. Moreover, we note that the sign of the thermoelectric power S=($\Delta V / \Delta T$) is a positive. This result indicates that the majority carrier for electrical conductivity in the superionic phase is protonic.

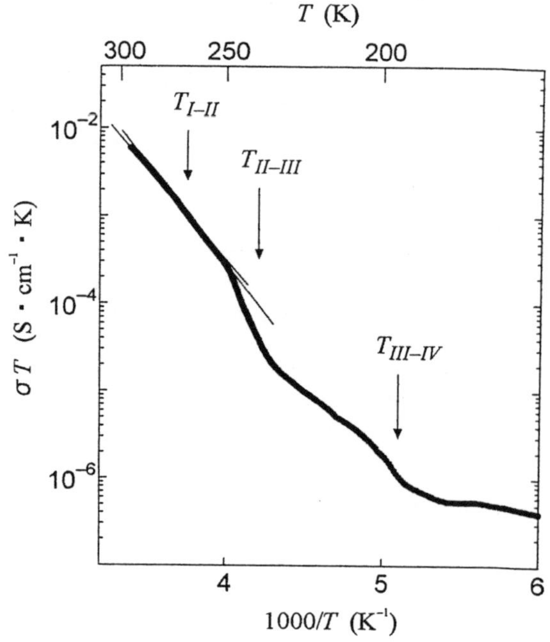

Figure 2. $\sigma T - 1/T$ characteristic

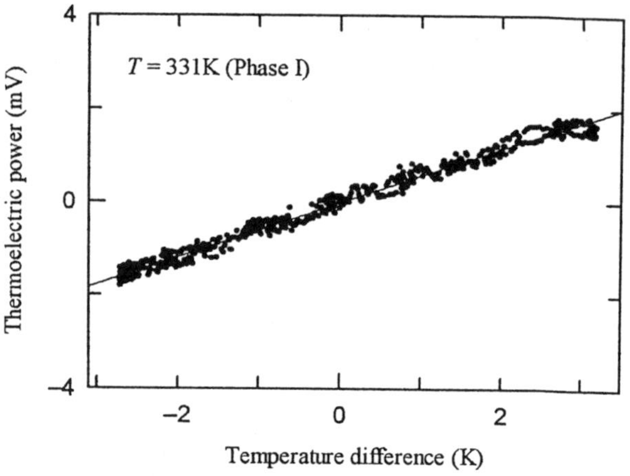

Figure 3. $\Delta V - \Delta T$ characteristic

Figure 4 shows the ^{1}H-NMR absorption lines for various temperatures. As shown in Fig. 4, the ^{1}H-NMR absorption lines show the broad ones in the phases III, IV and V, and become sharp above $T_{\text{II-III}}$. The shape of the NMR absorption lines in the phases III, IV and V are described well with the Gaussian curve which corresponds to the NMR line for rigid lattice. On the other hand, the NMR absorption lines observed above $T_{II\text{-}III}$ are described with the functional form mixed with the Gaussian curve and the Lorenzian curve. It is well-known that the NMR absorption line with the Lorenzian curve is caused by the existence of mobile protons. Therefore, this result indicates that mobile protons exist above $T_{II\text{-}III}$.

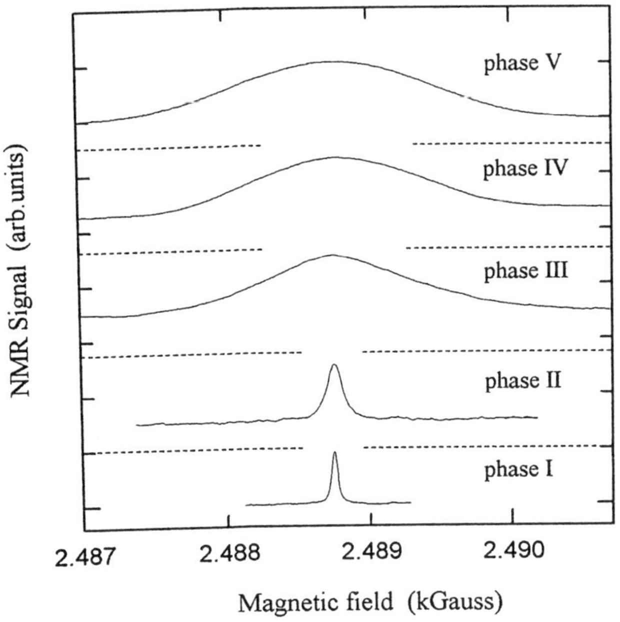

Figure 4. ^{1}H-NMR absorption lines for various temperatures

It is also noted that the ^{1}H-NMR line width becomes narrow above $T_{II\text{-}III}$ drastically. In order to examine this narrowing of ^{1}H-NMR absorption line in detail, we show the temperature dependence of the second moment M_2 in Fig. 5. The second moment is closely related with the line width of the ^{1}H-NMR absorption line. The second moment M_2 is calculated directly from the measured NMR absorption lines for various temperatures using the following equation,

$$M_2 = \frac{\int_{-\infty}^{\infty} (H - H_0)^2 f(H - H_0) dH}{\int_{-\infty}^{\infty} f(H - H_0) dH}, \quad (2)$$

where H_0 is the resonance magnetic field for ^{1}H-NMR and H the external magnetic field. The function $f(H)$ means the measured NMR absorption line. It is evident that the second moment M_2 begins to decrease above about 190 K with increasing temperature and that becomes below 0.01 G^2 in the superionic phases I and II, while M_2 becomes about 0.47 G^2 below about 190 K. The decrease of M_2 results from the motional narrowing of the ^{1}H-NMR absorption line, because the mobile proton leads to an averaging of the magnetic dipole-dipole interaction of proton. This result indicates that the hopping motion of protons begins above about 190 K and that in the superionic phases protons move with the hopping rate which is fast enough to narrow the ^{1}H-NMR absorption line.

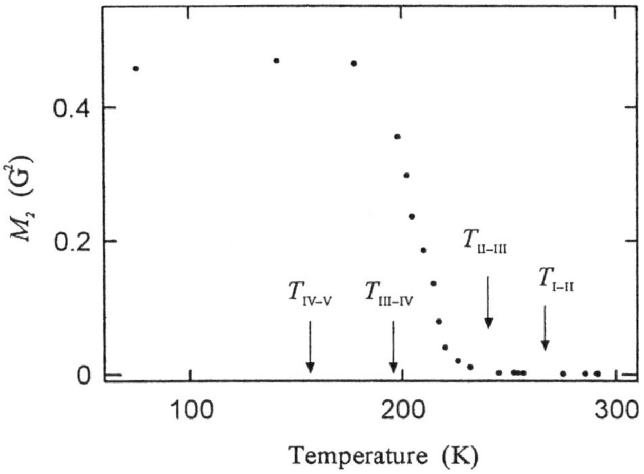

Figure 5. Temperature dependence of second moment M_2

As described above, the oxygen closely related with the hydrogen bond in the SO_4 tetrahedrons moves three equivalent sites. Considering this fact, the hopping motion of protons is caused by the breaking and rearrangement of the hydrogen bonds via the hopping motion of the oxygen between three equivalent sites. Moreover, it should be noted that the onset of the decrease of M_2 appears below the superionic phase transition. It is known that the hopping motion of proton is observed even below the superionic phase for the isomorphic $Rb_3H(SeO_4)_2$ crystal, accompanied by the growth of the micro-domain structure[3, 9]. Therefore, this decrease of M_2 observed below $T_{II\text{-}III}$ is caused by the precursor effect of the hopping motion of proton in the superionic phase. Furthermore, analysis of the second moment makes it possible to determine the correlation frequency ν_c for the hopping motion of proton narrowing the NMR absorption line. The analysis was carried out using the expression[14]

$$\nu_c = \frac{\gamma}{2\pi} \frac{\alpha\sqrt{M_2}}{\tan\left(\frac{\pi}{2}\frac{M_2 - M_2^H}{M_2^L - M_2^H}\right)}, \quad (3)$$

which describes the temperature dependence of ν_c obtained from various values of M_2. Here, M_2^L and M_2^H are the second moments before and after narrowing. The constant α is $(8 \ln 2)^{-1}$. The correlation time is assumed to obey the Arrhenius relation

$$\nu_c = \nu_0 \exp(-E_a / kT) \quad (4)$$

so that we can obtain the correlation frequency for high-temperature limit ν_0 and the activation energy E_a for the hopping motion of proton producing the motional narrowing.

Figure 6 shows the temperature dependence of the correlation frequency ν_c obtained for the hopping motion of proton.

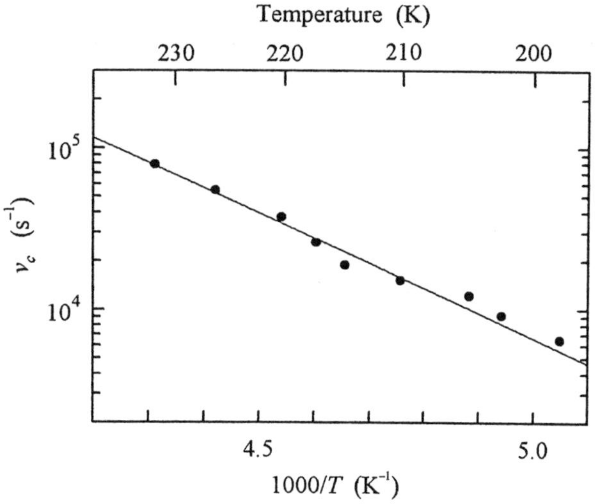

Figure 6. Temperature dependence of the correlation frequency ν_c

It is evident from Fig. 6 that $\log\nu_c$ is proportional to $1/T$. From this result, we find $\nu_0 = 7.7\times10^{10}$ sec^{-1} and E_a = 0.33 eV. It is known that the activation energies of the $M_3H(XO_4)_2$ type compounds in the super-ionic phase become 0.26~0.39 eV in the *a-b* plane of the hexagonal system in the super-ionic phase and become 0.40~0.61 eV along the *c*-axis[2]. As described above, the activation energy obtained from the measurement of electrical conductivity along the *c*-axis for $Tl_3H(SO_4)_2$ is 0.46 eV in the phase II. In NMR measurement, we used the powder sample and therefore obtain the activation energy averaged on all directions. Considering this fact, the activation energy 0.33 eV for the hopping motion of protons corresponds to that obtained from the electrical conductivity measurement. From these results it is deduced that the hopping motion of protons, which is caused by the breaking and rearrangement of the hydrogen bonds, leads to the electrical conductivity in the superionic phase.

4. Summary

We have carried out the thermoelectric power, electrical conductivity and ^{1}H-NMR measurements and have investigated the role of proton in electrical conductivity on the $Tl_3H(SO_4)_2$ crystal. We have obtained from the thermoelectric power $S = (\Delta V / \Delta T)$ that the majority carrier for electrical conductivity in the superionic phase is a proton. Moreover it is also found from the analyses of the second moment of ^{1}H-NMR absorption line that the activation energy for the hopping motion of protons in the superionic phase (phase II) corresponds to that observed in the electrical conductivity measurement. These results indicate that the electrical conductivity of the $Tl_3H(SO_4)_2$ crystal observed in the superionic phase results from the hopping motion of protons accompanied by the breaking and rearrangement of the hydrogen bonds.

Acknowledgment

This work was supported by a GRANT-IN-AID of the Promotion and Mutual Aid Corporation for Private Schools of Japan.

References

1 Y. Matsuo, K. Takahashi, K. Hisada and S. Ikehata, *J. Phys. Soc. Jpn.* **68** 2965 (1999).
2 A. Pawlowski, Cz. Pawlaczyk and B. Hilczer, *Solid State Ionics* **81** 17 (1990).
3 Y. Matsuo, K. Takahashi and S. Ikehata, *J. Phys. Soc. Jpn.* **70** 2934 (2001).
4 C. Abramic, J. Dolinsek and R. Blinc, *Phys. Rev.* B **42** 442 (1990).
5 A. Bohn, R. Melzer, T. Sonntag, R. E. Lechner, G. Schuck and K. Langer, Solid State Ionics. 77 111 (1995).

6 B. V. Merinov, Solid State Ionics. **84** 89 (1996).
7 M. Komukae, K. Sakata, T. Osaka and Y. Makita, *J. Phys. Soc. Jpn.* 63 1009 (1994).
8 T. Ito and H. Kamimura, *J. Phys. Soc. Jpn.* **67** 1999 (1998).
9 H. Kamimura and S. Watanabe, *Philosophical Magazine* B **81** 1011 (2001).
10 S. M. Haile, D. A. Boysen, C. R. I. Chisholm and R. B. Merle, *Nature* **410** 910 (2001).
11 T. Norby, Nature **410** 877 (2001)
12 Y. Matsuo, K. Takahashi and S. Ikehata, *Solid State commn.*. **120** 85 (2001).
13 Y. Matsuo, S. Kawachi, Y. Shimizu and S. Ikehata, *Acta Cryst.* ***C*58** i92 (2002).
14 A. Abragam, *The Principles of Nuclear Magnetism* (The Clarendon Press, 1961) Chap. 10

STRUCTURAL AND COMPOSITIONAL ANALYSIS OF LaF_3 THIN FILMS SUITABLE FOR OXYGEN SENSOR

S.SELVASEKARAPANDIAN, M.VIJAYAKUMAR,

Solid State and Radiation Physics Laboratory, Department of Physics, Bharathiar University, Coimbatore – 641 046, INDIA.
Email: sekarapandian@yahoo.com

SHINOBU FUJIHARA, SHINNOSUKE KOJI

Department of Applied Chemistry, Faculty of science and Technology, Keio University, 3-14-1, Hiyoshi, Kohoku-Ku, Yokohama 223-8522, JAPAN.

Abstract

The lanthanum fluoride film has been prepared by thermal evaporation method. The XRD pattern shows the polycrystalline nature and hexagonal structure of the LaF_3 film. The unit cell parameters are calculated and are found to be in good agreement with standard values. The grain size has been calculated using the Debye scherrers formula and is found to be 53 nm. The XPS spectra recorded in the binding energy range 500-850 eV shows three intense peaks corresponding to lanthanum ($La3d_{5/2}$), fluorine ($F1s_{1/2}$) and oxygen (O1s) at binding energies around 836, 684 and 530 eV respectively. The oxygen peak corresponding to the surface of the film is due to the adsorption of oxygen at LaF_3 film surface. The inner layers of the film consist of multiple oxygen peaks due to the presence of peroxide (HO^{2-}) and superoxide ions (O_2^-). This impurities leads to the formation of lanthanum oxyfluoride and F deficiencies, which are favorable for the oxygen sensing mechanisms. The optical spectrum shows the low porosity and high transparent nature of the film.

Key words: ionic conductors, lanthanum fluoride, xps analysis, oxygen sensor, optical spectrum

1. Introduction

Lanthanum fluoride (LaF_3) is an excellent F^- ionic conductor among other rare earth fluorides [1,2]. LaF_3 based chemical sensors has potential application in sensing the fluorine, oxygen and carbon monoxide because of its high chemical stability and ionic conductivity [3,4]. The dissolved oxygen sensing property of the LaF_3 material has been utilized to construct the biosensors with suitable enzymes as auxiliary electrode [5]. The fast response and good sensitivity of these sensors rely on the F^- ionic conductivity. The increase in F^- ionic conductivity due to the existence of metastable oxygen species, specifically superoxide (O_2^-) and peroxide ions (HO_2^- or O_2^{2-}) has reported previously [6]. N.Miura etal. [7] reported the construction of oxygen sensor with sputtered LaF_3 thin film with high response rate due to the water vapor treatment, which provides the superoxide and peroxide ions to LaF_3. Our recent study shows that the thermal evaporation method provides LaF_3 thin film with desired peroxide and superoxide ions without any external water vapour treatments [8]. However the response time and ionic conductivity of these films strongly depends on the structure and F^- vacancies. Hence it is essential to characterize the structure and composition of LaF_3 thin films to identify its suitability for the usage of chemical sensors. In the present study LaF_3 thin film prepared by thermal evaporation has been analyzed by XRD, XPS & Transmittance spectra and the results are reported.

2. Experimental

The high purity lanthanum fluoride (99.9%), has been used as a starting material. "Tungsten helical source" has been used for thermal evaporation, and glass slides as substrate material. The film is coated under a vacuum pressure of around 2×10^{-5} Torr using HINDHIVAC 12 AD coating unit. The substrate has been kept at 473 K during evaporation and thickness and evaporation rate are monitored by in-situ digital thickness monitor. Crystalline phase of the film has been identified with a glancing-angle X-ray diffractometer (Rigaku) using CuK_α radiation (40kV-150mA; scan speed 2□ / min). Chemical bonding states of the constituent elements in the film were examined by X-ray photoelectron spectroscopy (XPS) (model JPS9000, JEOL) using Mg K_α radiation (10 kV – 10 mA). Peak positions were calibrated by C_{1s} position at 284 eV. Compositional analysis of the LaF_3 film has been done by calculating the ratio of peak heights in the XPS spectra using relative sensitivity coefficients. Optical transmittance spectrum has been recorded by using JASCO 540 double beam spectrophotometer in the wavelength range of 400 – 2500 nm. All measurements were made in laboratory air at room temperature.

3. Results and Discussions

3.1 Structural analysis

The crystalline nature and structural parameters of the LaF_3 films were identified by the X-ray diffraction analysis. The glancing angle X-ray diffraction pattern is recorded and shown in Fig.1. The presence of well defined peaks reveals the polycrystalline nature of the films. The values are found to be in good agreement with bulk LaF_3 material [9]. The peaks observed around 24.16°, 24.86°, 43.62° and 50.623° are due to (002), (110), (300) and (302) reflections respectively. These are characteristics of LaF_3 film in hexagonal symmetry.

Micro-structural parameters such as grain size (D), micro-strain (ε) and dislocation density (δ) are calculated from the analysis of (300) peaks using the below equations [10],

$$D = \frac{0.94\lambda}{\beta_{2\theta}\cos\theta} \quad (1)$$

$$\delta = 1/D^2 \quad (2)$$

$$\varepsilon = \left(\frac{\lambda}{D\cos\theta} - \beta\right)x\frac{1}{\tan\theta} \quad (3)$$

Where, λ is the wavelength of the X-ray and $\beta_{2\theta}$ is the full width at half maximum of the corresponding peak of the XRD pattern. The dislocation density and particle size are calculated to be 2.160 X 10 ^{18} and 7.5 nm respectively.

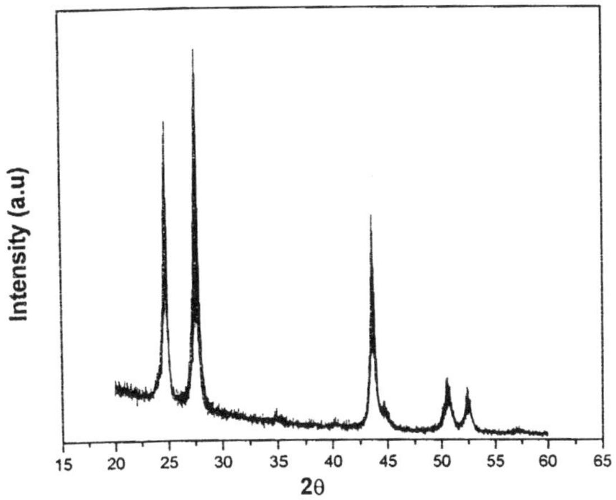

Fig.1 XRD pattern of the LaF_3 thin film

3.2. XPS analysis

X-ray photoelectron (XPS) spectra for the surface layer of lanthanum fluoride film have been shown in Fig.2. The spectra recorded in the binding energy (BE) range 100-850 eV show three intense peaks corresponding to lanthanum ($La3d_{5/2}$), fluorine ($F1s_{1/2}$) and oxygen (O1s) at binding energies around 836, 684 and 531 eV respectively [11]. However the impurity peaks like silicon and carbon are also traced out. The chemical state of lanthanum in LaF_3 films can be characterized by analyzing the energy position, chemical shift and FWHM of the two core level binding energy peaks of lanthanum ($La3d_{5/2}$), fluorine ($F1s_{1/2}$) and oxygen (O1s). The presence of oxygen peak in the surface of the film is due to the adsorbed oxygen. This may be due to the presence of oxygen during coating procedure. The vacuum level during the evaporation is around 8 x 10^{-5} Torr, hence there will be possible oxygen inside the chamber which reacts with LaF_3 vapor, leading to the oxygen impurities in the films.

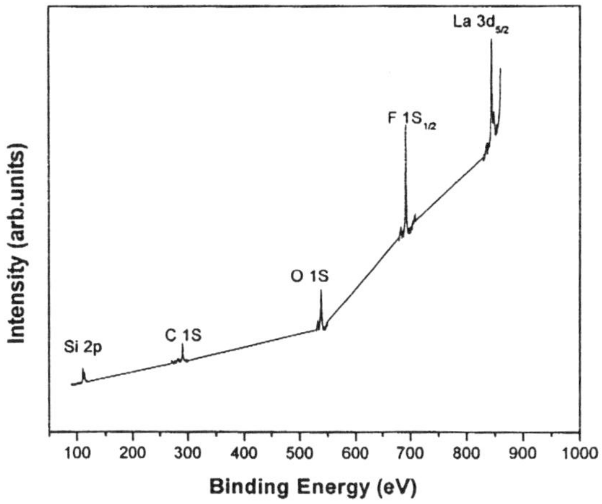

Fig.2 XPS analysis of LaF_3 film

Fig.3.a-c, shows depth profile XPS spectra of lanthanum, fluorine and oxygen respectively along the direction of the film. The oxygen peak in the depth profile XPS spectra shown in Fig.3.c, reveals the presence of oxygen throughout the LaF_3 film. The oxygen spectra for surface layer of the LaF_3 show a single oxygen peak at 531.5 eV, which is due to the adsorbed oxygen. The lanthanum fluoride films were etched by Ar^+ ion up to 2000 s to analyze compositions in direction of film thickness. Interestingly two peaks at 531.5 and 528.7 eV have been observed for oxygen spectra in the inner layers of the film, which reveals that the oxygen has two binding state in LaF_3 film. The presence of additional peak at the lower energy side may be due to the presence of oxide impurities such as superoxide and peroxide ions. Presence of these ions leads to the formation of lanthanum oxyfluoride in the lanthanum fluoride films. This has been confirmed from the chemical shift of the fluorine peak towards higher binding energy in the XPS spectra (shown in Fig.3.b) with increase in oxygen peak intensity in the inner layer of the film. It has been reported that the LaF_3 reacts chemically react with oxide ions to form lanthanum oxyfluoride (LaOF) with binding energy of F1s will move toward higher value [12]. Further the reaction of rare earth fluoride with oxide ions and formation of oxyflouride has already been reported [13]. The formation of the lanthanum oxyfluorides in the sol-gel prepared lanthanum fluoride has also been reported elsewhere [11].

Due to the similarity in size between oxide ion and fluoride ion, oxide ion must substitute or incorporate for fluoride ion or vacant site in the structure of lanthanum fluoride. Hence the formation of lanthanum oxyfluoride causes the F^- vacancies and free fluoride ions, which gives rise to the F^- ionic conductivity and hence high response rate of the sensor. This is represented in Kroger-Vink notation as,

$$O_2^{-} + F_F^{x} = O^{x}_{F'} + F^{-} \quad (4)$$

$$OH^{-} + F_F^{X} = OH_F^{X} + F^{-} \quad (5)$$

The [O]/[F] ratio is the critical parameter which is the ratio of oxygen content with fluorine content in the LaF_3 films. N. Miura et.al, [12] reported the presence of oxide impurities in the sputtered lanthanum fluoride films with [O]/[F] ratio of 0.22, which gives the response rate as 5 minutes. The [O]/[F] ratio has been increased to 0.31 by the water vapor treatment at 90°C for 1 hour of sputtered LaF_3 films which shows high response rate of 0.5 minutes. In the present work, the thermally evaporated LaF_3 film yields the average [O]/[F] ratio as 0.35, even without water vapor treatment.

3.3. Transmittance analysis

Optical characteristics of the films are strongly influenced by the thickness of the film and the deposition method. In the present work thickness of the films are monitored by in-situ digital thickness monitor and the evaporation rate has been controlled so as to get thickness of 800 nm for all the films. The optical absorption and transmittance studies are useful for the identification of band gap, refractive index, extinction coefficient etc. Fig.4 shows the optical transmittance spectra of the LaF_3 film. In the present study the porosity of the film may be calculated by extending the Lorentz-Lorentz formula [15]

$$1-p=\frac{(n^2+2)(n^2{}_m-1)}{(n^2-1)(n^2{}_m+2)} \quad (6)$$

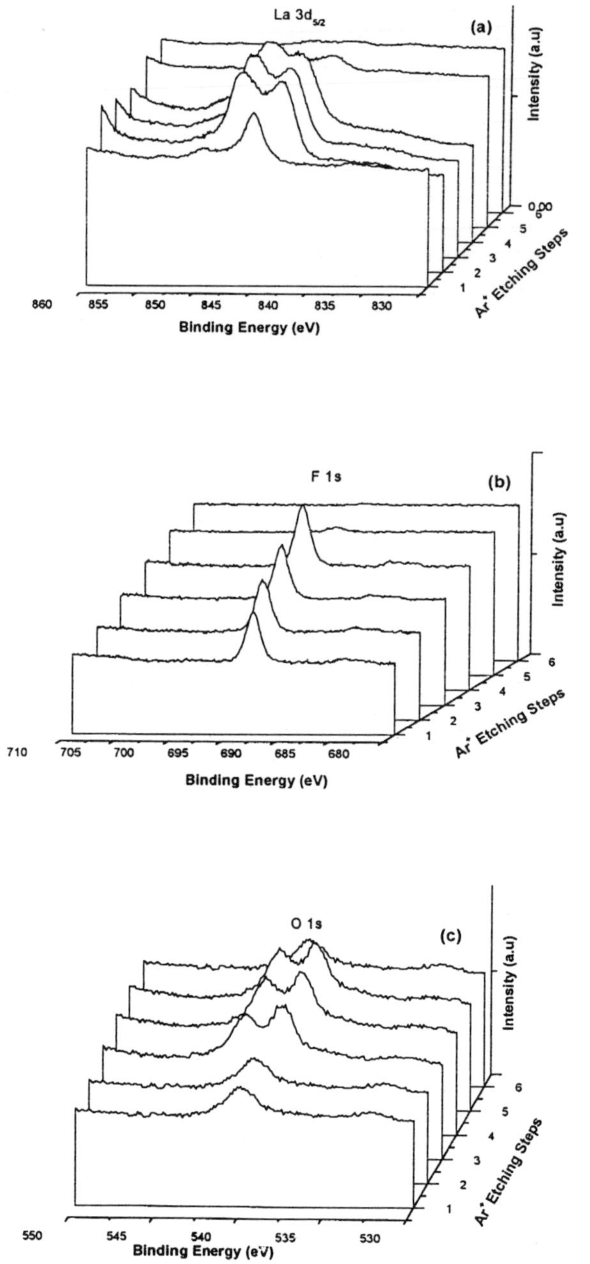

Fig.3 a-c, Depth profile composition analysis of LaF_3 Film

where p is porosity and n and n_m are the theoretical and measured refractive indices respectively. The theoretical refractive index of the LaF_3 material is 1.55. Hence the porosity of the film has been calculated by using Eq.6 and found to be 0.8. This low porosity reveals that the film is uniform and free from pin holes which will make adverse effects in sensor fabrications.

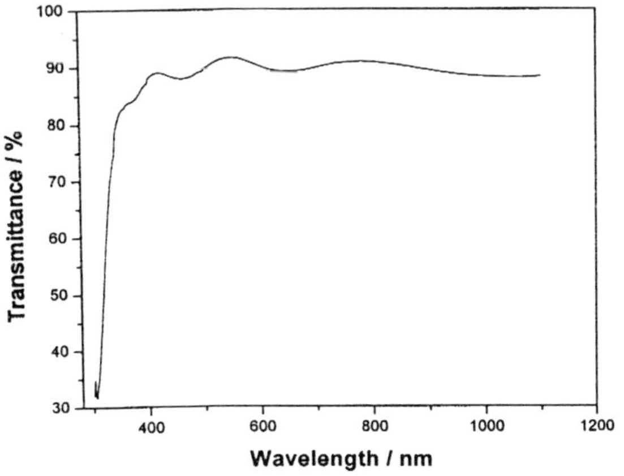

Fig.4 Optical transmittance spectra of LaF_3 film

4. Conclusion

The LaF_3 film has been coated by using thermal evaporation method. The formation of lanthanum oxyfluorides due to the presence of oxygen impurities has been detected by XPS analysis. The average [O]/[F] ratio has been found to be 0.32 which is desirable for the oxygen sensor construction. The low porosity of the film calculated from the transmittance spectra reveals its suitability for the use in the oxygen sensors.

References

1. A.Roos, A.F.Aalders, J.Schoonman, A.F.M. Arts, H.W.DeWijin *Solid State Ionics* **571** 9-10 (1983).
2. J.Schoonman, G.Oversluizen, K.E.D. Wapennar *Solid State Ionics* **1** 211 (1980).
3. J.Szeponik, W.Mortiz, *Sensors and Actuators* **B 2** (1990) 243.
4. W.Mortiz, L.Muller, *Analyst* **116** 589 (1991).
5. N.Miura, N.Matayoshi and N.Yamazoe, Jpn. *J.Appl.Phys.* **28** L1480 (1989).
6. S.Selvasekarapandian, M.Vijayakumar, Shinobu Fujihara, Shinnosuke Koji, *Physica* **B 337/1-4** 52 (2003).
7. N.Miura, J.Hisamoto, N.Yamazoe, S.Kuwata, J.Salardenne, *Sensors and Actuators* **16** 301 (1989).
8. M.Vijayakumar, S.Selvasekarapandian, Shinobu Fujihara, Shinnosuke Koji, *Appl. Surf. Sci.* **222** 125 (2004)
9. M.Tada, S.Fujihara, T.Kimura, *J. Mater. Res.* **14** 1610 (1999).
10. B.D.Cullity, *"Elements of X-ray diffraction"*, (Addison-Wesley, Reading, MA, 1978)
11. S.Fujihara, C.Mochizuki, T.Kimura, *J.Non-Cryst.Solids* **244** 267 (1999).
12. M.Ryzhkov, J.Electron Spectrons. *Relat. Phenom.* **21** 193 (1980).
13. T.Balaji, S.Buddhudu, *Spectrosc. Lett.* **26** 113 (1993).
14. N.Miura, J.Hisamoto, N.Yamazoe, S.Kuwata, *Appl.Surf.Sci.* **33** 1253 (1988).
15. M.Tada, S.Fujihara, and T.Kimura, *J.Mater.Res.* **14** 1610 (1999).

FAST MIGRATION PHENOMENON OF Tl^+ IONS DIFFUSING IN A KCl CRYSTAL THROUGH THE INTERFACE WITH LIQUID TlCl AND SOLID KCl

WEI YU[1] AND ATSUHIRO FUJII[2]

[1] *Graduate School of Science and Technology, Kumamoto University*
[2] *Shockwave and Condensed Matter Research Center, Kumamoto University*
Kurokami, Kumamoto, 860-8555, Japan

A fast migration phenomenon has been observed for thallous (Tl^+) ions diffusing in KCl single crystals through the interface of liquid TlCl and KCl single crystal. Diffusion coefficients are much larger than those in the diffusion processes through the interface of TlCl vapor or solid TlCl and KCl single crystals. A possible model is proposed to explain these diffusion phenomena. A hopping model is extended in our model where Tl^+ ions are assumed to move in a vacancy rank by one jump. The gradient of the chemical potential of Tl^+ ions in a KCl crystal is thought to be an origin for the fast migration phenomenon.

1. Introduction

Diffusion of impurity ions into crystalline solid through the interface of liquid phase, vapor phase and solid phase is one of the most important subjects in materials research as it is involved in many fundamental material processes, such as epitaxy, corrosion and thermal annealing. The fabrication of electronic devices depends also on the underlying diffusion mechanisms.

Several investigations have been reported on the diffusion of thallous ions into potassium chloride single crystals.[1-3] In these investigations, the diffusion coefficient is experimentally determined from the diffusion of radioactive tracer ions and the diffusion processes are occurring with solid and vapor diffusion sources. In other words, the diffusion of Tl^+ ions into KCl single crystals, through the interface between solid TlCl and KCl single crystals (S-S) or vapor TlCl and KCl single crystals (V-S), was investigated.

In this paper, we report an experimental study of the diffusion processes of Tl^+ ions in KCl single crystals through three kinds interfaces, liquid TlCl and KCl single crystal (case L-S), case S-S and V-S, by optical methods.

The main results of our study are that (a) through the interface of case L-S, we found that the diffusion coefficients of these diffusion processes are very much larger than that in the case S-S or V-S at the same temperature, which means that the fast migration phenomenon has been observed in the Tl^+ ions diffusion processes in KCl single crystal, and (b) the activation energy of these diffusion processes is much larger than that in KCl single crystal through the interface in the case of S-S or V-S.

The organization of the paper is as follows. In Sec. 2, we report the experimental process and the results. In Sec. 3, we evaluate the restriction (limit) of the classical hoping model and propose a possible mechanism.

2. Experiments and Results

The results of Tl^+ ion diffusion into KCl single crystal through the interface of KCl single crystal and one of three phases of TlCl are reported in this section.

2.1 *Sample Preparations*

2.1.1 *Solid-solid diffusion*

The KCl single crystal was cleaved to pieces with the size of about $10\times10\times5$ mm^3 by a razor blade and polished by lapping film of 8μm and 3μm together with ethanol. The TlCl solid diffusion source is evaporated as a thin film on one side of the KCl specimen. Then this specimen is placed in a quartz capsule together with the TlCl powder. The capsule is evacuated to $3{\sim}5\times10^{-4}$ Pa and sealed in vacuum. The sealed capsule is set in an electric furnace and heated at a given temperature in the range from 250°C to 400°C for the required time. The temperature of the diffusion processe was controlled by an automatic regulator with tolerance of ±0.5°C. In this way the specimen was obtained from the diffusion process in the case of S-S diffusion.

2.1.2 *Vapor-solid diffusion*

In this case, the KCl single crystal was cleaved with a size of about $12\times12\times8$ mm^3 and polished with a lapping film in ethanol. Diffusion of thallium ions was carried out in an evacuated quartz capsule. Pure thallium chloride powder was used as the diffusion source. The KCl specimen was placed in a quartz capsule together with the diffusion source of thallium ion. The capsule was evacuated to $3\sim5\times10^{-4}$ Pa and sealed in vacuum. The sealed capsule was set in an electric furnace and heated at a given temperature in the range from 360°C to 550°C for the required time.

After the diffusion, the capsule was removed from the furnace and cooled rapidly, so that evaporated thallium chloride was deposited not on the crystal but on the surface of the quartz capsule.

2.1.3 *Liquid-solid diffusion*

In this experiment, a KCl single crystal was cleaved and mechanically polished to about $10\times10\times8$ mm^3. Diffusion of Tl^+ ions was also carried out in an evacuated quartz capsule. The pure TlCl powder was used as the diffusion source. The processed KCl single crystal sample was placed in a quartz capsule together with the TlCl powder. The capsule was evacuated to $3\sim5\times10^{-4}$ Pa and then filled with N_2 gas to 3×10^4 Pa. The sealed capsule was set in an electric furnace and heated at a given temperature in the range from 430°C (melting point of TlCl crystal) to 520°C for the required time. Thus the specimen for optical measurements was obtained from the diffusion process in a TlCl-KCl system through the interface of liquid TlCl and solid KCl.

After every diffusion process mentioned above, thin layers were removed from the surfaces of the KCl crystal except the main diffusion surface. This procedure prevents the effect of Tl^+ ion diffusion through the side surfaces of the crystal on the optical absorption, which are explained in the next section.

2.2 *Measurements*

The diffusion coefficient of Tl^+ ions diffusing into KCl single crystal was evaluated by measuring the optical density of the characteristic absorption band with a multi-channel spectrometer (model PMA-50) at room temperature. Then the activation energies of these processes were determined from the temperature dependence of the diffusion coefficient.

2.2.1 *Estimation of relative concentration by optical methods*

KCl single crystals without additives are transparent in the visible and near-ultraviolet region. When a small amount of thallous halide is added, a KCl:Tl solid solution can be obtained. The incorporation of the Tl^+ ions leads to new characteristic absorption bands, usually called Tl^+ centers. According to the result suggested by Smakula, [4] the product of optical density times band width for these absorption bands is proportional to the concentration of thallous ions, which have been incorporated into KCl single crystals. One of these characteristic absorption bands, labeled A, at 247nm, was used as a tracer to estimate the concentration of Tl^+ ion in our work. [5,6]

To demonstrate how the concentration of Tl^+ ion is related to the optical density (*O.D*) in our experiment, we consider the Smakula's formula and definition of mean optical density.

For evaluation of the concentration distribution of Tl^+ ion in KCl, the optical density of the A band was measured at room temperature after repeatedly taking off a layer of several to 10-20μm from the diffusion surface by polishing it on lapping film with ethanol. The results are shown in Fig. 1.

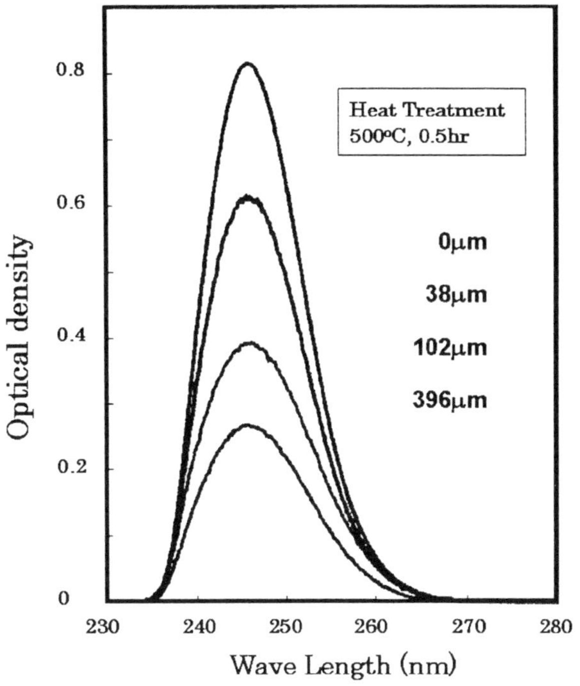

Fig.1 Characteristic absorption band of Tl^+ ions in a KCl crystal. The listed values of length are the thickness of the polished off thin layer of the crystal surface

The mean optical density of the A band of removed thin layer can be defined and determined by

$$\frac{\Delta O.D}{\Delta x} = \frac{O.D_1 - O.D_2}{\Delta x}, \qquad (1)$$

where x is the distance from the surface of the crystal to the middle of a thin layer, $O.D_1$ is the measured optical density before polishing, $O.D_2$ is the measured optical density after polishing, and Δx is the thickness of the removed layer.

This procedure was repeated to a depth of about 100-500μm from the surface. The corresponding values of the thickness are read on a micrometer with direct reading to 1μm.

By Smakula's formula and definition of the optical density, we get

$$\frac{\Delta O.D}{\Delta x} = kC(x.t), \qquad (2)$$

where k is the proportionality constant, and $C(x,t)$ is the concentration of the Tl^+ ions. The concentration $C(x,t)$ can be determined from the experimental data $\Delta O.D$ and Δx.

2.1.2 *Estimation of diffusion coefficient*

The migration of Tl^+ ions into the KCl crystal during the diffusion process is described by a simple one-dimensional Fick's equation

$$\frac{\partial C(x,t)}{\partial t} = \frac{\partial}{\partial x}\left(D\frac{\partial C(x,t)}{\partial x}\right), \qquad (3)$$

where C is concentration of the diffusing ions, x is the diffusion distance, and D is the diffusion coefficient. If D does not depend on position, the diffusion equation becomes

$$\frac{\partial C(x,t)}{\partial t} = D\left(\frac{\partial^2 C(x,t)}{\partial x}\right), \qquad (4)$$

where t is the time of heat treatment.

The analytical solution of this equation for the boundary condition in our experiment of case L-S and V-S is [7]

$$C(x,t) = C_0 erfc\left(\frac{x}{2\sqrt{Dt}}\right), \qquad (5)$$

in which C_0 is the constant concentration at the surface. This result gives

$$\frac{dC(x,t)}{dx} = -\frac{C_0}{\sqrt{\pi Dt}}\exp\left(-\frac{x^2}{4Dt}\right). \qquad (6)$$

A convenient expression to determine D from the macroscopic measurable quantities can be obtained by combining equation (2) and (6),

$$\ln\left(-\frac{\Delta}{\Delta x}\left(\frac{\Delta O.D}{\Delta x}\right)\right) = \ln\left(\frac{kC_0}{\sqrt{\pi Dt}}\right) - \frac{x^2}{4Dt}. \qquad (7)$$

The diffusion coefficient D can be obtained from the plot of the logarithm of the derivative of the concentration $C(x,t)$ of ions as a function of the square of distance, x^2, of the layer from the surface of the diffusion sample. The diffusion coefficient can be determined by estimating the slope of this straight line.

On the other hand, the solution of equation (4) for the boundary condition in the case S-S mentioned above is [8]

$$C(x,t) = \left(\frac{C_0}{\sqrt{\pi Dt}}\right)\exp\left(-\frac{x^2}{4Dt}\right), \qquad (8)$$

here $C(x,t)$ is the concentration of Tl^+ ions at x after an annealing period t and C_0 is the initial concentration at the surface. The mean optical density of a removed layer is proportional to the concentration at approximately the center of the layer, and thus a plot of the logarithm of the mean optical density against the square of the distance should give a straight line with a slope of $-1/4Dt$. Then the diffusion coefficient D can be obtained from this slope.

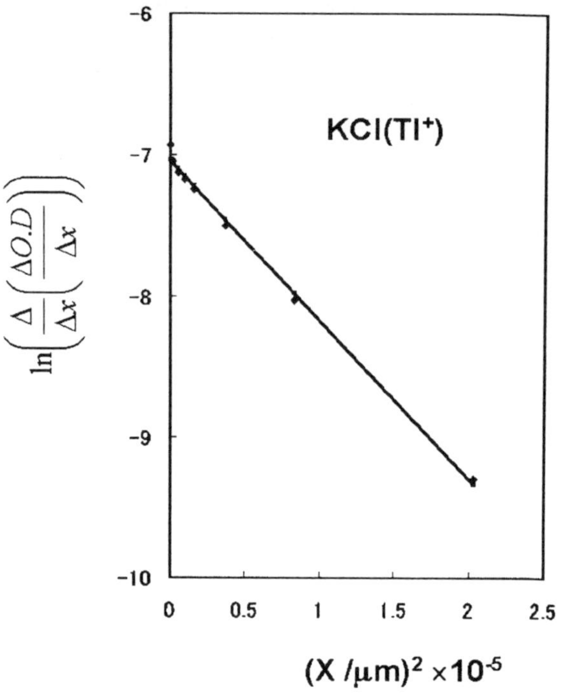

Fig.2. Logarithm of the differential of mean optical density of Tl^+ center vs square of penetration depth after heat-treated at 500°C for 12 minutes in KCl (case L-S).

Figure 2 shows a typical example, which was plotted as the logarithm of the derivative of the Tl^+ ion concentration versus the squared penetration depth. This figure shows the penetration of Tl^+ ions diffusing into the KCl crystal by heating at 500°C for 12 minutes through the liquid-solid interface.

2.1.3 *Estimation of activation energies*

The activation energy of diffusion process can be determined from the temperature dependence of the diffusion coefficient.

Over a wide range of temperatures, experimentally measured diffusion coefficients often fit the relation.

$$D = D_0 \exp\left(-\frac{E}{k_B T}\right), \qquad (9)$$

where D_0 is a constant, k_B is the Boltzmann constant, T is the absolute temperature and E is the activation energy of the diffusion process. In above formula, both E and the pre-exponential factor D_0 are independent of temperature.

In our work, the diffusion coefficients of Tl^+ ion in KCl crystal as a function of temperature are represented by the fully drawn curves in Fig.3, where the logarithms of the experimental values of diffusion coefficient D versus the inverse of the absolute temperatures are plotted

2.3 *Results*

From the slope of the straight line in Fig. 3, the activation energies of the diffusion of Tl^+ ions into KCl single crystal are derived as 1.08eV, 1.09eV and 4.36eV for the cases of solid-solid, vapor-solid and liquid-solid interfaces. There are two characteristic trends, observed in Fig. 3. First, the values of diffusion coefficient D in the case of L-S are very large compared to those in the case V-S and the case S-S at the same temperature. For example, the ratio $D_{(L\text{-}S)}/D_{(V\text{-}S)}$ is about 10^4 at 500°C. Secondly, The activation energy of Tl^+ ion in the case L-S is much larger than that of Tl^+ ions in the case V-S and the case S-S.

3. Discussion

In the case of the vapor-solid interface, the diffusion coefficient of Tl^+ ion and the activation energy E for the diffusion was derived as 7.01×10^{-8} cm^2/s at 520℃ and 1.09eV, respectively.

Tamai [9] studied thermal diffusion of Tl^+ ion in KBr single crystal by heating in Tl metal vapor from 520℃ to 700℃. The diffusion coefficient of Tl^+ ion at 520℃ and the activation energy E for the diffusion were give to be 1×10^{-11} cm^2/s and 1.98eV, respectively.

It is well known that the case V-S and S-S can be explained by a classical hopping model,[1-3,9] which is generally accepted for alkali halides of the NaCl-type. According to this model, the activation energy E of an elementary diffusion process can be expressed as

$$E = \frac{E_f}{2} + E_j, \qquad (10)$$

where E is the activation energy of the diffusion process, which can be measured experimentally. E_f is the energy required to form a Schottky vacancy-pair and E_j is the motion energy that is required to move an atom to a vacancy in the nearest neighbor site.

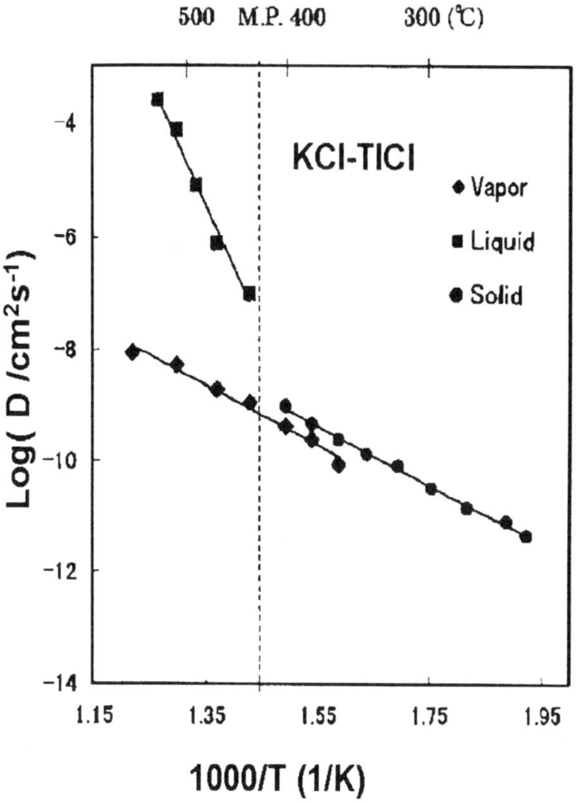

Fig.3 Diffusion coefficient D vs reciprocal temperature 1/T in the cases of V-S(◆), L-S(■) and S-S(●).

Rittner, et al.[10] reported the theoretical and experiment value of vacancy-pair formation energy E_f as 1.86eV and 2.22~2.3eV in KCl crystal, respectively. And Henderson [11] reported the theoretical value as 1.75~2.3eV. So we can resonably assume that the value is about 2eV. Then the motion energy E_j of a Tl^+ ion required to move to the vacancy in a nearest neighbor site can be derived as about 0.1eV at case S-S and V-S in our experiment.

Comparing the motion energy of K^+ ion in KCl, which is 0.7eV, [11] the E_j of the Tl^+ ion is small. We can understand that this difference is caused by the difference of the mass and the volume of K^+ and Tl^+ ions.

In contrast to the systems mentioned above, Ag^+ ions in an AgI crystal of α-phase have been known to be extremely mobile. Mobility is comparable to that of an ion in a molten salt. Such a material as AgI is named superionic conductor. The activation energy for the diffusion and the diffusion coefficient of Ag^+ ion in an α-AgI crystal are 0.048eV and the order of 10^{-5} cm^2/s.[12]

On other hand, in the case of the liquid-solid interface (case L-S), the activation energy for the diffusion of Tl^+ ions in KCl crystals is measured as 4.36eV. This value is much larger than that of the Ag^+ ion in an α-AgI crystal and 2 times or more larger than that of the Tl^+ ion in alkali halide crystals which were heated in TlCl vapor or the Tl metal vapor (case V-S and S-S). The diffusion coefficients of Tl^+ ions at 500°C in this case is nearly equal to that of Ag^+ ion in an α-AgI crystal and much larger than that of Tl^+ ion in alkali halide crystals with interface cases V-S and S-S.

We believe that the phenomenon of Tl^+ ion diffusion in KCl single crystals in case L-S can not be explained by the same model as that used in cases S-S and V-S.

3.1 *Hoping model*

The atomistic theory of diffusion is used to describe the processes by which an atom gets from one part of a crystal to another. The lattice sites in a crystal are generally taken as the fixed locations of atoms making up the crystal. The study of specific heats clearly demonstrates that atoms oscillate around these lattice sites, which are their equilibrium position. These oscillations lead to a finite probability that an atom will move from its lattice site to another position in the crystal. There are several mechanisms by which atoms can move from one site in the crystal. One of these classical hopping models, called the Schottky defect model is generally accepted for the alkali halides crystal of the NaCl-type.

As is well know, in thermal equilibrium, any crystal at a temperature above absolute zero contains a certain number of vacant lattice sites, and this number is generally a function of temperature. These vacancies provide an easy path for diffusion. When a lattice atom moves into an adjacent vacant site, this process is called the hopping diffusion mechanism. The atom jump in the hopping mechanism is the jump of an atom into a neighboring vacancy. The site previously occupied by the atom then is vacant, so that the atom and vacancy merely exchange their positions in effect. Each atom moves through the crystal by making a series of exchanges with the various vacancies, which are in its vicinity from time to time.

The transfer of mass and charge occurring in alkali halide crystals can be described by means of an ionic migration process. The ionic migration processes of the transport phenomena are related to the macroscopic quantities measured in a laboratory by means of the phenomenological law of steady-state diffusion and the theory of random walk. The steady-state diffusion equation, Fick's second law, can be expressed as

$$\frac{\partial C}{\partial t} = \nabla \cdot D \nabla C , \qquad (11)$$

where C is the concentration of diffusing ions, D is the diffusion coefficient. An expression for the mean-square displacement (σ) of the diffusing ions can be obtained from the general solution of Eq. (11). The result is

$$\sigma^2 = 6Dt , \qquad (12)$$

where t is the time.

The mean-square displacement of the diffusing ions can also be calculated from an analysis of the motion of the individual ions. If an ion makes Γ random jumps per unit time, and if all the jumps are of the same length β, the random walk theory is applicable and gives the result that the mean-square displacement is equal to the product of the total number of jumps times the square of the length of each jump,

$$\sigma^2 = \beta^2 \Gamma t . \qquad (13)$$

An expression for D, a macroscopic measurable quantity, in terms of the microscopic behavior of the individual ions, or vacancies, in the crystal is obtained by combining Eqs. (12) and (13), i.e.,

$$D = \frac{1}{6} \beta^2 \Gamma . \qquad (14)$$

An expression for the jump probability for a vacancy in ionic crystals has been derived using several different approaches, e.g., absolute rate theory, many-body theory of equilibrium statistics, and the dynamical theory of diffusion. All of these derivations have resulted in an expression of the following form:

$$\upsilon = \upsilon_0 \exp\left(-\frac{E_j}{kT} \right) , \qquad (15)$$

where v is jump frequency of an elementary jump process with which the atom and a neighboring vacancy exchange their positions. The E_j, motion energy, is the height of the free-energy barrier that an ion must surmount in order to pass to the adjacent vacancy. The ν_0, hopping frequency, is an effective vibration frequency of the ion in the initial site.

If the number of nearest neighboring sites around an atom are Z, the mole fraction of Schottky defects is N_v and the probability per unit time that a vacancy will jump from one position to another is ν, then the jump frequency of diffusion process in a specific ionic crystal lattice, Γ, is

$$\Gamma = ZN_v\upsilon . \qquad (16)$$

So that the expression for the diffusion coefficient of the individual ion, or vacancy, becomes

$$D = \frac{1}{6}\beta^2 ZN_v\upsilon_0 \exp\left(-\frac{E_j}{kT}\right). \qquad (17)$$

The diffusion coefficient is determined experimentally from the diffusion of tracer ions. Hence, the diffusion equations must be modified to include a correlation factor f that accounts for the nonrandom motion of the tracer ions. So, the diffusion coefficient of the tracer ion, or impurity ion, is equal to the product of the correlation factor times the diffusion coefficient of the individual ions, or vacancies, in the same crystal lattice, as follows,

$$D = \frac{1}{6}f\beta^2 ZN_v\upsilon_0 \exp\left(-\frac{E_j}{kT}\right). \qquad (18)$$

For the diffusion coefficients of impurity ions (Tl^+ ions in our experiment) in two diffusion processes, the cases of liquid-solid and vapor-solid, applying the result of formula (18),

$$D_{L\to S} = \frac{1}{6}f\beta_{L\to S}{}^2 Z_{L\to S}N_{vL\to S}\upsilon_0 \exp\left(-\frac{E_{jL\to S}}{kT}\right) \qquad (19)$$

and

$$D_{V\to S} = \frac{1}{6}f\beta_{V\to S}{}^2 Z_{V\to S}N_{vV\to S}\upsilon_0 \exp\left(-\frac{E_{jV\to S}}{kT}\right) \qquad (20)$$

are derived. The marks $L\to S$ and $V\to S$ mean the condition of diffusion interface. Comparing the Eqs. (19) and (20), we get

$$\frac{D_{L\to S}}{D_{V\to S}} = \frac{\beta_{L\to S}{}^2 Z_{L\to S}N_{vL\to S}}{\beta_{V\to S}{}^2 Z_{V\to S}N_{vV\to S}} \cdot \exp\left(-\frac{1}{kT}\left(E_{jL\to S} - E_{jV\to S}\right)\right). \qquad (21)$$

The value of term in left-hand side, the ratio $D_{(L-S)}/D_{(V-S)}$, can be estimated with our experimental results. For example, it is about 10^4 at 500°C. In the right-hand side, according to classical hopping model, the crystal structure and imperfection distribution is independent of diffusion interface, the relations can be expressed as $\beta_{L-S}=\beta_{V-S}$, $Z_{L-S}=Z_{V-S}$, $E_{jL-S}=E_{jV-S}$ and $N_{vL-S}=N_{vV-S}$. Then the value of term in right side of equation (21) must equal 1.

Because the difference of the values of terms in left-hand side and right-hand side can not be explained by classical hopping model, we have to suggest the onset of a new possible mechanism to explain such a deviation from the behavior of the classical hopping model.

3.2 *A possible mechanism*

To understand the results of above formula (21), we try to extend the meaning of the hopping model. Firstly, when there is a vacancy-rank (which consists of neighboring vacancies in succession) in the nearest neighboring site of a Tl^+ ion, we can assume that the Tl^+ ion can move in a rank from one end to the other by one step. With this assumption, the formation energy (E_{fL-S}) of a vacancy-rank is nearly equal to the product of the number of vacancies and the formation energy (E_{fV-S}) of the monovacancy. Secondly, we assume that the effective jump energy in a vacancy-rank is nearly equal to the jump energy of monovacancy, $E_{jL-S}\approx E_{jV-S}$; there is no potential barrier for a Tl^+ ion to move inside the vacancy-rank. According to this new possible mechanism, named as the Multi Jump Model (MJM), equation (21) can be changed to

$$\frac{D_{L\to S}}{D_{V\to S}} = \left(\frac{Z_{L\to S}}{Z_{V\to S}}\right)\left(\frac{N_{vL\to S}}{N_{vV\to S}}\right)\left(\frac{\beta_{L\to S}{}^2}{\beta_{V\to S}{}^2}\right). \qquad (22)$$

The parentheses on the right-hand side of Eq. (22) emphasize the fact that the ratio is proportional to the product of the three terms, which can be obtained by individual investigation.

According to our experimental results shown in Fig.3, the values of D in the case of L-S are very large compared to those in the case V-S and the case S-S. For example, the ratio $D_{(L-S)}/D_{(V-S)}$ is about 10^4 at 500°C. Using this value, we get

$$\left(\frac{Z_{L\to S}}{Z_{V\to S}}\right)\left(\frac{N_{vL\to S}}{N_{vV\to S}}\right)\left(\frac{\beta_{L\to S}{}^2}{\beta_{V\to S}{}^2}\right) \approx 10^4 . \qquad (23)$$

From Eq. (10), $E_{jV-S}=E_{V-S}-E_{fV-S}/2$. In the case of V-S of our experiment, $E_{jV-S}=1.08$eV, and the average value of formation energy of Schottky pair in KCl crystal is 2.0±0.1eV.[10] So the $E_{fV-S}/2=1$eV and the $E_{jV-S}=0.1$eV, respectively. As mentioned before, the ratio E_{L-S}/E_{V-S} is about 4. Therefore from these values and Eq. (10) we can estimate the ratio E_{fL-S}/E_{fV-S} is nearly equal to 4.2. This means that the vacancy-rank consists of 4 neighboring vacancies in succession. With this result, the average length of vacancy-rank is about four lattice intervals. The value β_{L-S} is assumed to be about 4 times the amount of nearest site interval. This value is the same as 4 times lattice constant of KCl crystal. So the term $(\beta_{L-S}/\beta_{V-S})^2$ can be calculated to be about 16.

On the other hand, Z_{V-S} is the number of the nearest neighbor sites in the V-S case and is equal to 12 for the NaCl-type crystal lattice. Z_{L-S} is the number of sites around the vacancy-rank in the L-S case and is assumed as the total number of 1st nearest neighbor sites up to 4th neighbor sites. Then Z_{L-S} is 54 and the ratio of Z_{L-S}/ Z_{V-S} is 4.5.

From formula (23) and the above results, we can calculate the ratio of N_{vL-S}/N_{vV-S} is about 10^2. This result means that the number of vacancies (the vacancy concentration) in the L-S case is at least 10^2 times as large as that in the V-S case.

Using this model, the vacancy concentration in the L-S case can be deduced to be much larger than those in the V-S case over the temperature range of 430~520°C. For example, their ratio is about 10^2 at 500°C. These vacancies of high concentration are considered to result in the long distance migration via 4 vacancies.

This result may be supported partly with the phase diagram of the system KCl-TlCl.[13] Molten TlCl and KCl crystals can be dissolved with each other. Because the diffusion rate in the liquid is much larger than in the solid, the diffusion process is not in an equilibrium state in our experiment. The large difference of dissolution of the Tl^+ ion in a KCl crystal and the K^+ ion in molten TlCl can be confirmed in this case. We believe that this large difference in dissolution caused the high concentration differential to occur.

A high chemical concentration gradient in the molten TlCl and KCl crystals may induce the large difference in dissolution. In this case, the driving force for the diffusion process must be derived by a consideration of thermodynamic quantities, such as chemical potential gradients in nonideal solutions.[14]

References

1. R. J. Tiernan and B. J. Wuensch, *J. Chen. Phys.*, **55** (1971) 4996.
2. A. Glasner and R. Reisfeld, *J. Phys. Chem. Solids*, **81** (1961) 345.
3. E. R. Dobrovinskaya and N. M. Podorzhanskaya, *Ukr. Fiz. Zhorn.*, **11** (1966) 227.
4. A. Smakula, *Z. Phys.* **59** (1930) 603, and W. B. Fowler, *"Physics of Color Centers"*, (Academic Press, 1968).
5. W. Yu and A. Fujii, *Phys. Rep. Kumamoto Univ.*, **10** (1998) 209.
6. W. Yu and A. Fujii, *Inter. J. Mod. Phys.*, **B16** (2002) 108.
7. See, for example, J. Crank, *"The Mathematics of Diffusion"*, (Clarendon Press, Oxford, 1975). 32
8. See, for example, J. Crank, *"The Mathematics of Diffusion"*, (Clarendon Press, Oxford, 1975). 13
9. T.Tamai, *J. Phys. Soc. Japan*, **16** (1961) 2463.
10. E.S.Rittner, *J. Chem. Phys.*, **17** (1959) 198.
11. R. Henderson, *Prog. Mat. Science*, **10** (1963) 151
12. W. Jost and J. Nolting, *Z. Phys. Chem.*, **7** (1956) 383.
13. R.S.Roth, M.A.Clevinger and D.Mckenna, *"PHASE DIAGRAMS FOR CERAMISTS 1969 SUPPLEMENT"* edited by THE AMERICAN CERAMIC SOCIETY INC., (1969)291
14. N.B.Hannay, *"Solid-state chemistry"*, (Prentice-Hall, Inc., Englewood Cliffs, New Jersey, 1967) Chapter 5.

PRESSURE DEPENDENCE OF THE EFFECTIVE CHARGE IN COPPER HALIDES

MASARU ANIYA and FUYUKI SHIMOJO
Department of Physics, Faculty of Science, Kumamoto University, Kumamoto 860 8555, Japan

After showing how the effective charges are related to the superionic properties, the pressure dependence of the localized effective charges in copper halides is focused, where interesting material dependence has been reported. Namely, the localized effective charges in CuI, CuBr and CuCl increases, remains almost constant and decreases with pressure, respectively. By using the results of an *ab initio* simulation, it is found that in CuI, the Cu d-electrons contribute largely to the localized effective charge. For CuCl this contribution is diminished when compared with CuI.

1. Introduction

At ambient temperature and pressure, the copper halides CuCl, CuBr and CuI are some of the most ionic zinc-blende structure compounds [1]. They undergo a number of structural phase transitions under pressure [2,3], exhibit ionic conductive phases at high temperatures [4], and exhibit many other unusual physical properties mainly related to lattice dynamics.

A quantity of fundamental importance in the study of the dynamics of crystalline lattices is the effective charge of the ions. Since this quantity is expected to be intimately related with the superionic conducting properties, in recent years, many experimental and theoretical works have been devoted on this subject [5-12].

In the present paper, after showing how the effective charges are related to the superionic properties, we focus on the pressure dependence of the localized effective charge, where interesting material dependence among CuCl, CuBr and CuI has been reported [12].

2. The Effective Charges of Superionic Materials

A large part of papers dealing with effective charge concerns the behavior of transverse or Born effective charge e_T^* [13-17]. For diatomic crystals, e_T^* measures the macroscopic polarization induced by a relative displacement of the cation and anion sublattices. This quantity is thus, implicitly related to the bond polarity of the material [16].

Experimentally, the transverse effective charge is obtained from [17],

$$e_T^* = \left[\frac{\varepsilon_\infty \mu V}{4\pi}\left(\omega_{LO}^2 - \omega_{TO}^2\right)\right]^{1/2}, \qquad (1)$$

where ε_∞ is the high frequency dielectric constant, μ is the reduced mass of the ions, and V is the volume per formula unit. ω_{LO} and ω_{TO} are the longitudinal and transverse phonon frequencies, respectively.

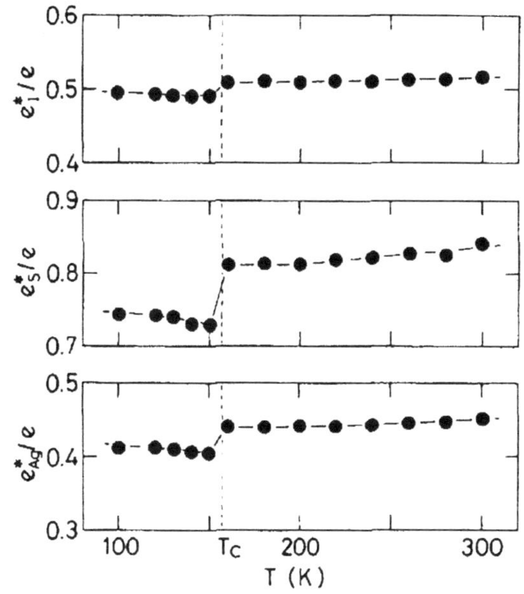

Figure 1. Temperature dependence of the transverse effective charge in Ag_3SI [9].

By studying the temperature dependence of e_T^*, important insight to understand the superionic behavior is obtained. For instance, it has been shown for AgI [9], Ag_3SI [9,10], Ag_3SBr [5] and CuI [8] that the transverse effective charge decreases with temperature in the non-superionic phase, whereas in the superionic phase it increases. As an example, the temperature dependence of e_T^* for Ag_3SI is shown in Figure 1 [9]. The characteristic behavior of the temperature dependence of

e_T^* has been interpreted by using the bond fluctuation model of superionic conductors [18-20]. This model suggests that the change of bonding that occurs locally and fluctuates in time is the key mechanism for the understanding of superionic properties. The increase of e_T^* with temperature in the superionic phase is interpreted to arise from the increased number of atoms bonded ionically. The bond fluctuation model has been corroborated theoretically by using a pseudopotential method [19], and more recently, by an *ab initio* molecular dynamics simulations [21].

3. Pressure Dependence of the Effective Charge

The transverse effective charge e_T^* is related to the bond polarity of the material. By applying pressure, the overlap of wave functions increases, leading to an increase of bond metallicity and a decrease in ionicity. Indeed, the e_T^* measured experimentally for a number of III-V zinc-blende compounds decreases in magnitude under the pressure [15].

The transverse effective charge includes contributions from charge localized near the ion sites, as well as charge distributed through the unit cell [22]. It has been pointed out by Burstein and others [22] that in semiconductors, the effective charge due to valence electrons which have extended nature, is not effective to produce the Lorentz field as in the case of good ionic crystals such as alkali halides. They showed that only localized charges give rise to dipolar forces that contribute to the TO-phonon frequency. Burstein suggested that there were two components to the macroscopic or transverse effective charge, such that

$$e_T^* = e_l^* + e_{nl}^* , \tag{2}$$

where e_l^* and e_{nl}^* are the localized and non-localized charges, respectively. It has been also shown that for zinc-blende and wurtzite structured crystals, the localized effective charge correlates with the Phillips ionicity f_i as,

$$e_l^* = Z_{eff} f_i , \tag{3}$$

where Z_{eff} is the effective chemical valence [22].

It is well known that compounds which have Phillips ionicity near the critical value f_c =0.785, exhibit high ionic conductivity. Therefore, it may be interesting to study the behavior of localized effective charge.

The pressure dependence of the localized effective charge in copper halides has been studied previously [12], and the result is shown in Figure 2. The behavior shown in Figure 2 has been obtained by using elastic constant data [23]. The same method has been used also in the study of localized effective charges and lattice vibrational properties of zinc-blende structure compounds [24].

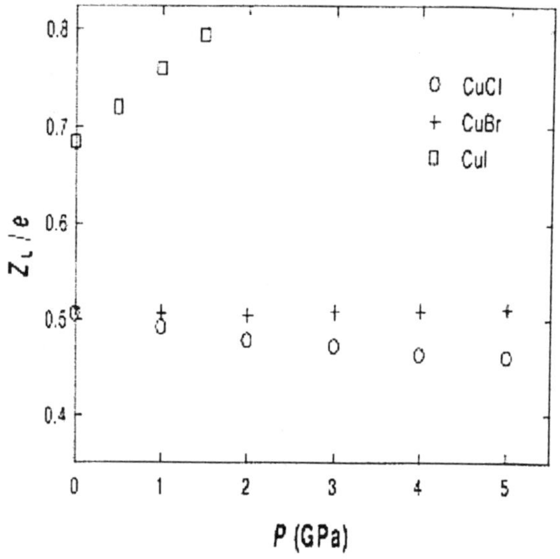

Figure 2. Pressure dependence of the localized effective charge in copper halides [12].

Figure 2 exhibits a very interesting material dependence. The localized effective charge increases, remains almost constant and decreases with pressure in CuI, CuBr and CuCl, respectively. By making connection with the result of previous section, there is a possibility that such differences might be related with the difference in the ionic transport properties.

Concerning the ionic transport in copper halides, some suggestions have been done on the high pressure phase, based on structural studies [25]. However, as far as the authors are informed, no direct ionic conductivity measurements under pressure exist in the zinc-blende structured phase.

Although showing a correlation with the Phillips ionicity, the localized effective charge e_l^* is a model dependent quantity. Therefore, it is difficult to understand the pressure dependence shown in Fig. 2 from a microscopic point of view. In the following, we present the result of an *ab initio* simulation study which may help to gain further insights to understand the obtained material dependence shown in Figure 2.

In the study, static and dynamic calculations were performed. In the static case, the atomic positions are frozen in the lattice sites, whereas in the dynamic case, the atomic motions are included. The electronic states were calculated within the framework of the density functional theory, in which the generalized gradient approximation was used for the exchange-correlation energy [26]. The energy functional was minimized using an iterative scheme based on the preconditioned conjugate-gradient method [27]. The ultrasoft pseudopotentials were used [28]. The electronic wave functions and the charge density were expanded in plane waves with cutoff energies of 17 Ry and 120 Ry, respectively. The system size was 64 (32 Cu + 32 halogens) atoms in a cubic supercell with periodic boundary conditions. For the valence electrons, we considered Cu: $3d^{10}4s^14p^0$, Cl: $3s^23p^53d^0$ and I: $5s^25p^55d^0$. In the dynamic case, the Nosé-Hoover thermostat technique was used [29,30], in which the temperature was set to be 300 K, and the equations of motion were solved via the velocity Verlet algorithm with a time step $\Delta t = 2$ fs.

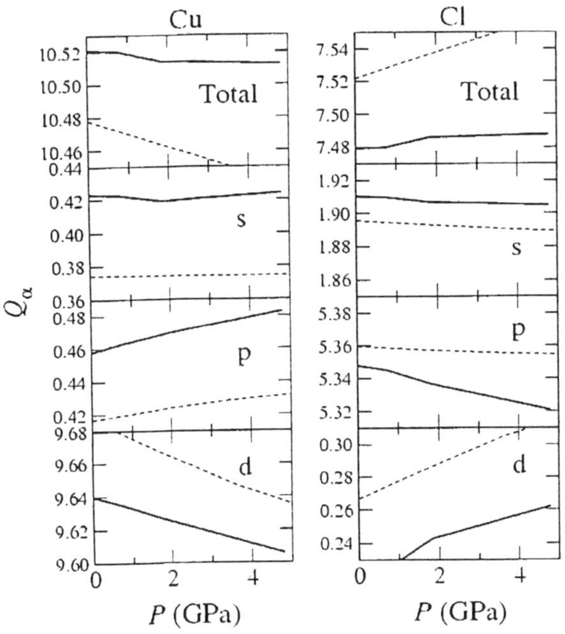

Figure 3. Pressure dependence of the static (broken line) and dynamic (full line) charge on Cu and Cl in CuCl. The decomposition of the total charge on different orbitals is also shown.

The results of the calculation for CuCl and CuI are shown in Figure 3 and Figure 4, respectively. The vertical axes in these figures represent the electronic charge, or in other words, the number of valence electrons present on the ions. We can note that in CuCl, the total charge on Cu (Cl) decreases (increases) with pressure. On the other hand, in CuI the total charge on Cu (I) increases (decreases) with pressure. In other words, in CuCl the charge transfer from Cu to Cl increases with pressure, whereas in CuI the charge transfer from Cu to I decreases with pressure. This behavior of the total charge is opposite to the trend of the localized effective charge shown in Figure 2. In Figure 3 and Figure 4, we can also note the effects of the temperature on the charges through the difference in the static and dynamic charges. The decomposition of the total charge on different orbitals are also shown.

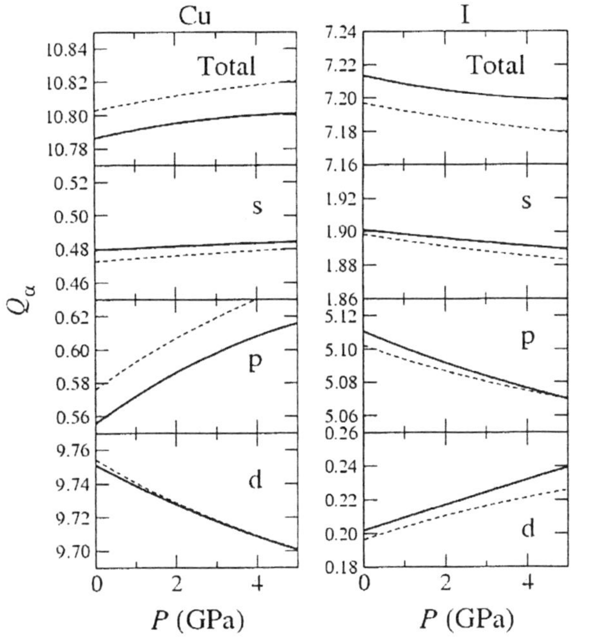

Figure 4. Pressure dependence of the static (broken line) and dynamic (full line) charge on Cu and I in CuI. The decomposition of the total charge on different orbitals is also shown.

The valence band of copper halides is formed by strongly hybridized d-electrons of copper and p-electrons of halogens [31]. At ambient pressure, the fraction of density of states with p-symmetry at the valence band is estimated to be approximately 25% in CuCl, 35% in CuBr and 50% in CuI [31]. This fact suggests that in CuI, the d-electrons are more localized than in the other two compounds. Indeed, the small difference between the static and dynamic charges shown in Figure 4 supports this view. Therefore, we may say that the localized effective charge e_l^* in CuI is mainly due to Cu d-electrons. An additional support to this view is that the number of Cu d-electrons in CuI decreases with pressure, which means that the localized effective charge on Cu increases with pressure as shown in Figure 2.

On the other hand, the situation in CuCl is very different. In this compound, the Cu d-electrons contribute largely to the density of states of the valence band. This is reflected in the large difference between the static and dynamic charges of Cu d-electrons shown in Figure 3. Therefore, in CuCl the contribution of Cu d-electrons to the localized effective charge diminishes. The d-electrons in CuCl hybridize strongly with the atomic orbitals of Cl and play an active role in the dynamic process of the atoms.

4. Conclusion

A quantity of fundamental importance in the study of the dynamics of crystalline lattices is the effective charge of the ions. An interesting material dependence was reported for the pressure dependence of the localized effective charge by one of the authors [12]. It was found that the localized effective charges in CuI, CuBr and CuCl increases, remains almost constant and decreases with pressure, respectively. In the present paper, by using the result of an *ab initio* simulation, the origin of the material dependence of the localized effective charge has been investigated. It is found that in CuI, the Cu d-electrons contribute largely to the localized effective charge. On the other hand, in CuCl this contribution is diminished when compared with CuI.

Acknowledgments

This work was supported in part by a Grant-in-Aid for Scientific Research from the Japan Society for the Promotion of Science (No. 15560586).

References

[1] J.C.Phillips, Rev.Mod.Phys. **42**, 317 (1970).
[2] E. Rapoport and C.W.F.T.Pistorius, Phys.Rev. **172**, 838 (1968).
[3] S.Hull and D.A.Keen, Phys.Rev. **B 50**, 5868 (1994).
[4] T.Jow and J.B.Wagner, J.Electrochem.Soc. **125**, 613 (1978).
[5] K.Wakamura, Solid State Commun. **82**, 705 (1992).
[6] K.Wakamura, J.Phys.Chem.Solids **59**, 591 (1998).
[7] T.Tomoyose, J.Phys.Soc.Jpn. **64**, 1616 (1995).
[8] M.Aniya, Mem.Fac.Gen.Educ., Kumamoto Univ., Natur.Sci. **31**, 17 (1996).
[9] M.Aniya and K.Wakamura, Physica **B 219&220**, 463 (1996).
[10] M.Aniya and K.Wakamura, Solid State Ionics **86-88**, 183 (1996).
[11] T.Tomoyose, A.Fukuchi and M.Aniya, J.Phys.Soc. Jpn. **65**, 3692 (1996).
[12] M.Aniya, Solid State Ionics **121**, 281 (1999).
[13] P.Lawaetz, Phys.Rev.Lett. **26**, 697 (1971).
[14] S.Katayama and H.Kawamura, Solid State Commun. **21**, 521 (1977).
[15] K.Aoki, E.Anastassakis and M.Cardona Phys.Rev. **B 30**, 681(1984).
[16] T.Sengstag, N.Binggeli and A.Baldereschi, Phys. Rev. **B 52**, R8613 (1995).
[17] F.J.Manjón, J.Serrano, I.Loa, K.Syassen, C.T.Lin and M.Cardona, Phys. Rev. **B 64**, 064301 (2001).
[18] M.Aniya, Solid State Ionics **50**, 125 (1992).
[19] M.Aniya, J.Phys.Soc.Jpn. **61**, 4474 (1992).
[20] M.Aniya, Rec.Res.Devel.Phys.Chem.Solids **1**, 99 (2002).
[21] F.Shimojo and M.Aniya, J.Phys.Soc.Jpn. **72**, (2003) 2702.
[22] G.Lucovsky, R.M.Martin and E.Burstein, Phys.Rev. **B 4**, 1367 (1971).
[23] R.K.Singh and D.C.Gupta, Phys.Rev. **B 40**, 11278 (1989).
[24] H.Neumann, Cryst.Res.Technol. **20**, 773 (1985).
[25] S.Hull, D.A.Keen, W.Hayes and N.J.G.Gardner, J.Phys.Cond.Matt. **10**, 10941 (1998).
[26] J.P.Perdew, K.Burke and M.Ernzerhof, Phys. Rev. Lett. **77**, 3865 (1996).
[27] F.Shimojo, Y.Zempo, K.Hoshino and M.Watabe, Phys.Rev. **B 52**, 9320 (1995).
[28] D.Vanderbilt, Phys. Rev. **B 41**, 7892 (1990).
[29] S.Nosé, Mol.Phys. **52**, 255 (1984).
[30] W.G.Hoover, Phys. Rev. **A 31**, 1695 (1985).
[31] A.Goldmann, Phys.Stat.Solidi **(b) 81**, 9 (1977).

AB INITIO MOLECULAR-DYNAMICS SIMULATIONS OF SUPERIONIC PHASES OF Cu HALIDES AND Ag CHALCOGENIDES

FUYUKI SHIMOJO and MASARU ANIYA

Department of Physics, Kumamoto University, Kurokami 2-39-1, Kumamoto 860-8555, Japan
Fax: +81-96(342)3488, E-mail: shimojo@kumamoto-u.ac.jp

KOZO HOSHINO

Faculty of Integrated Arts and Sciences, Hiroshima University, Higashi-Hiroshima 739-8521, Japan

The dynamic properties of mobile ions in the superionic conductors CuI and Ag_2Se are studied by *ab initio* molecular-dynamics simulations. The superionic behavior of these materials is successfully reproduced by our simulations. To investigate the bonding nature of these materials, the atomic charge of each ion and the bond order between mobile and immobile ions are calculated by the population analysis as a function of time. It is shown that, in CuI, the covalent bonding around the Cu ions weakens when they diffuse in the octahedron cage, and the ionicities for the Cu ions at the octahedral sites are larger than those for the Cu ions at the tetrahedral sites. For Ag_2Se, we demonstrate that the bond order for the mobile Ag ions becomes smaller and their atomic charges become larger accompanied with the diffusive motion. These results reveal that the high ionic conductivity is the outcome of the fluctuation of bonding properties in the materials, and agree well with the predictions of the bond fluctuation model proposed for the conduction mechanism of the mobile ions in the superionic conductors.

1. Introduction

Cuprous halides and silver chalcogenides are typical superionic conductors. The Cu and Ag ions migrate between sites in the immobile halides and chalcogenides ions, respectively. Although the dynamic properties of mobile ions in these materials have been extensively investigated both experimentally and theoretically, there remain problems as for the correlation between the disordered ionic structure and the electronic states. To clarify the microscopic mechanism of high ionic conduction, it is essentially important to consider the electronic states around the mobile ions as a function of time [1-3].

In this study, we investigate the atomic dynamics and the electronic states in the superionic phases of CuI and Ag_2Se by *ab initio* molecular-dynamics (MD) simulations. We focus on the time evolution of electronic states around mobile ions by utilizing the population analysis, and discuss the fluctuation of chemical bonding accompanied with the diffusive motion.

In CuI, the mobile Cu ions stay at two possible sites, tetrahedral and octahedral sites, in the fcc sublattice formed by the immobile I ions. Although the tetrahedral sites are preferable for the Cu ions to stay, the Cu ions must pass through an octahedral site when they migrate between two tetrahedral sites. On the other hand, the immobile Se ions in Ag_2Se form the bcc lattice, which consists of the tetrahedron cages only. The mobile Ag ions migrate directly between two tetrahedral sites, and their diffusion mechanism could be different from that for Cu ions in CuI. It is, therefore, worth while comparing the diffusion mechanism of the mobile ions in these materials.

2. Method of Calculations

The electronic structure calculations were performed within the framework of the density functional theory, in which the generalized gradient approximation [4] was used for the exchange-correlation energy. We used the ultrasoft pseudopotentials [5]. The electronic wavefunctions were expanded in the plane-wave basis set. The energy functional was minimized using an iterative scheme based on the preconditioned conjugate-gradient method [6,7]. The atomic forces were obtained by the Hellmann-Feynman theorem. We used systems of 64 (32Cu+32I) atoms for CuI and 162 (108Ag+54Se) atoms for Ag_2Se in cubic supercells with periodic boundary conditions. Using the Nosé-Hoover thermostat technique [8,9], the equations of motion were solved via the velocity Verlet algorithm with a time step Δt = 2-2.4 fs. The temperature was set to be 700 K for CuI and 500

K for Ag_2Se. The Γ point was used for the Brillouin zone sampling.

3. Results and Discussions

3.1. *Mean Square Displacement*

Figures 1 and 2 show the mean square displacements (MSD's) for CuI and Ag_2Se, respectively. The solid and dashed lines display the MSD's for the anion and cation, respectively, in these materials. The MSD's for Cu and Ag ions have finite slope, while those for I and Se ions remains almost constant at large time, which clearly shows that our *ab initio* MD simulations successfully reproduce the superionic behavior of CuI and Ag_2Se. The diffusion coefficients for the mobile ions obtained from the slope of the MSD's are in reasonable agreement with the experiments.

3.2. *Electronic Density of States*

The electronic densities of states (DOS's) in CuI and Ag_2Se are shown in Figs. 3 and 4, respectively. The origin of the energy is taken to be the Fermi level (E_F = 0). In the partial DOS's for I and Se ions, the electronic states around -13 eV are s-like in character, and those between -7 and 0 eV are p-like. The large peaks in the partial DOS for Cu and Ag ions originate from 3d and 4d electronic states, respectively. It is seen that the d electronic states of Cu and Ag ions hybridize with the p

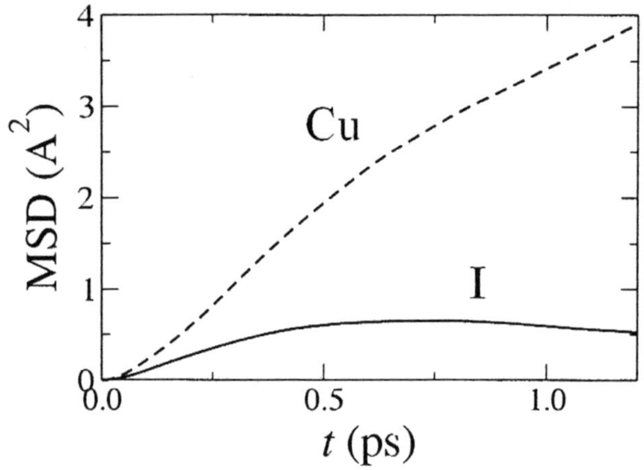

Figure 1. The mean square displacements (MSD's) in CuI. The solid and dashed lines show the MSD's for I and Cu ions, respectively.

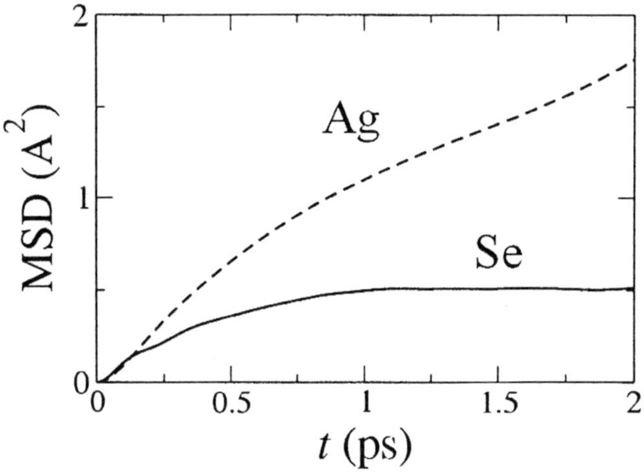

Figure 2. The mean square displacements (MSD's) in Ag_2Se. The solid and dashed lines show the MSD's for Se and Ag ions, respectively.

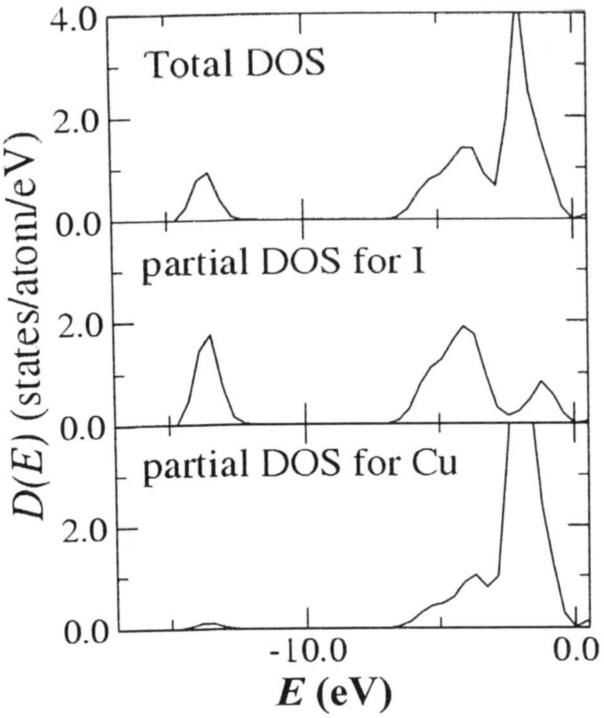

Figure 3. The electronic density of states in CuI.

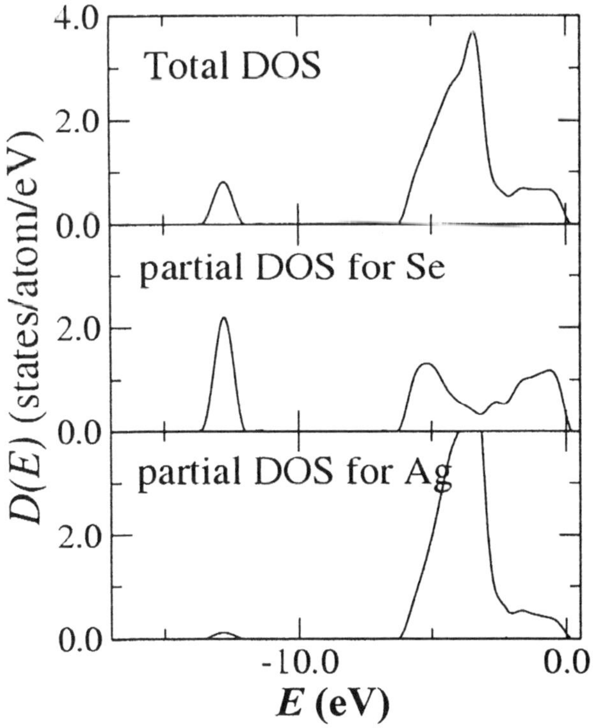

Figure 4. The electronic density of states in Ag_2Se.

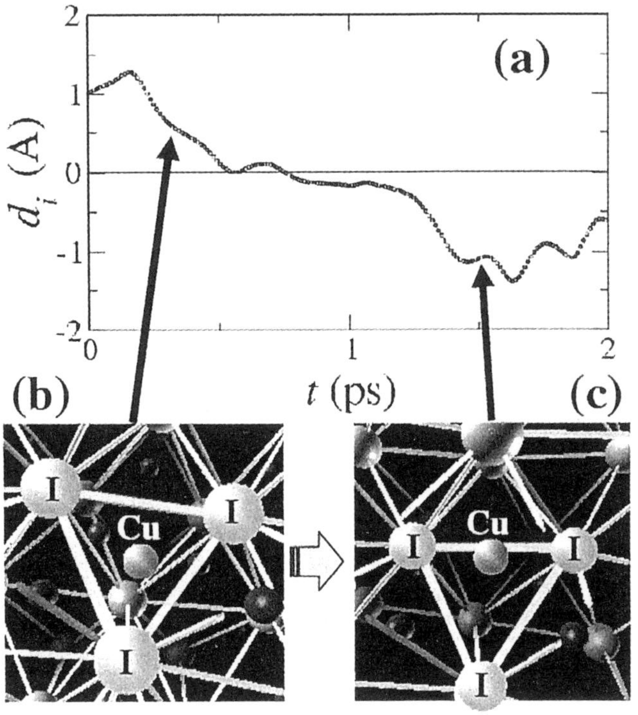

Figure 5. (a) The time evolution of the distance d_i between the Cu ion and the triangle formed by three I ions. (b) The atomic configuration at t=0.3 ps around the Cu ion inside the tetrahedron. The focused triangle is formed by three I ions denoted as 'I'. (c) The atomic configuration at t=1.5 ps around the Cu ion inside the octahedron.

states of the immobile ions with an energy range from -7 to 0 eV. Such p-d hybridization is expected to be an important role for high ionic conductivities.

3.3. *Diffusive Motion of Mobile Ions*

Figure 5 shows a typical example of diffusive motion of Cu ions in CuI. At t=0 ps, the focused Cu ion stays at a tetrahedral site, and it goes on a neighbor octahedral site through a triangle formed by three I ions denoted as 'I' in Figs. 5(b) and 5(c). The focused Cu ion is denoted as 'Cu' in the figures. Figure 5(a) shows the time evolution of the distance d_i between the Cu ion and the triangle. d_i is defined to be positive and negative, when the Cu ion is on the inside and outside of the tetrahedron, respectively. As shown in Fig. 5(a), d_i changes its sign around t=0.6-0.7 ps, which means that the Cu ion moves its position toward the neighbor octahedral site.

A typical example of diffusive motion of Ag ions in Ag_2Se is shown in Fig. 6. The focused Ag ion is denoted as 'Ag' in the Figs. 6(b) and 6(c). Figure 6(a) shows the time evolution of the distance d_i between the Ag ion and the triangle formed by three Se ions denoted as 'Se'. d_i is defined to be positive and negative, when the Ag ion is on the inside and outside of the tetrahedron stayed at t=0, respectively. Since d_i changes its sign around t=0.5 ps as shown in Fig. 6(a), we see that the Ag ion migrates to the neighbor tetrahedral site at this time.

3.4. *Population Analysis*

To investigate the bonding nature of these materials in detail, we utilized the population analysis [10] by expanding the electronic wavefunctions in atomic orbital basis sets. The population analysis describes intuitively the electronic structure of not only crystals but also disordered systems in terms of chemical concepts [11]. Utilizing the formulation of the analysis for the ultrasoft pseudopotentials [12], we calculated the bond order O_i and the atomic charge Q_i for i-th ion as a function of time.

Figure 7(a) and 7(b) show the time evolutions of the Q_i and O_i of the focused Cu ion in CuI, respectively. The time t in the horizontal axis corresponds to that for Fig. 5(a). The solid line in each figure shows the average values of all Cu ions. From Fig. 7 (a), it is seen

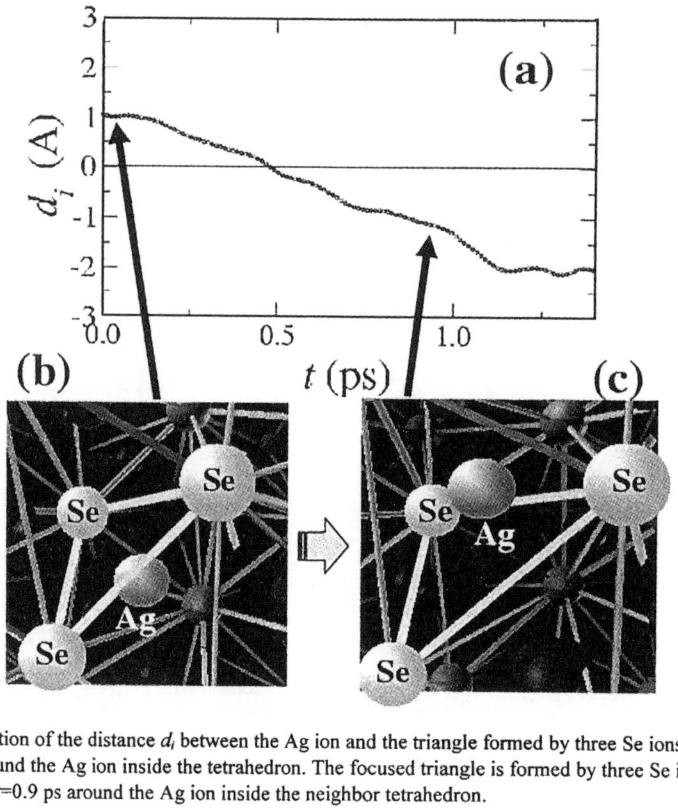

Figure 6. (a) The time evolution of the distance d_i between the Ag ion and the triangle formed by three Se ions. (b) The atomic configuration at t=0.1 ps around the Ag ion inside the tetrahedron. The focused triangle is formed by three Se ions denoted as 'Se'. (c) The atomic configuration at t=0.9 ps around the Ag ion inside the neighbor tetrahedron.

that the Q_i has the largest value at about t=1.5 ps, when the Cu ion migrates in the octahedron, while the values of Q_i are 0.2 or less, when the Cu ion is at the tetrahedral site. These results suggest that the ionicities for the Cu ions at the octahedral sites are larger than those for the Cu ions at the tetrahedral sites. This observation corresponds to the time evolution of the O_i shown in Fig. 7(b). We see that the O_i for the Cu ion in the octahedral sites (at about t=1.5 ps) has smaller values than those in the tetrahedral sites, which indicates that the covalent bonding around the Cu ion weakens when it passes through the octahedral site during the migration between neighboring tetrahedral sites. Such fluctuation of bonding properties can be confirmed by viewing the time evolution of the spatial distribution of the electronic-charge density. It should be noted that around t=0.5 ps the values of O_i are smaller than the average values even when the Cu ion is in the tetrahedral site. It is also seen that O_i around t=1 and 2 ps has larger values, because the distances between the Cu ion and neighboring I ions become smaller when the Cu ion passes through the triangles formed by three I ions.

Figure 8(a) and 8(b) show the time evolutions of the Q_i and O_i of the focused Ag ion in Ag_2Se, respectively. The time t in the horizontal axis corresponds to that for Fig. 6(a). The solid line in each figure shows the average values of all Ag ions. As shown in Fig. 6(a), the Ag ion passes through the triangle formed by three Se ions around t=0.5 ps, and migrates from a tetrahedral site to its neighbor tetrahedral site. We see from Fig. 8(a) that the Q_i has larger values than the averaged values while the Ag ion diffuses around t=0.5 ps. It is also seen from Fig. 8(b) that the O_i has smaller values than the averaged values around t=0.5 ps. These results suggest that the covalent bonding around Ag ions weaken and their ionicities become larger when Ag ions migrate between neighboring tetrahedral sites, although there is no octahedral site in Ag_2Se. Thus, such fluctuation of bonding properties accompanied with the diffusive motion of the mobile ions would be common to Cu halides and Ag chalcogenides.

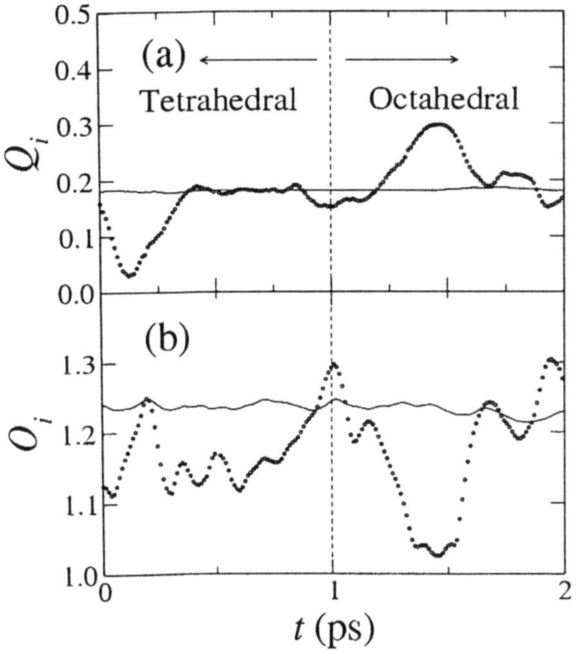

Figure 7. The time evolution of (a) the atomic charge Q_i and (b) the bond order O_i of the focused Cu ion in CuI. The time t in the horizontal axis corresponds to that for Fig. 5(a). The solid lines show the average values of all Cu ions.

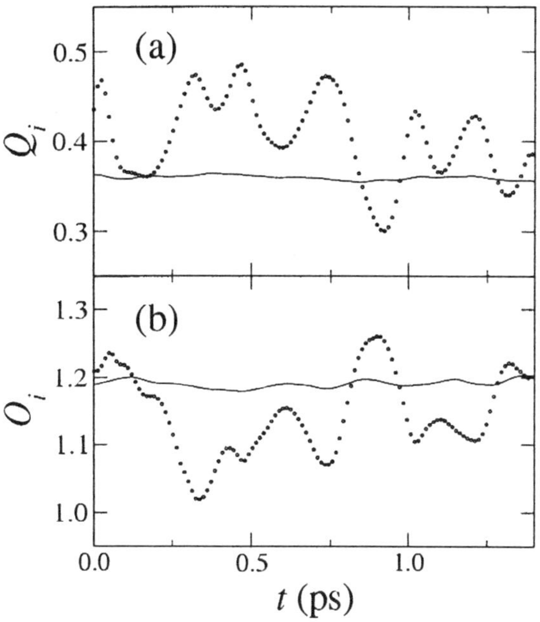

Figure 8. The time evolution of (a) the atomic charge Q_i and (b) the bond order O_i of the focused Ag ion in Ag_2Se. The time t in the horizontal axis corresponds to that for Fig. 6(a). The solid lines show the average values of all Ag ions.

4. Summary

We have investigated the structural and electronic properties of the superionic conductors CuI and Ag_2Se by means of *ab initio* molecular-dynamics simulations. The mean square displacements show that our simulations have successfully reproduced the superionic behavior of these materials. Utilizing the population analysis, we have detailed the time evolution of the electronic states around mobile ions, and have shown that the bonding properties of the mobile ions vary accompanied with the diffusive motion. These results of our MD simulations agree well with the predictions of the bond fluctuation model proposed by one of the authors [1-3].

References

1. M. Aniya, *Solid State Ionics* **50**, 125 (1992)
2. M. Aniya, *J. Phys. Soc. Jpn.* **61**, 4474 (1992)
3. M. Aniya, *Solid State Ionic Materials*, eds. B. V. R. Chowdari *et al.*, (World Scientific, Singapore, 1994) pp.223.
4. J. P. Perdew, K. Burke, and M. Ernzerhof, *Phys. Rev. Lett.* **77**, 3865 (1996)
5. D. Vanderbilt, *Phys. Rev. B* **41**, 7892 (1990)
6. G. Kresse and J. Hafner, *Phys. Rev. B* **49**, 14251 (1994)
7. F. Shimojo, Y. Zempo, K. Hoshino, and M. Watabe, *Phys. Rev. B* **52**, 9320 (1995)
8. S. Nosé, *Mol. Phys.* **52**, 255 (1984)
9. W. G. Hoover, *Phys. Rev. A* **31**, 1695 (1985)
10. R. S. Mulliken, *J. Chem. Phys.* **23**, 1833 (1955)
11. D. Sanchez-Portal, E. Artacho, and J. M. Soler, *J. Phys.: Condens. Matter* **8**, 3895 (1996)
12. F. Shimojo, K. Hoshino, and Y. Zempo, *J. Phys. Soc. Jpn.* **72**, 2822 (2003)

MODELING CONDUCTION PATHWAYS IN IONIC CONDUCTORS

S. ADAMS AND A. PREUSSER

GZG, Abt. Kristallographie, Universität Göttingen, Goldschmidtstr. 1, 37077 Göttingen, Germany
FAX: +49 (551) 39 9521 Email: sadams@gwdg.de

J. SWENSON

Department of Applied Physics, Chalmers University of Technology, 412 96 Göteborg, Sweden

A modified bond-valence approach proved to be useful for studying the interplay between the microscopic structure and the transport properties of solid electrolytes. Combining this approach with reverse Monte Carlo (RMC) modeling or molecular dynamics (MD) simulations provides a deeper understanding of ion transport, especially in amorphous or highly disordered crystalline solid electrolytes. Here we discuss the requirements for a consistent determination of bond- softness sensitive bond-valence parameters as well as their application to analyze the transport mechanisms in various crystalline and amorphous solid electrolytes. Local structure models for crystalline systems may be identified with the crystal structure or derived from a combination of crystallographic information with simulations. RMC models for a wide variety of ion conducting glasses have been analyzed to identify ion transport pathways, i.e. regions of sufficiently low bond-valence mismatch. The strong correlation between the volume fraction of the percolating pathway cluster and the transport properties yields a prediction of both the absolute value and activation energy of the dc ionic conductivities directly from the structural models. The effect of temperature on the conductivity is related to the dependence of the pathway volume fraction on the bond valence mismatch threshold. The application of this procedure to mixed alkali glasses reveals that the extreme drop of the ionic conductivity when a fraction of the mobile ions is substituted by another type of mobile ions can be mainly attributed to the blocking of conduction pathways by unlike cations. The efficiency of this blocking is rationalized by the reduced dimensionality of the pathways on the length scale of ion hops. The variability of the pathways in time is studied by analyzing the evolution of the bond valence pathways in MD trajectories.

1. Bond Valence Concept

The concept of bond valence is widely used in crystal chemical considerations, e.g. to assess equilibrium positions of atoms in crystal structures from empirical relationship between bond length R_{A-X} and bond valence $s_{A-X} = \exp\left[(R_0 - R_{A-X}) / b\right]$ as sites where the bond valence sum $V(A) = \Sigma_X\, s_{A-X}$, approaches the formal valence V_{ideal} of the central atom A. It appears straightforward to postulate that for solid electrolytes an analogous condition should apply not only to equilibrium sites but to all accessible sites for mobile ions A^{+}. Low-energy transport pathways for the motion of ions between these sites should then correspond to pathways along which the valence sum deviation $|V(A) - V_{ideal}(A)|$ remains as low as possible.

Established bond-valence (BV) parameter tables are typically based on the postulate that the BV sum V of a central atom is fully determined by interactions counterions in its first coordination shell (see e.g. Brown[1]). As this convention obstructs an independent refinement of the two BV parameters R_0 and b from reference crystal structures, a universal value of b = 0.37Å is commonly assumed, which essentially fixes the shape of the interatomic potential irrespective of the polarizabilities of the interacting particles. Therefore the conventional BV calculations provide a reliable tool to check whether a certain site in the crystal structure is a plausible equilibrium site for a certain type of ions, but the value of the bond valence mismatch calculated from these parameters cannot be expected to yield sensible estimates for the site energy of the often irregularly coordinated non- equilibrium sites along an ion transport pathway. Particularly, bond valence mismatches for a cation in different types of anion coordinations cannot be compared to each other.

The application of the BV approach to model the ion transport in solid electrolytes obviously requires more accurate estimates of their site energy and thus an adaptation of the BV parameter *b* (that determines the shape of the bond length - valence mismatch "potential") to the "softness" of the bond for the respective atom pair.

The inclusion of interactions with higher coordination shells in the refinement of bond valence parameters for our *softBV* parameter set[2,3] enabled us to determine BV parameters that systematically account for the *bond softness*. Utilizing differences between the *absolute softnesses*[4] of the interacting particles as an independent measure for the bond softness, a close correlation between these freely refined b values and the bond softness has been observed.[2]

2. Pathway Models in Crystalline Solid Electrolytes

Presuming that sites with a high bond valence mismatch for a mobile ion are energetically unfavorable, it appears straightforward to identify those pathways between equilibrium sites along which the bond valence mismatch remains as low as possible with probable ion transport pathways. Isosurfaces of constant bond valence mismatch then visualize regions that the mobile ion can reach with an activation energy that corresponds to the chosen value of the BV mismatch threshold. The lowest mismatch value for which the isosurface becomes infinite should correspond to the activation energy. The isosurface for this valence mismatch threshold represents a probable pathway for the *dc* ionic conduction (provided that this pathway contains both occupied sites and vacancies).

For crystalline ionic conductors local structure models as a basis for the BV calculations are available from crystal structure determinations. For structures with a low degree of disorder the crystal structure data may be approximately identified with the local structures, while the average structure models for solid electrolytes with a high degree of disorder in the immobile sublattice have to be converted into a representative model of the instantaneous local structure, e.g. by molecular dynamics simulations.[6]

As an example Fig. 1 shows a bond valence model of the Ag^+ transport pathways in the structure of the silver iodide silver oxyacid salt Ag_3IMoO_4 (AgI – Ag_2MoO_4) that we recently determined from X-ray powder diffraction data by Reverse Monte Carlo modeling and a subsequent Rietveld refinement.[5] It should be noted that none of the crystal structures of

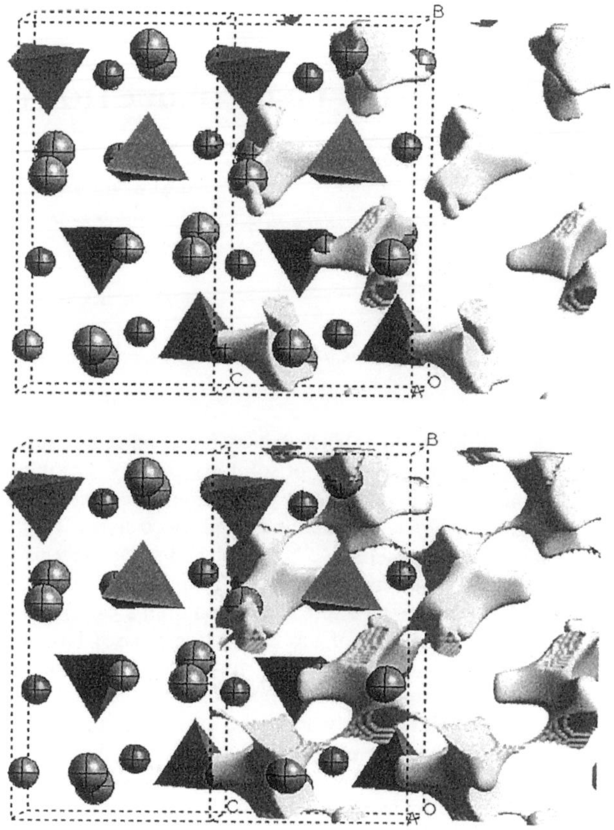

Fig. 1: Crystal structure of Ag_3IMoO_4 [5] and isosurface of constant Ag bond valence sum mismatch as a model for the Ag ion transport pathways. The left-hand side shows one unit cell of the crystal structure (MoO_4 as tetrahedra, iodide (silver) ions as small (large) ellipsoids.), the right-hand side part displays the bond valence map and the central part their superposition. Upper row: low energy pathways for local Ag^+ motion for $\Delta V = 0.10$ valence units; lower row: infinite pathways for $\Delta V = 0.28$ valence units).

AgI- AgM_xO_y compounds that we had determined (AgI-Ag_2MoO_4, AgI-$AgPO_3$, AgI-$Ag4P_2O_7$, 2AgI-$3Ag_2CrO_4$, $Ag_8I_4V_2O_7$) contains any exclusively iodide coordinated Ag sites. Thus for the crystalline state the earlier postulate that the ion transport in AgI- AgM_xO_y compounds should follow AgI-like pathways needs to be revised. This also sheds further doubt on the idea that clearly separated AgI-like regions in the corresponding AgI- AgM_xO_y glasses should be responsible for the ionic conductivity. Nevertheless, in these systems the mobility of an Ag^+ ion at a certain site is indisputably linked to the relative contribution of Ag-I bonds to the Ag bond valence sum. In the bond valence model this is accounted for by the higher value of the bond valence parameter *b* for Ag-I bonds.

Fig. 2 shows for a variety of Ag^+ ion conducting crystalline phases that the bond-valence mismatch

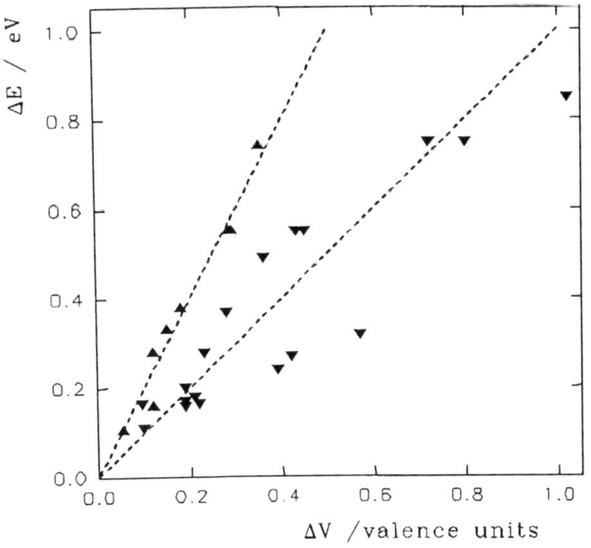

Fig. 2: Correlation between the experimental activation energy for the ionic conduction in various Ag^+ ion conductors and the threshold value of the bond valence mismatch, for which an isosurface contains an infinite pathway. As a guide to the eye broken lines indicate $\Delta E = \Delta V$ und $\Delta E = 2\Delta V$. The symbol ▲ (▼) refers to compounds with a high (low) degree of disorder in the mobile cation sublattice.

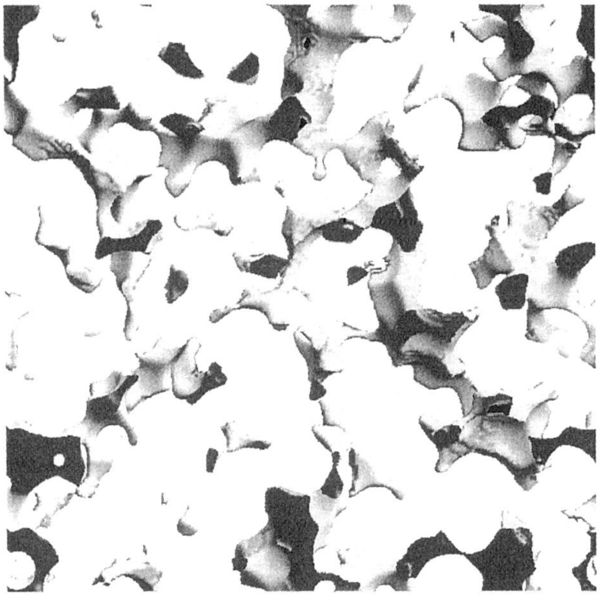

Fig. 3: Isosurface of constant Ag bond valence mismatch in a $(42.7\ \text{Å})^3$ RMC structure model of the glass AgI-$AgPO_3$ (bond valence mismatch threshold 0.1 valence units; thickness of displayed slice: 10 Å).

barrier, which a mobile ion has to cross in the rate-limiting elementary transport step, is linked to the experimental activation energy for this process.[5,7]

3. Structure Conductivity Correlation in Glasses

While the ion transport processes in crystalline solid electrolytes may be traced back to a few fundamental mechanisms, a generally accepted model of ion transport in fast ion conducting glasses is still lacking. This is largely due to the fact that the information contained in diffraction or spectroscopy data of amorphous systems is not sufficiently detailed to yield a unique structure model. The reverse Monte Carlo (RMC) technique, however, at least permits to produce models that reproduce characteristic features of the structure and fully harmonize with all experimentally available information on the structure of the system (e.g. x-ray, neutron, EXAFS).[8] Hence, these RMC models of the instantaneous local structure may be used as a basis for the analysis of preferable ion transport pathways by the bond valence method. To further enhance the chemical plausibility of the RMC refinements, we also included soft constraints of the bond valence mismatch for the mobile ions. The main effect of this constraint is that it helps to keep the number of nearly non-bonded particles sufficiently low.[9] It should be emphasized that RMC models are not unique structure solutions and thus the extraction of any information on the properties of the system from such models requires a statistical analysis of the entire model. In contrast to the crystalline phases, a detailed analysis of a few migration barriers in the model cannot provide representative information on the characteristics of the ion transport in glasses.

Three-dimensional maps of sites with a low BV mismatch for the mobile ion ("accessible sites") are calculated from RMC models of silver or alkali ion conducting glasses. (*cf.* a slice of the pathway model for the glass AgI-$AgPO_3$ in Fig. 3).[10,11] Clusters of adjacent accessible sites represent pathways for *ac* ion conduction, while the *dc* conductivity is determined by the ionic motion within the percolating pathway cluster. Due to the non-statistical distribution of the accessible volume (as well as of the mobile cations) a volume fraction of accessible sites of only a few percent is sufficient to ensure the existence of such a percolating pathway cluster.

From these conduction pathway models the effect of the type of coordinating anions on the cation mobility can be studied. As mentioned above for the related crystalline solid electrolytes, long-range ion transport pathways in the halide doped oxide glasses almost entirely consist of sites with a mixed oxide-

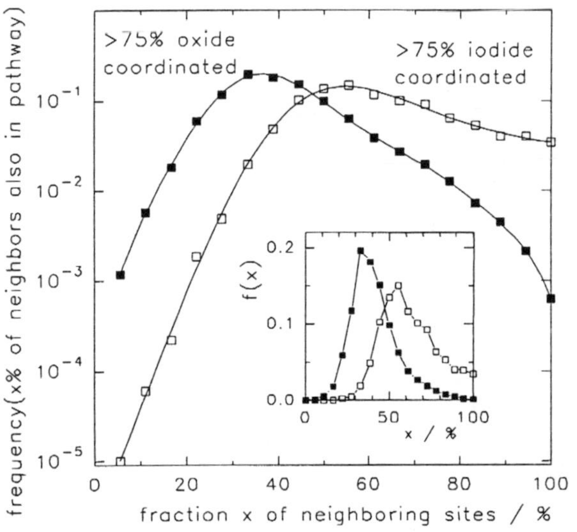

Fig. 4: Distribution of the number of adjacent (face or edge sharing) volume elements of a Ag site in the infinite conduction pathway for the RMC structure model of glassy $(AgI)_{0.5}$-$(AgPO_3)_{0.5}$. Open (full) symbols show the distribution for sites where the Ag bond valence sum originates mainly from bonds to iodide (oxide) anions. The logarithmic scale chosen for the relative frequency f(x) of a certain number of neighbors within the pathway in the main diagram highlights the asymmetry of the distributions. The inset displays the same data on a direct scale.

halide coordination and it is impossible to construct infinite pathways from exclusively halide coordinated sites only. Nevertheless the higher mobility of cations on halide sites can be estimated from comparing the local dimensionality of the pathways for cations on sites that are mainly coordinated by one type of anions. As illustrated in Fig. 4. for the RMC model of the AgI – $AgPO_3$ glass, Ag^+ ions on mainly iodide coordinated sites are more mobile than those on mainly oxide coordinated sites as the local pathway dimensionality is higher, *i.e.* there are on average more neighboring accessible targets for the mobile ion.

Moreover the bond valence analysis provides a simple tool to estimate the ionic conductivity of a glass system directly from its structure. For a variety of glasses with a single mobile cation we found that for a given mismatch threshold the cube root of the pathway volume fraction F is linearly related to both the experimentally observed absolute conductivity σ_{exp} and of the conduction process and to its activation energy E_σ.[9-11]

$$a_1' \cdot \sqrt[3]{F} + b_1' = \log(\sigma_{exp} T) \quad , \qquad (1a)$$

$$a_2' \cdot \sqrt[3]{F} + b_2' = -\frac{E_\sigma}{k_B T} \quad , \qquad (1b)$$

where k_B and T symbolize the Boltzmann constant and the absolute temperature, while a_i', b_i' represent empirical constants that depend on the type of the mobile ion, the chosen valence mismatch threshold value for the determination of F and obviously on the temperature. Both the thermal velocity of mobile ions and the conversion factor between bond valence and bond energy depend on the mass M of the mobile ion. These dependencies may be compensated by an appropriate scaling of Eqs. (1a) and (1b).[12] Thereby the values of the room temperature conductivity σ_{exp} as well as of E_σ can be predicted for a wide variety of ion conducting glasses with different types of mobile ions (Ag^+, Li^+, Na^+, Rb^+) from the RMC model of the respective glass structure by determining the volume fraction F of the infinite cluster of accessible sites via the correlations

$$a_1 \cdot \sqrt[3]{F\sqrt{M}} + b_1 = \log(\sigma_{exp} T \cdot \sqrt{M}) \quad , \qquad (2a)$$

$$a_2 \cdot \sqrt[3]{F\sqrt{M}} + b_2 = -\frac{E_\sigma}{k_B T} \quad . \qquad (2b)$$

Here, the empirical constants a_i, b_i in contrast to Eq. (1) are independent of the type of mobile ions. Choosing a fixed valence mismatch threshold of $\Delta V = 0.2$ valence units, a regression for room temperature RMC models of 20 glass systems that contained oxide and / or halide (iodide , chloride) anions yielded $a_1 = 18.3$ amu$^{-1/6}$ and $b_1 = -14.7$ (if data are given in atomic mass units *amu* for M, in S·cm^{-1} for σ_{exp} and log is understood as $\log_{10}$). For predictions of the activation energy E_σ the same set of experimental reference systems leads to $a_2 = 46.7$ amu$^{-1/6}$ and $b_2 = -49.9$. Using the procedure described above, the mixed alkali effect[13-15] (i.e. the drastic drop of the conductivity caused by the replacement of one type of mobile ion by another type of mobile ions) can be reproduced as shown in Fig. 5 for the typical mixed alkali glass system $Li_xRb_{1-x}PO_3$. According to the RMC structure model the two types of alkali ions in a mixed alkali glass are randomly mixed within the regions between the network of the immobile (poly-)anions. Nevertheless the two types of cations have distinctly different pathways. The low dimensionality d of these pathways on the length scales of thermal motion or individual hops (typically in the range of $1.5 < d < 2$) leads to a highly effective blocking of ***A*** ions even by a low number of adjacent immobile ***B*** ions and vice versa. Compared to this dominant blocking effect the reduced adaptation of the

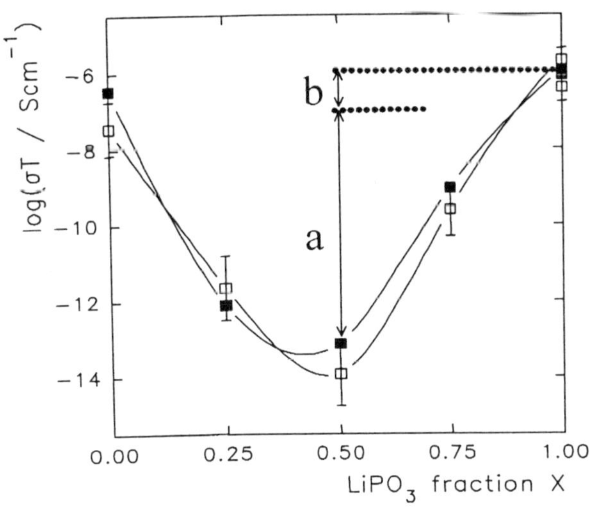

Fig. 5: Variation of the conductivity in the mixed alkali glass system $Li_xRb_{1-x}PO_3$ ($0 < x < 1$, T=300 K). Open symbols: experimental data[16]; filled symbols: predictions from using equation (2a). The vertical arrow (a) represents the predicted effect of the blocking of Li^+ pathways by immobile Rb^+; the arrow (b) indicates the minor effect of a reduced adaptation of the phosphate glass matrix to Li^+ ions. (redrawn after Swenson & Adams[17]).

glass matrix to either type of mobile ions in the mixed alkali system only plays a minor role. This can be investigated by comparing the Li^+ pathway volume fractions in the RMC models of $Li_{0.5}Rb_{0.5}PO_3$ and $LiPO_3$ after formally switching off the blocking effects of Rb^+ ions in the bond valence analysis. If the blocking effect is disregarded the slightly reduced volume fraction of Li pathways in $Li_{0.5}Rb_{0.5}PO_3$ should give rise to a conductivity drop by less than one order of magnitude (*cf.* arrows in Fig. 5).

4. Dynamic models

The relevance of pathways in static snapshots obviously depends on the time period over which the ion transport pathway (or more exactly a pathway network with constant characteristic features) exists. Thus we have started to investigate the time evolution of BV pathway networks from molecular dynamics simulation trajectories in alkali silicate glasses.[18] The amorphization of the structure model has been achieved in the usual way by heating the model to 6000 K, gradually cooling it back through the glass transition temperature (with a cooling rate of 10 K/ps) and annealing it at room temperature. First results for the system Na_2O-$2SiO_2$ (864 atoms, NVT) using two different two-body force fields (msi glass force field 1.01[19] and BKS force field[20,21]) suggest that the backbone of the pathway remains essentially unchanged over the time scale of the MD simulation (cf. Fig. 6) and that the glass structures resulting from the BKS force field exhibit more realistic pathway volume fractions.

Provided that the rigid matrix ensures a long-time stability of the pathway network, peculiarities of the ion transport in glasses can be effectively studied by simulating the motion as a random walk through the complex- shaped network of BV pathways. Such a simplistic approach permits to map the pathways of a mobile ion over time scales that are inaccessible to MD simulations.

The application of the bond valence analysis to molecular dynamics simulation trajectories permits also to investigate the effect of temperature (pressure, etc.) on the conductivity of ion conducting glasses. Higher

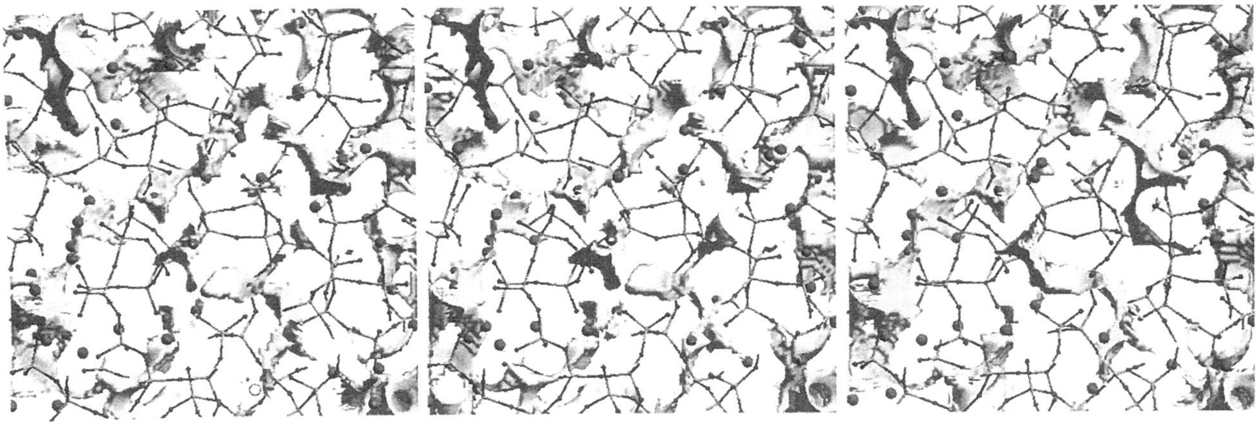

t_0 *t_0 + 25ps* *t_0 + 50ps*

Fig. 6: Isosurfaces of constant Na bond-valence mismatch for three different time steps from a MD simulation trajectory for Na_2O-$2SiO_2$ (NVT, 300K, 864 atoms, BKS force field). Large spheres: Na; light (dark) small spheres: Si (O); thickness of the displayed slice: 8 Å

temperatures favor ionic conductivity in various ways. As the valence mismatch threshold corresponds to the activation energy barriers that a mobile ion may cross with a certain probability, the appropriate valence mismatch threshold should increase with temperature. In the case of glassy Na_2O-$2SiO_2$ sufficiently below its glass transition temperature, MD simulations at 300-900K indicate that this is the prevailing effect and other factors like the increase of the pathway volume fraction with temperature (taking into account the reduced density) for a fixed valence mismatch threshold or the reduction of the time stability of the network exhibit no significant influence on the predicted conductivity of the glass system and its standard deviation.

Since the correlation between the activation energy and the bond valence mismatch depends on the mass of the mobile ions, it is to be expected that the correlation between the valence mismatch threshold and the temperature depends on the (reduced) mass of the mobile ion.

Acknowledgments

Financial support by the Deutsche Forschungsgemeinschaft is gratefully acknowledged.

References

1. I. D. Brown, The Chemical Bond in Inorganic Chemistry - The bond valence model; Oxford University Press (2002).
2. S. Adams, *Acta Crystallographica* **B57**, 278 (2001).
3. S. Adams, *softBV* parameter database, Universität Göttingen, Göttingen, Germany (2003). *http://kristall.uni-mki.gwdg.de/softBV/.*
4. R. G. Parr and R. G. Pearson, *J. Am. Chem. Soc.* **105**, 1503 (1983).
5. A. Preusser, *Thesis*, Göttingen (2002).
6. S. Adams and J. Swenson, *Solid State Ionics* **154/155**, 151 (2002).
7. S. Adams and J. Maier, *Solid State Ionics* **105**, 67 (1998).
8. R. L. McGreevy and L. Pusztai, *Mol. Simul.* **1**, 359 (1988).
9. J. Swenson and S. Adams, *Phys. Rev.* **B64**, 024204 (2001).
10. S. Adams and J. Swenson, *Phys. Rev. Lett.* **84**, 4144 (2000).
11. S. Adams and J. Swenson, *Phys. Rev.* **B63**, 054201 (2000).
12. S. Adams and J. Swenson, *Phys. Chem. Chem. Phys.* **4**, 3179 (2002).
13. J. O. Isard, *J. Non-Cryst. Solids* **1**, 235 (1969).
14. D. E. Day, *J. Non-Cryst. Solids* **21**, 343 (1976).
15. M. D. Ingram, *Glass Sci. Technol.* **67**, 151 (1994).
16. C. Karlsson, A. Mandanici, A. Matic, J. Swenson and L. Börjesson, *Phys. Rev.* **B68**, 064202 (2003)
17. J. Swenson and S. Adams, *Phys. Rev. Lett.* **90**, 155507 (2003).
18. S. Adams and J. Swenson, *Solid State Ionics*, in the press.
19. Glass force field gff_1.01, Molecular Simulations.
20. B. W. H. van Beest, G. J. Kramer and R. A. van Santen, *Phys. Rev. Lett.* **64**, 1955 (1990).
21. X. Yuan, A. N. Cormack, *J. Non-Cryst. Solids* **319**, 31 (2003).

ION CONDUCTION PATH AND LOW-TEMPERATURE FORM: ARGYRODITE-TYPE SUPERIONIC CONDUCTORS

M. ONODA, H. WADA, A. SATO AND M. ISHII
Advanced Materials Laboratory, National Institute for Materials Science, Namiki, Tsukuba, Ibaraki, 305-0044, Japan

The structures of the orthorhombic room-temperature phase of Cu_8GeS_6 (phase II) and the monoclinic low-temperature phase of Ag_7TaS_6 (phase II) have been successfully refined based on X-ray diffraction data from 12-fold twinned (Cu_8GeS_6 II) and 24-fold twinned (Ag_7TaS_6 II) crystals. Respectively among 6 major and 6 minor twin domains of Cu_8GeS_6 II, or among 12 major and 12 minor twin domains of Ag_7TaS_6 II, the argyrodite-type frameworks, GeS_6 or TaS_6, can be superposed to each other in principle, and only Cu-Cu or Ag-Ag network directions differ. At higher temperature, the crystals were considered to be 2-fold twinned crystals of superionic-conductor phase I with a space group $F\bar{4}3m$. On cooling, each domain transforms into 6 domains of orthorhombic Cu_8GeS_6 II or 12 domains of monoclinic Ag_7TaS_6 II. Superposed projections along 6 directions of the structure of Cu_8GeS_6 II and along 12 directions of the structure of Ag_7TaS_6 II seem to show approximate expressions for Cu-ion and Ag-ion conduction paths in superionic-conductor phases, Cu_8GeS_6 I and Ag_7TaS_6 I.

1. Introduction

A high-temperature form (phase I) of argyrodite-type Cu_8GeS_6 is cubic with a space group $F\bar{4}3m$ and a=9.9Å above about 330K [1-5]. The room-temperature phase (phase II) is orthorhombic with a space group $Pmn2_1$ and A=7.04, B=6.97, C=9.87Å [5,6]. Though X-ray diffraction data from a single crystal of Cu_8GeS_6 II seem to be from a twined rhombohedral crystal with a=9.9Å and α=90°, the orthorhombic structure of phase II has been refined successfully on the assumption that the X-ray data come from a 12-fold twinned crystal, with 6 major twin domains and 6 minor twin domains [7].

A new ternary sulfide Ag_7TaS_6 has been synthesized and its high Ag ion conductivity at ambient temperature was found [8, 9]. The compound was identified as a cubic phase of the argyrodite-family with the space group $F\bar{4}3m$, with lattice constant a=10.5Å. Low-temperature powder X-ray diffraction study revealed two kinds of low-temperature phase, *i.e.* phase II (below approx. 280K) and phase III (below approx. 170K) [10-13]. The crystal structure of Ag_7TaS_6 II was described based on a space group Pn with the lattice constants A=7.45, B=7.40, C=10.54Å and β=90.07° from a Rietveld analysis of a low-temperature powder X-ray diffraction pattern [11,12]. However, X-ray diffraction data from a single crystal of Ag_7TaS_6 II measured at 223K seem to be from a twined pseudo-cubic crystal with a=10.50Å. In a similar manner to the analysis of Cu_8GeS_6, low-temperature X-ray diffraction data from a single crystal have been examined on the assumption that the crystal was 24-fold twinned, with 12 major domains and 12 minor domains.

In the present work, the results of symmetry consideration and structure refinement of Cu_8GeS_6 II and Ag_7TaS_6 II are described. Next, superposed projections along 6 directions of the structure of Cu_8GeS_6 II and along 12 directions of the structure of Ag_7TaS_6 II are reported in connection with phase transitions of argyrodite-type superionic conductors.

2. Structure Refinement of Cu_8GeS_6 II and Ag_7TaS_6 II using intensity data from multiple-twinned crystals

The structure of the orthorhombic room-temperature phase of Cu_8GeS_6, phase II, has been refined on the basis of X-ray diffraction data from a 12-fold twinned crystal applying a six-dimensional twin refinement technique [14]. For 1804 unique reflections [$I>2\sigma(I)$] measured using Mo$K\alpha$ radiation, R_F was 0.083 with 77 structure parameters and 12 scale factors [7]. Twin operations of six major twin domains 1-6 are x,y,z, y,z,x, z,x,y, x,-y,-z, -y,-z,x and -z,x,-y based on a pseudo-cubic cell (a=9.9 Å). The minor twin domains 7-12 are respectively related to the major domains 1-6 by a two-fold rotation on [1 1 1], and they may have their origins in a minor domain of the twinned crystal of the high-temperature phase (phase I) of Cu_8GeS_6. The symmetry operations, the unit cell and other crystal data are (0,0,0; 1/2,1/2,0)+ x,y,z; y,x,z;

1/4-x,3/4-y,1/2+z; 3/4-y,1/4-x,1/2+z; $a=b=9.907$, $c=9.870$Å, $\alpha=\beta=90°$, $\gamma=90.642°$, Z=4. In the argyrodite-type framework GeS_6, a tetrahedrally closed- packed S sublattice is present. Cu ions are in distorted S-S tetrahedrons and in S-S triangle faces shared by two distorted tetrahedrons, and they are coordinated by some each other and seem to form a Cu-Cu network. Among 6 major twin domains and among 6 minor twin domains, the GeS_6 frameworks can be superposed to each other in principle, and only Cu-Cu network directions differ.

Intensity data collection from a single crystal of Ag_7TaS_6, prepared at 1123K followed by slow cooling to 873K, was performed at 223K with graphite monochromatized MoKα radiation on an Enraf-Nonius CAD4 diffractometer using a hexagonal setting: $a_h=b_h=14.85=\sqrt{2}\times10.50$, $c_h=18.19=\sqrt{3}\times10.50$ Å. The reflections seem to be caused by a twinned rhombohedral or pseudo-cubic crystal, with two reflection conditions – $h_h+k_h+l_h=3n$ or $h_h-k_h+l_h=3n$. In a similar manner to the analysis of Cu_8GeS_6 [7], the reflections were re-indexed respectively based on the lattice relations $\mathbf{a}_h=\mathbf{b}-\mathbf{c}$, $\mathbf{b}_h=\mathbf{c}-\mathbf{a}$, $\mathbf{c}_h=\mathbf{a}+\mathbf{b}+\mathbf{c}$ and the relations $\mathbf{a}_h=\mathbf{c}'-\mathbf{b}'$, $\mathbf{b}_h=\mathbf{a}'-\mathbf{c}'$, $\mathbf{c}_h=\mathbf{a}'+\mathbf{b}'+\mathbf{c}'$, and each reflection is expressed by $\mathbf{q}=h\mathbf{a}^*+k\mathbf{b}^*+l\mathbf{c}^*+h'\mathbf{a}'^*+k'\mathbf{b}'^*+l'\mathbf{c}'^*$. The observed 1230 reflections [$I>2\sigma(I)$] are of three groups. The first

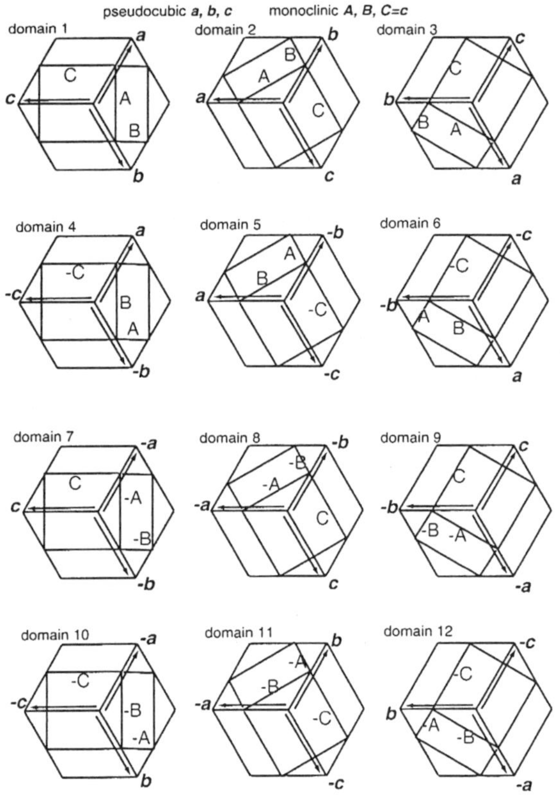

Figure 1. Twelve major domains of low-temperature phase, phase II, of Ag_7TaS_6.

group (323 reflections) is assignd by both $hkl000$ and $000h'k'l'$, the second group (590 reflections) assigned by $hkl000$, and the third group (317 reflections) assigned by $000h'k'l'$. The data were examined on the assumption that the crystal was 24-fold twinned monoclinic Ag_7TaS_6 II, with 12 major twin domains (Fig. 1) and 12 minor twin domains respectively, related to the major twin domains by a two-fold rotation on [1 1 1]. The initial structural model is obtained from previous work [11,12]. New bases $\mathbf{a}=\mathbf{A}+\mathbf{B}$, $\mathbf{b}=\mathbf{B}-\mathbf{A}$, $\mathbf{c}=\mathbf{C}$ are selected. The lattice constants are $a=b=c=10.500$Å, and a centering translation (1/2, 1/2, 0)+ is generated. Atomic parameters x, y, z based on $\mathbf{a}$, $\mathbf{b}$ and $\mathbf{c}$ are calculated from the monoclinic coordinates X, Y, Z using the relations x=(X+Y)/2, y=(Y-X)/2 and z=Z. Symmetry operations are also converted into the formula based on C-lattice with $\mathbf{a}$, $\mathbf{b}$ and $\mathbf{c}$: (1/2, 1/2, 0)+ x,y,z; 3/4-y,1/4-x,1/2+z. For the refinement using both $hkl000$ and $000h'k'l'$ reflections, symmetry operations are expressed in the (3+3)-dimensional formalism in a manner described in the literature [14,7]. Refinement was performed on the basis of |F| through FMLSM [14,15]. Besides structural parameters, 24 scale factors were considered as parameters. The agreement was fairly good for 1230 reflections with 49 structural parameters (40 positional parameters, 8 isotropic thermal parameters for Ta and Ag and one common isotropic thermal parameter for S) and 24 scale factors; $R_F=0.124$ and $wR_F=0.138$. The final parameters will be reported in another publication [16].

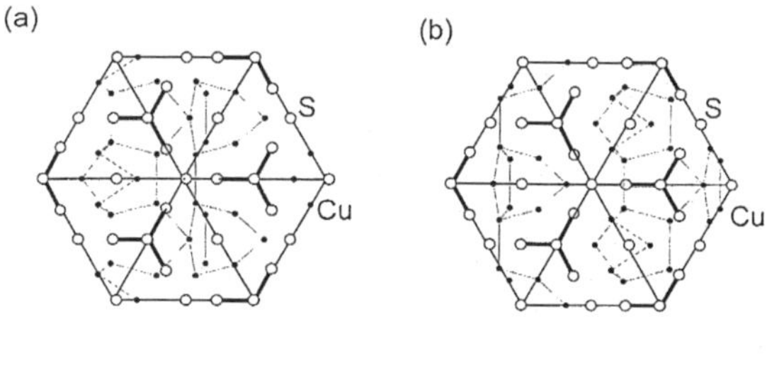

Figure 2. Projections of (a) the model of Cu_8GeS_6 II and (b) the model parallel-shifted by $\mathbf{a}/2+\mathbf{c}/2$, along pseudo-cubic [-1 -1 -1] and bounded by $0.7<x+y+z<1.03$. Small solid, medium open and large open circles represent Cu, Ge and S respectively.

3. Superposed projections along six directions of the structure of Cu_8GeS_6 II

The projection, along pseudo-cubic [-1 -1 -1] and bounded by $0.7<x+y+z<1.03$, of the model of Cu_8GeS_6 II is shown in Fig. 2(a) with that of the model parallel-shifted by $\mathbf{a}/2+\mathbf{c}/2$ (Fig. 2(b)). Both projections, Fig. 2(a) and 2(b), are from the same crystal structure, because the latter is obtained by parallel translation of the former. The model of Cu_8GeS_6 II in a pseudo-cubic

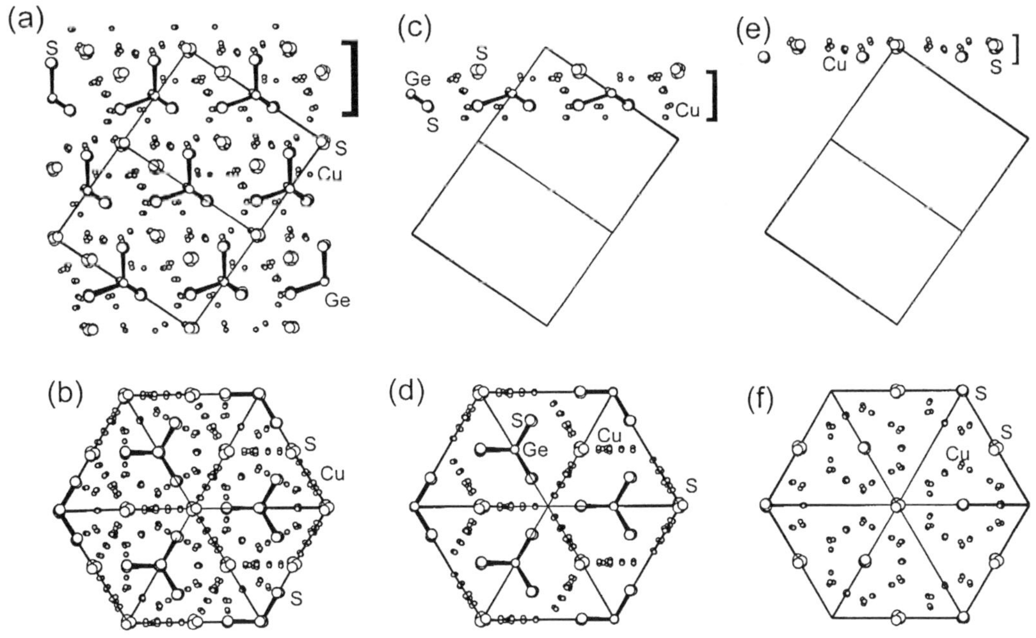

Figure 3. Superposed projections of the model of Cu_8GeS_6 II along [-1 1 0] (upper) and [-1 -1 -1] (lower) based on a pseudo cubic cell (a=9.9Å). Bounded areas are (a) $0.0<x+y+z<1.1$, (b) $0.7<x+y+z<1.03$, (c)–(d) $0.7< x+y+z<0.94$, (e)–(f) $0.94 <x+y+z<1.03$ and projections along six directions which correspond to six twin operations are superposed on to each other (see text).

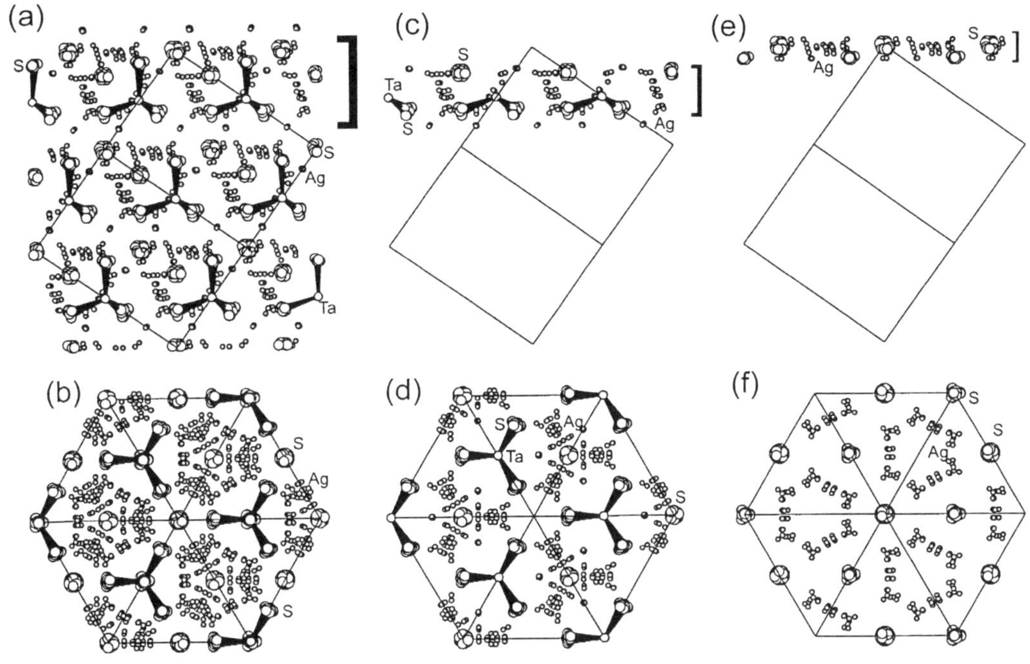

Figure 4. Superposed projections of the model of Ag_7TaS_6 II along [-1 1 0] (upper) and [-1 -1 -1] (lower) based on a pseudo cubic cell (a=10.5Å). Bounded areas are (a) $0.0<x+y+z<1.1$, (b) $0.7<x+y+z<1.03$, (c)–(d) $0.7<x+y+z <0.94$, (e)–(f) $0.94 <x+y+z<1.03$ and projections along 12 directions which correspond to 12 twin operations (Fig. 1) are superposed on to each other (see text).

cell and that parallel-shifted by **a**/2+**c**/2 have been projected along [-1 1 0] (left) and [-1 -1 -1] (right) in Fig. 3, bounded by (a) 0.0<x+y+z<1.1, (b)0.7< x+y+z<1.03, (c)-(d) 0.7<x+y+z<0.94, (e)-(f) 0.94<x+y+z<1.03. Next, projections along five directions with equivalent bounded areas are superposed to each figure of Fig. 3. after changes from **a, b, c** to **b, c, a** ; **c, a, b** ; **a, -b, -c** ; **-b, -c, a** ; **-c, a, -b.**

4. Superposed projections along twelve directions of the structure of Ag_7TaS_6 II

Twin operations of superposed projections of twelve major twin domains 1-12 of the low-temperature phase Ag_7TaS_6 II based on a pseudo-cubic cell (a=10.5 Å) are shown in Fig. 1. The origin of the minor twin domains 13-24 is a minor domain of the twinned crystal of the room-temperature phase (phase I) of Ag_7TaS_6.

Quite similarly to the case of Cu_8GeS_6 II, the model in a pseudo-cubic cell and that parallel-shifted by **a**/2+**c**/2 have been projected along [-1 1 0] (left) and [-1 -1 -1] (right) in Fig. 4(a)–4(f). Then, projections along 11 directions with equivalent bounded areas are superposed to each figure of Fig. 4. after changes from **a, b, c** to **b, c, a** ; **c, a, b** ; **a, -b, -c** ; **-b, -c, a** ; **-c, a, -b**; **-a, -b, c** ; **-b, c, -a** ; **c, -a, -b**; **-a, b, -c** ; **b, -c, -a** ; **-c, -a, b.**

5. Discussion

Satisfactory or fairly good convergence was obtained in structure refinement of Cu_8GeS_6 II or Ag_7TaS_6 II using intensity data from the multiple twinned crystal. At higher temperature, the crystal used in data collection is considered to be a 2-fold twinned crystal of Cu_8GeS_6 I or Ag_7TaS_6 I with a space group F $\bar{4}$3m. In each domain of Cu_8GeS_6 I or Ag_7TaS_6 I, the framework composed of GeS_4 or TaS_4 tetrahedra and free sulfurs is considered to be continuous substantially. On cooling, each domain of phase I transforms into 6-fold domains of orthorhombic Cu_8GeS_6 II and 12-fold domains of monoclinic Ag_7TaS_6 II due to low symmetry arrangement of Cu ions and Ag ions, while the argyrodite-type framework shows few distortions.

In the model of Cu_8GeS_6 II or Ag_7TaS_6 II, Cu or Ag ions are coordinated by some each of other Cu or Ag ions, and they seem to form a Cu-Cu or Ag-Ag network. In superposition of 6 kinds of projections of Cu_8GeS_6 II or 12 kinds of projections of Ag_7TaS_6 II with basically a common argyrodite-type framework, namely in Fig. 3(a)-3(f) or Fig. 4(a)-4(f), Ag or Cu sites are adjoined to one another within the fields which are distant from Ge or Ta ions and not too close to S ions. In other words, superposed projections of Cu_8GeS_6 II and Ag_7TaS_6 II seem to show approximate expressions for Cu-ion and Ag-ion conduction paths in superionic-conductor phases Cu_8GeS_6 I and Ag_7TaS_6 I. In superionic-conductor phases with F $\bar{4}$3m, mobile ions may drift with deformed three-dimensional network-like geometry among S ions of argyrodite-type framework. On cooling through the phase-transition temperature, Cu or Ag ions relax into a network-like arrangement with orthorhombic or monoclinic symmetry. The structure of phase I may be sectioned off into multiple domains with their various orientations of the low-temperature form, and the crystal may be converted into an aggregate of crystals of Cu_8GeS_6 II or Ag_7TaS_6 II.

References

1. H. Hahn, H. Schulze & L. Sechster, *Naturwissenschaften* **52**, 451 (1965).
2. W.F. Kuhs, R. Nische & K. Scheunemann, *Mat. Res. Bull.* **14**, 241 (1979).
3. M. Khanafer, O. Gorochov & J. Rivet, *Mat. Res. Bull.* **9**, 1543 (1974).
4. N. Wang & W. Viane, *Neues Jahrb. Mineral. Monatsh.*, **10**, 442 (1974).
5. Goetz (1988). Inst. f. Krist. Der Aachen, JCPDS Grant-in Aid Report; JCPDS 39-1202,40-1190.
6. H. Müller and B. D. Serot, *Phys. Rev.* **C52**, 2072 (1995).
7. M. Ishii, M. Onoda & K. Shibata, *Solid State Ionics* **121**, 11 (1999).
8. H. Wada and M. Onoda, *J. Less-Common Metals* **175**, 209 (1991).
9. H. Wada, *J. Alloys Compounds* **178**, 315 (1992).
10. M. Onoda , H. Wada, K. Yukino and M. Ishii, *Solid State Ionics* **79**, 75 (1995).
11. M. Onoda, H. Wada, and M. Ishii, *Solid State Ionics* **86-88**, 217 (1996).
12. M. Onoda , H. Wada, P. Pattison, A. Yamamoto, M. Ishii and G. Chapuis, *Mol. Cryst. Liq. Cryst.* **341**, 879 (2000).
13. H. Wada, A. Sato, M. Onoda, S. Adams, M. Tansho and M. Ishii, *Solid State Ionics* **154-155**, 723 (2002).
14. K. Kato, *Z. Krist.*, **212** (1997), 423.
15. K. Kato, *Acta Cryst.*, **A50** (1994) 351.
16. M. Onoda, H. Wada, A. Sato and M. Ishii, *J. Alloys Compounds*, (2004) in press.

GLASSY AND POLYMERIC IONIC CONDUCTORS: STATISTICAL MODELING AND MONTE CARLO SIMULATIONS

O. DÜRR AND W. DIETERICH
Fachbereich Physik, Universit¨at Konstanz, D–78457 Konstanz, Germany
E-mail: wolfgang.dieterich@uni-konstanz.de

Stochastic semi-microscopic models for glassy and polymer ion conductors provide a framework for understanding a variety of experimental transport properties in these materials. In the case of glasses we discuss effects of counter ions and propose a mechanism for constant dielectric loss in terms of a dipolar lattice gas. The behavior of glasses is put into contrast to polymeric systems, where ions are diffusing through a fluctuating host medium. A lattice model based on specific cation-polymer interactions allows us to deduce diffusion properties as a function of temperature, pressure and salt content, consistent with experimental observations on polyethylene-oxide(PEO)/salt complexes. Directions for future research, based on an extension of the above models and also by further coarse-graining are pointed out.

1 Introduction

Because of their variability in composition, electrical transport and mechanical properties, amorphous ionic conductors have become of prime importance in the search for solid electrolyte materials which are suitable for practical applications. There exist two classes of amorphous ionic conductors with distinct properties:

i) In *glassy materials,* e.g. in alkali doped silicate or borate glasses, ions diffuse in an essentially rigid, disordered energy-landscape containing preferred sites which are created during the process of cooling the melt to the glassy state. Temperature dependent dc-conductivities generally follow the Arrhenius-law. The ac conductivity shows pronounced dispersion, normally characterized by the Jonscher power-law regime in the megahertz range [1, 2] and by the nearly constant loss (NCL)-regime at higher frequencies [3, 4, 5].

ii) In *polymer electrolytes*, such as PEO-salt complexes, dissociation of salt is a consequence of the attraction of cations by polar groups in the polymer chains. Motions of chain segments provide pathways which rearrange in time and allow ions to diffuse. Free volume theories and dynamic percolation ideas have been advanced as phenomenological descriptions of the diffusion mechanism[6], which is distinctly different from the hopping mechanism pertaining to glasses. This difference is reflected both in dc- and ac-conductivities. In polymer electrolytes, as a function of temperature generally follows the empirical Vogel-Tammann-Fulcher (VTF) law. Dispersive transport effects in polymers are much less important thanin glasses.

Fundamental differences between both groups of materials are apparent also from composition-dependent properties. In polymer electrolytes, ¾*dc* rises by adding salt but at the same time the mobility of charge carriers decreases because cations induce transient cross-links within polymer chains so that the network rigidity gets enhanced. Therefore, when the salt content is increased, the conductivity levels off or passes a maximum. More dramatic concentration effects are known for network glasses, where the activation energy decreases with increasing ion concentration. One way to understand this is by assuming preferred sites in the network with a density nearly equal to the density of mobile ions and by a reduction in the effective barrier between two such sites, when their distance gets reduced. Glasses containing two species of ions show pronounced mixed alkali effect [7, 1, 8, 9, 10] not known for polymers.

In this communication we summarize very briefly some recent results on the ion dynamics in glasses and polymers, obtained from studies of coarse-grained stochastic models. Our emphasis with this approach is on the understanding of important features which can be observed experimentally in many materials. For details we refer the reader to the original literature cited below.

2. Counterion effects and NCL-response in glasses

A conceptually transparent model for ion diffusion in glasses can be designed when we consider the situation in alkali-doped network glasses, like $(Na_2O)_x\ (SiO_2)_{1-x}$. There we assume that preferred sites for alkali ions exist on account of negatively charged, immobile counterions, which were introduced through the doping process [11]. The simplest model is to start from a random distribution of counterions. Cations diffuse in the

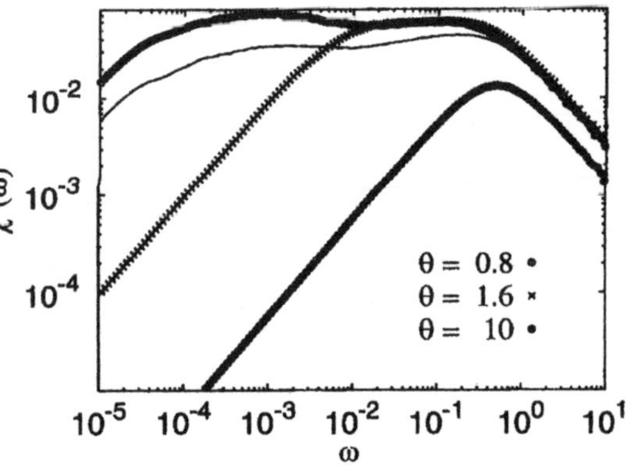

Fig.1. Dielectric loss spectrum $\chi''(\omega)$ of a dipolar lattice gas with concentration $c = 1/64$ of dipolar centers, showing a gradual transition from Debye to NCL-behavior as the reduced temperature $\theta = k_B T = V_{dip\text{-}dip}$ is lowered. The continuous line represents the self-part of the susceptibility at $\theta = 0.8$.

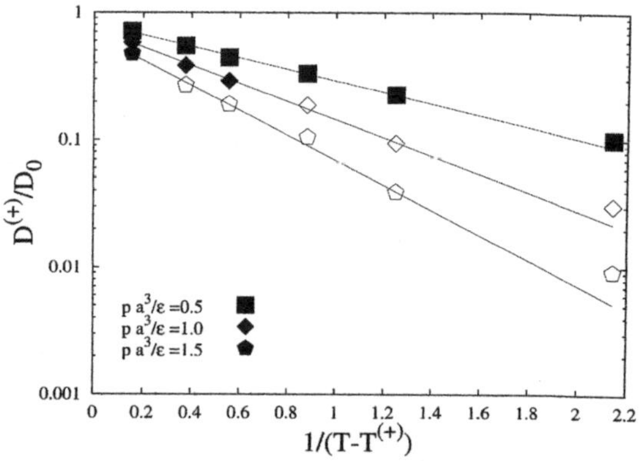

Fig.2. Ionic diffusion constants, normalized by the free particle diffusion constant D_0 versus inverse distance in temperature from the VTF-temperature $T^{(+)}$. The ion concentration is $x = 0.16$. Different data sets refer to different reduced pressures and thus display the influence of pressure on the energetic parameter $E^{(+)}$ in the VTF-law, equation (1). Full symbols: simulated data points, light symbols: extrapolation for fixed T with the help of equation (2).

associated energy landscape which can be described as a superposition of randomly placed Coulombic traps. Discretising this model, which includes also cation-cation interactions, we arrive at a special form of a Coulomb lattice gas with correlated structural disorder. Concentration-dependent transport properties, Haven-ratios, conductivity dispersion and correlation functions describing nuclear magnetic resonance and ultrasonic attenuation have been worked out by Monte-Carlo simulation and give a reasonable description of experiments. Such studies and related models are reviewed in [12], see also [13].

A particular limit of that model, where ions remain bound to a set of lattice sites next to an associated counterion, leads to the "random dipolar lattice gas". This model provides a mechanism for NCL-behavior, a phenomenon common to many disordered materials, but not yet understood in a satisfactory way. Irrespective of ions, the random dipolar lattice gas should apply to any system containing localized, charged defects which undergo reorientational motions. Fig.1 shows a series of dielectric loss spectra, obtained by kinetic Monte Carlo simulation. The ultra-slow decay of the polarization correlation function, which can be observed in the simulations at low temperatures, is attributed to rare events of large fluctuations in the Coulomb field acting on one particular dipolar center. In fact, these simulations indicate the existence of extended time scales for fluctuations which are sufficiently large to change the minimum energy orientation of that particular dipole and hence to introduce a relaxation step. Since dipole-dipole interactions decay with distance r as $V_{dip\text{-}dip} \sim r^{-3}$, the most important contribution to that Coulomb field arises from the closest neighbor. The distribution of relaxation times referring to the correlated motion of nearest-neighbor pairs can indeed be shown to determine an ultra-slow decay of the polarization correlation function [12].

3. Lattice model of ion conducting polymers

For PEO-type electrolytes we discuss a simple model of lattice chains [14], where beads occupy a sequence of nearest-neighbor points on a simple cubic lattice of spacing a. We distinguish between C-beads and X-beads in sequences $C(XCC)_n$, where X corresponds to an oxygen atom and C to a hydrocarbon group. The total length of the chain is $r = 3n + 1$. Beads on nearest-neighbor positions interact with a common repulsion $\varepsilon > 0$, which drives the system from a fluid to an amorphous state. For simplicity we deal only with one species of "ions", represented as point-like particles, which attract X-beads, again with strength ε. Elementary moves of the chains follow the generalized Verdier-Stockmayer algorithm, including king jump, end jump and crank shaft moves, while ions simply perform nearest-neighbor hops.

Monte Carlo simulations of that model at constant volume are combined with an approximate equation of state, derived from the quasi-chemical approximation (QCA). This procedure allows us to get access to the pressure p as an independent variable. That aspect is important because we wish to describe phenomena under varying ion content $x = N_I/N_X$, where N_I denotes the number of ions and N_X the total number of X-beads.

Calculated ion and polymer chain diffusion constants at constant pressure are well represented by the VTF-law,

$$D^{(\alpha)}(T,x) = D_\infty^{(\alpha)} \exp\left(-\frac{E^{(\alpha)}(x)}{k_B\left(T - T^{(\alpha)}(x)\right)}\right), \tag{1}$$

with $\alpha = +$ (ion) and $\alpha = P$ (polymer). $E^{(\alpha)}(x)$ is a characteristic energy and $T^{(\alpha)}(x)$ the VTF-temperature. Conversely, at constant temperature one finds an exponential decrease of $D^{(+)}$ with increasing pressure,

$$D^{(+)}(p,T) = D_0^{(+)}(T)e^{-p/p_0(T)} \tag{2}$$

Both of these findings qualitatively agree with measurements [6, 15]. Fig.2 shows a plot of log $D^{(+)}$ versus $(T\text{-}T^{(+)})^{-1}$ at constant x for different pressures. The linearity of the different sets of data verifies the form (1). At the same time Fig. 2 indicates that both the energetic parameter $E^{(+)}$ and the prefactor in eq. (1) as a function of pressure, can be represented as

$$E^{(+)} = E_0 + Ap; \quad D_\infty^{(+)} = \text{const.e}^{-Bp} \tag{3}$$

where E_0, A and B are constants. Combination of these relations with (2) and (1) describes the combined T-and p-dependence of $D^{(+)}$. Measurements on PEO-based alkali ionomers [15] revealed a variation of the conductivity with T and p qualitatively similar to Fig. 2. These measurements, however, also covered temperatures close to $T^{(+)}$ which are inaccessible by our present simulations.

Another issue of great importance for transport in polymer electrolytes is the dependence of VTF-temperatures $T^{(\alpha)}(x)$ on the ion concentration x. Fig. 3 shows results for both $T^{(+)}(x)$ and $T^{(P)}(x)$ at fixed pressure. As long as $x \leq 0.5$ which implies that all ions have a chance to bind to X-beads, both quantities coincide, but decouple when $x \geq 1$. A more detailed analysis of dilute systems with $x \leq 0.5$ shows that the majority of ions in fact binds to two or more X-beads and thus provides crosslinks within a chain or between chains. Crosslinked chains have lower configurational entropy. In this context, DiMarzio has employed the idea of an ideal transition to an amorphous state when the chain configurational entropy vanishes [17]. Such a transition is normally associated with the VTF-temperature. Despite of the fact that the DiMarzio concept cannot be regarded as generally applicable [18], we tentatively follow this idea and calculate the configurational entropy $S_c(x,T)$ by means of the QCA. A characteristic temperature $T_c(x)$ is then extracted, where

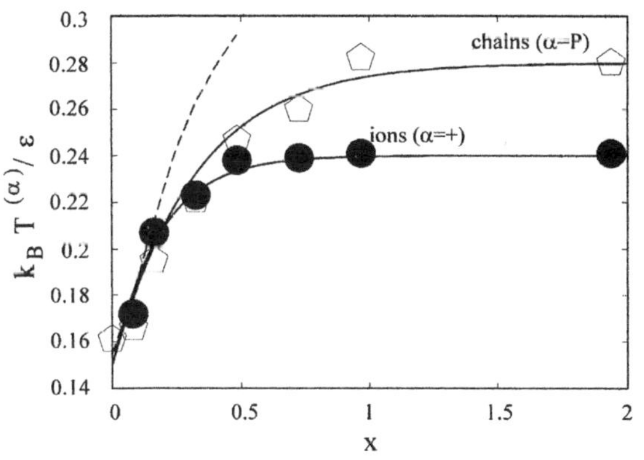

Fig.3. VTF-temperatures $T^{(+)}$ and $T^{(P)}(x)$ for ions and chains as a function of ion concentration x at fixed pressure $pa^3/\varepsilon = 0.35$. Full lines represent an interpolation. Also shown (dashed line) is the temperature $T_c(x)$ defined by the condition of a vanishing configurational entropy, as predicted by the QCA.

S_c vanishes. In this way we obtain the dashed curve in Fig. 3, which is in remarkable agreement with the rise of $T^{(\alpha)}(x)$ for small x. Further analysis of the entropy for chains with different length and under varying pressure seems to be an interesting task.

4. Outlook

A generalization of the counterion model of section 2 to mixed (AB) cation glasses and a description of mixed alkali effects seems possible when we invoke the idea of two different energy landscapes seen by A or B-ions [19]. Within the counterion model it may be sufficient to distinguish between two classes of counterions, which preferentially bind either A or B-ions. Consider a glass that initially contains only A-ions as mobile species. When A-ions are progressively exchanged by B-ions such that the total ion concentration is fixed, the distances between A-type centers will increase. From that we expect a reduction in the mobility of A-ions, one of the key features in mixed alkali effects.

For polymer ion conductors one would like to construct an even more coarse-grained theory for ion diffusion, which eliminates irrelevant degrees of freedom of the polymer host. A successful step in this direction has been achieved recently by a quantitative mapping of systems of athermal chains and point-like particles onto dynamical percolation theory [16]. Extension of this procedure to more realistic polymer electrolyte models will be the subject of future studies.

References

1. M.D. Ingram in *Materials Science and Technology*, Vol. **9**, edited by J. Zarzycki (VCV Publications, Weinheim, New York, 1991), p. 715.
2. J.C. Dyre and T. B. Schroeder, *Rev. Mod. Phys.* **72**, 873 (2000).
3. A.S. Nowick, B.S. Lin and A.V. Vaysleyb, *J. Non-Cryst. Solids* **172-174**, 1243 (1994).
4. K.L. Ngai, *J. Chem. Phys.* **110**, 10576 (1999).
5. T. Ishii and T. Abe, *J. Phys. Soc. Jpn.* **69**, 2549 (2000).
6. M. Ratner in: J.R. McCallum and C.A. Vincent (Eds.), *Polymer Electrolyte Reviews–1*, (Elsevier, New York, 1989), pp. 173–236, 7.
7. D. E. Day, *J. Non-Cryst. Solids* **21**, 343 (1976).
8. A. W. Imre, S. Voss and H. Mehrer, *J. Non-Cryst. Solids* **333**, 231 (2004).
9. H. Sato and R.J. Kikuchi, *Phys. Rev.* **B28**, 648 (1983).
10. J. Swenson and S. Adams, *Phys. Rev. Lett.* **90**, 155507 (2003).
11. D. Knödler, P. Pendzig and W. Dieterich, *Solid State Ionics* **86–88**, 29 (1996).
12. For a review see W. Dieterich and P. Maass, *Chemical Physics* **284**, 439 (2002).
13. J. Reinisch and A. Heuer, *Phys. Rev.* **B66**, 063401 (2002).
14. O. Dürr, W. Dieterich and A. Nitzan, submitted to *J. Chem. Phys.*.
15. M. Duclot, F. Alloin, O. Brylev, J.Y. Sanchez and J.L. Souquet, *Solid State Ionics* **136–137**, 1153 (2000).
16. O. Dürr, T. Volz, W. Dieterich and A. Nitzan, *J. Chem. Phys.* **117**, 441 (2002).
17. E. A. DiMarzio, *Journal of Research of the National Bureau of Standards-A. Physics and Chemistry* Vol. **68A**, 6, 611 (1964).
18. H. P. Wittmann, *J. Chem. Phys.* **95**, 8449 (1991).
19. P. Maass, *J. Non-Cryst. Solids* **255**, 35 (1999).

UNIVERSALITIES OF ION-HOPPING IN RANDOM SYSTEMS

Tadao Ishii and Toshinori Abe
Faculty of Engineering, Okayama University, Okayama 700-8530, Japan

Theoretical studies of dynamic conductivity and dielectric function for hopping ions are made on the basis of the relaxation mode theory, which lead to universalities in random lattice systems. Especially attention is paid to quasi-Debye behavior.

Keywords *ion hopping, universalities, non-Debye response, quasi-Debye response, dynamic conductivity, dielectric functions, diffusion coefficients, incoherent scattering functions*

1. Introduction

Relaxation in random systems is considered to consist of many relaxation modes of wave number q and band ε [1,2]. In this report, we investigate "universality" which is inherent in a dynamical property of ion-hopping in random systems, on the basis of the relaxation mode theory.

2. Master Equations and Relaxation Mode Theory

Consider interacting lattice gas ions consisting of cations and anions on A and B sublattices, respectively, where immobile anions play a role only to provide a distribution of random energies: In this case, we have the master equation [2,3]

$$\begin{aligned}\frac{\partial p_n(t)}{\partial t} = \sum_\delta \Big[& (1-p_n(t))\,p_{n+\delta}(t)\,w_{n,n+\delta}(t) \\ & \times \prod_{j\neq n\; n+\delta\in A}\{1+f_{n,n+\delta;j}\,p_j(t)\}\prod_{\chi\in B} e^{\beta h_{n+\delta,\chi}} \\ & -(1-p_{n+\delta}(t))\,p_n(t)\,w_{n+\delta\, n}(t) \\ & \times \prod_{j\neq n\; n+\delta\in A}\{1+f_{n+\delta,n;j}\,p_j(t)\}\prod_{\chi\in B} e^{\beta h_{n,\chi}} \Big]\end{aligned} \qquad (1)$$

$$f_{n,n+\delta;j} = e^{\beta h_{n+\delta:j}} - 1$$

where $p_n(t)$ is the probability of finding a cation at site n and time t, the interaction energies are taken as $h_{n+\delta;j} > 0$ among like ions on A sublattice and $h_{n+\delta\,\chi} < 0$ for unlike ions on A and B sublattices, the attempt frequency ω_s is taken independent of sites, β is the inverse temperature $(k_B T)^{-1}$ and $w_{nm}(t)$ is a transition probability from sites n to m.

If we replace $p_j(t)$ by its thermal equilibrium value $p_j^{\,0}$ in the interaction terms among like ions, we can obtain the effective transition probability by

$$\Gamma_{nn+\delta} = w_{nm}(t) \prod_{j\neq n,\, n+\delta\in A}\{1+f_{n,n+\delta;j}\,p_j^{\,0}\}\prod_{\chi\in B} e^{\beta h_{n+\delta,\chi}} .$$

In a disordered state, we may further approximate it by $\Gamma_{nn+\delta} \sim w_{nm}(t)e^{\beta U_{n+\delta}}$, $U_{n+\delta} < 0$. In this case, its dynamical property for mobile cations is governed by the master equation given by [3,4]

$$\begin{aligned}\frac{\partial p_n(t)}{\partial t} = \sum_\delta \Big[& (1-p_n(t))\,\Gamma_{n,n+\delta}(t)\,p_{n+\delta}(t) \\ & -(1-p_{n+\delta}(t))\,\Gamma_{n+\delta,n}(t)\,p_n(t)\Big]\end{aligned} \qquad (2)$$

$$\Gamma_{nn+\delta}(t) = \omega_s \exp\left(-\beta U_{n,n+\delta}\right)$$

with an intrinsic activation energy $U_{n\,n+\delta} = U^s{}_{n\,n+\delta} - U_{n+\delta}$ where $U^s{}_{nm}$ is introduced as the saddle point energy between two nearest neighbor sites n and m [5,6]. In the presence of an external perturbation $V(t)$, we have

$$\frac{\partial|\psi(t)\rangle}{\partial t} = -H|\psi(t)\rangle - H'|\Phi\rangle, \quad H' = \beta H V(t) \qquad (3)$$

with

$$H_{nm} = \begin{cases}\sum_{m'\neq n} W_{m'n}\,, & n=m \\ -\sqrt{W_{nm}W_{mn}}\,, & n\neq m\end{cases} \qquad (4)$$

$$W_{mn} = (1-p_m^0)\,\Gamma_{mn}\,(1-p_n^0)^{-1}$$

for $|\psi(t)\rangle = |\Phi^{-1}\Delta p(t)\rangle$ being an excited state induced from $V(t)$ where we put $p(t) = p^0 + \Delta p(t)$, and $|\Phi\rangle$ is the thermal equilibrium state corresponding to the probability p^0. The dynamic conductivity of xx-component is generally given in the present system by

$$\begin{aligned}&\sigma(\omega) = \sigma(0) + \sigma_1(\omega) \\ &\sigma(0) = \beta e^2 z_\phi \sum_q \frac{E_{q\varepsilon=0}}{q^2}\,\delta_{q,0}\,, \quad z_\phi = \frac{\langle\Phi|\Phi\rangle}{\Omega} \\ &\sigma_1(\omega) = \beta e^2 z_\phi \sum_q \sum_{\varepsilon\neq 0} |\langle\phi|R|\psi_{q\varepsilon}\rangle|^2 \frac{-i\omega E_{q\varepsilon}}{E_{q\varepsilon} - i\omega}\end{aligned} \qquad (5)$$

with

$$\langle\Phi|\Phi\rangle = \sum_n p_n{}^0(1-p_n{}^0)$$

where e and Ω are the electronic charge and volume of the system, respectively, and R is the position operator of x-component. The probability in thermal equilibrium is expressed in the form: $p_n{}^0 = (1+e^{\beta(U_n-\mu)})^{-1}$ at site n and μ being the chemical potential. A set of eigenvalue $E_{q\varepsilon}$ and eigenfunction $\psi_{q\varepsilon}$ for a mode (q,ε) in eqs.(5) is given from eq.(3) in the absence of external perturbation such that

$$H\left|\psi_{q\varepsilon}\right\rangle = E_{q\varepsilon}\left|\psi_{q\varepsilon}\right\rangle . \qquad (6)$$

3. Exact Expression of Conductivity

Let us take a system consisting of M supercells, each supercell having $\ell \times \ell \times \ell$ sites, with a periodic boundary condition. Thus a κth site in nth supercell is specified by (n,κ)

3.1 Perturbation method

In order to obtain an exact and explicit conductivity, we utilize a small q- perturbation method following the quantum mechanics [7, 8]. The state in eq.(6) is written as

$$|\psi_{q\varepsilon}\rangle = \frac{1}{\sqrt{M}}\sum_{n\kappa} e_{q\varepsilon}(\kappa)\exp(i\boldsymbol{q}\cdot\boldsymbol{R})\,|n,\kappa\rangle \qquad (7)$$
$$\boldsymbol{R}\,|n,\kappa\rangle = \boldsymbol{R}_{n\kappa}\,|n,\kappa\rangle, \quad \boldsymbol{R}_{n\kappa} = \boldsymbol{R}_n + \boldsymbol{X}_\kappa .$$

Thus eq.(6) is rewritten as

$$\tilde{H}\,|e_{q\varepsilon}\rangle = E_{q\varepsilon}\,|e_{q\varepsilon}\rangle, \quad \tilde{H} = e^{-i\boldsymbol{q}\cdot\boldsymbol{R}}He^{i\boldsymbol{q}\cdot\boldsymbol{R}}$$
$$|e_{q\varepsilon}\rangle = \frac{1}{\sqrt{M}}\sum_{n',\kappa'} e_{q\varepsilon}(\kappa')\,|n',\kappa'\rangle \qquad (8)$$

with

$$\tilde{H} = H + qH_1 + q^2H_2 + \cdots$$
$$H_1 = i\hat{q}\cdot[H,\boldsymbol{R}], \quad H_2 = -\frac{1}{2}\hat{q}\cdot[[H,\boldsymbol{R}],\boldsymbol{R}]\cdot\hat{q} \qquad (9)$$

for a small q, in terms of the unit vector $\hat{q}$.

3.2 Conductivity expressed in H_1 and H_2

From eqs.(4)-(9), we obtain the conductivity

$$\sigma(0) = \beta e^2 z_\Phi \sum_q \sum_{\varepsilon=0} |\langle\phi\,|\,R\,|\,\psi_q(\varepsilon)\rangle|^2 E_{q\varepsilon}$$
$$= \beta e^2 z_\Phi \left\{ \langle e_{0\rho}\,|\,H_2\,|\,e_{0\rho}\rangle - \sum_{\varepsilon\neq 0}\frac{|\langle e_{0\rho}\,|\,H_1\,|\,e_{0\varepsilon}\rangle|^2}{E_\varepsilon} \right\} \qquad (10)$$

and

$$\sigma_1(\omega) = \beta e^2 z_\Phi \sum_{\varepsilon\neq 0} \left|\frac{\langle e_{0\rho}\,|\,H_1\,|\,e_{0\varepsilon}\rangle}{E_\varepsilon}\right|^2 \frac{-i\omega E_\varepsilon}{E_\varepsilon - i\omega} \qquad (11)$$

where we put $E_{0\varepsilon} = E_\varepsilon$.

3.3 Conductivity expressed in Γ

By using eqs.(6), (9) and the explicit expressions of H's in terms of Γ in eqs.(10) and (11), one obtains the alternative expressions as

$$\sigma(0) = \beta(ea)^2 z_\Phi \left\{ \sum_k{}' \sqrt{\Gamma_{\kappa\kappa+1}\Gamma_{\kappa+1\kappa}}\,{}_x e_{0\rho}(\kappa)e_{0\rho}(\kappa+1) - \sum_{\varepsilon\neq 0}\frac{1}{E_\varepsilon}\left|\sum_k{}' \begin{pmatrix} e_{0\rho}(\kappa+1)\sqrt{\Gamma_{\kappa+1\kappa}\Gamma_{\kappa\kappa+1}} \\ -e_{0,0}(\kappa-1)\sqrt{\Gamma_{\kappa-1,\kappa}\Gamma_{\kappa\kappa-1}} \end{pmatrix}_x e_{0\varepsilon}(\kappa)\right|^2 \right\} \qquad (12)$$

and

$$\sigma_1(\omega) = \beta(ea)^2 z_\Phi \sum_{\varepsilon\neq 0}\frac{-i\omega}{E_\varepsilon(E_\varepsilon - i\omega)} \times \left|\sum_k{}' \begin{pmatrix} e_{0\rho}(\kappa+1)\sqrt{\Gamma_{\kappa+1\kappa}\Gamma_{\kappa\kappa+1}} \\ -e_{0\rho}(\kappa-1)\sqrt{\Gamma_{\kappa-1\kappa}\Gamma_{\kappa\kappa-1}} \end{pmatrix}_x e_{0\varepsilon}(\kappa)\right|^2 . \qquad (13)$$

4. Non-Debye Conductivity in Random Lattices

In the following sections 4 through 7, we deal with a lattice with equipotential trough, which is a symmetric hopping. In this case, $W_{mn} = \Gamma_{mn} = \Gamma_{nm}$ and $z_\phi = n_0(1-c)$ where n_0 and c are the ion density a volume and a site, respectively.

4.1 Universality: double-power law

The anomalous non-Debye relaxation involved in hopping particles has been investigated in a wide range of materials [9]. Nowick, Lim and Vaysleyb [10] experimentally studied the dynamic conductivity $\sigma'(\omega)$ (real part of the conductivity $\sigma(\omega)$) varying with the frequency ω, to obtain

$$\sigma'(\omega) - \sigma(0) = A\omega^s + A'\omega^{s'}, \quad s \sim 0.6, \quad s' \sim 1 \qquad (14)$$

which is considered to be universal. The first frequency dependent term with coefficient A dominates the low-frequency and high-temperature region, often referred to as the Jonscher regime, and the second term with coefficient A' governs the high-frequency and low-

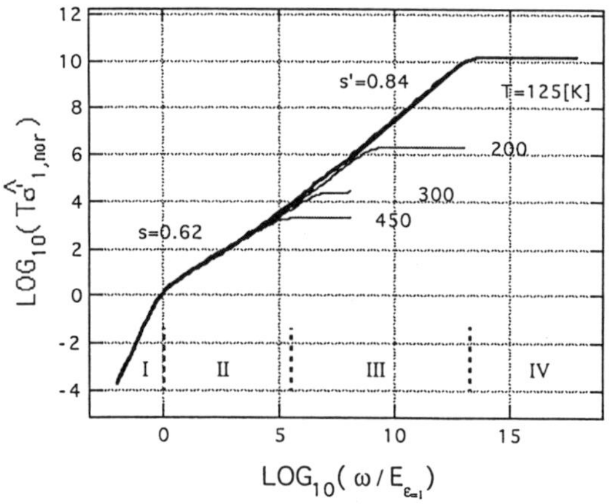

Fig.1 Frequency dependence of the scaled conductivities where frequency exponents s=0.62 and s'=0.84 in stages II and III, respectively, are at T=125[K]

temperature region, called nearly constant-loss regime.

we have theoretically proposed a mechanism of eq.(14) in a unified manner that stems from a characteristic distribution of the mode diffusion length $L_\varepsilon = \lim_{q\to 0}\sqrt{\ell}\left|\langle \phi | R | \psi_{q\varepsilon}\rangle\right|$ and density of states $D_\varepsilon = (dE_\varepsilon / d\varepsilon)^{-1}$ [2,11]. This discussion provides the anomalous conductivity including the double-power law of eq.(14), which is symbolically expressed in the form

$$\sigma'(\omega) - \sigma(0) = A\omega^2\big|_{\mathrm{I}} + A\omega^s\big|_{\mathrm{II}} + A'\omega^{s'}\big|_{\mathrm{III}} \qquad (15)$$

where the extended modes result in the exponent value $s \sim 0.6$ in stage II and the localized modes lead to $s' \to 1$ for $T \to 0$ in stage III (Fig.1 in 1D-random system). The normalized conductivity in Fig.1 is defined as

$$\hat{\sigma}'_{1\,nor}(\omega) = \frac{\sigma'_1(\omega)}{\sigma_{\varepsilon=1}}$$
$$\sigma_{\varepsilon=1} = \beta n_0(1-c)e^2 E_{q\to 0\,\varepsilon=1}\left|\langle \phi | R | \psi_{q\to 0\,\varepsilon=1}\rangle\right|^2. \qquad (16)$$

4.2 New universality: quasi-Debye law

Quite recently, we have obtained a quasi-Debye universality in 3D random systems [12]. This is concerned with a behavior in stage I of eq.(15) which is written anew as

$$\sigma'(\omega) - \sigma(0) = A''\omega^{s''}\big|_{\mathrm{I}} + A\omega^s\big|_{\mathrm{II}} + A'\omega^{s'}\big|_{\mathrm{III}}, \quad s'' \le 2 \quad (17)$$

where s'' is typically given by $s'' \sim 1.5$ (Fig.2): we should have ω^2-region for $0 < \omega < E_{0,\varepsilon=1}$ at any temperature in theory although numerically it has not been confirmed yet on account of computer accuracy so far. The calculation is carried out under a uniform distribution of the activation energies $0.1[eV] \le U \le 0.6[eV]$.

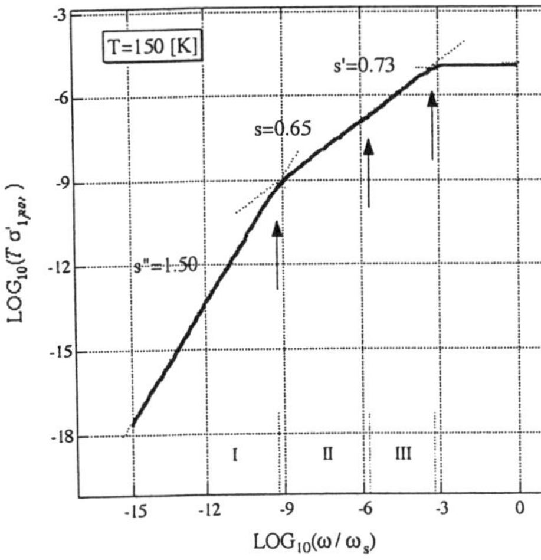

Fig.2 Quasi-Debye response (stage I) in the dynamic conductivity in 3D random systems.

The mechanism of the quasi-Debye regime is originated in the region of $dF_\varepsilon / dE_\varepsilon > 0$ ($F_\varepsilon = L_\varepsilon^2$), eq.(18), which implies that the mode diffusion length L_ε becomes larger with relaxation mode E_ε: this characteristic behavior is quite different from those of stages II and III. The product of F_ε and density of states D_ε behaves as

$$G(E_\varepsilon) = F_\varepsilon D_\varepsilon \sim K a^2 \gamma^{-1} \hat{E}_\varepsilon^{-n}, \quad \hat{E}_\varepsilon = E_\varepsilon / \gamma \qquad (18)$$

in two regions separated by $E_D = \tau_D^{-1}$: E_D is a crossover value between $dF_\varepsilon / dE_\varepsilon > 0$ and $dF_\varepsilon / dE_\varepsilon < 0$. The region $\omega\tau_D < 1$ is the quasi-Debye regime (Figs.2 and 3) being of interest at present. Here we call F_ε as the mean square displacement for mode ε (ε-MSD), and E_ε as relaxation eigenvalue (R-eigenvalue). Thus $F_\varepsilon D_\varepsilon$ signifies ε-MSD per R-eigenvalue.

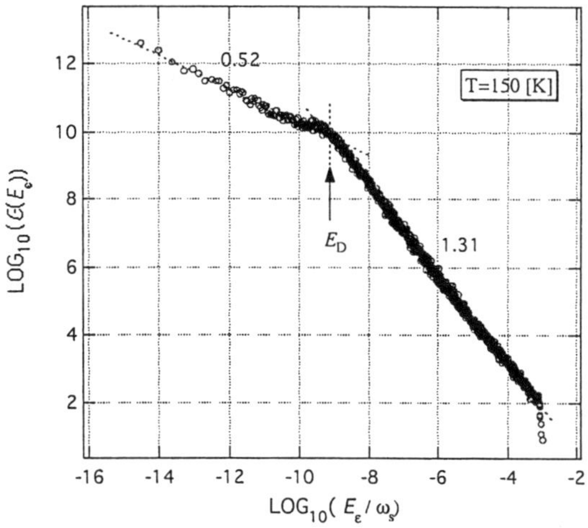

Fig.3 R-eigenvalue dependence of ε-MSD

In the quasi-Debye regime, we have the real and imaginary part of the approximate conductivity from eqs.(5) and (18) as

$$\sigma'_1(\omega) \sim Q''\gamma\,\hat{E}_D^{\,s''}\omega_n^{\,2}\int_0^1 dx \frac{x^{s''-1}}{x^2+\omega_n^2}$$
$$\sigma''_1(\omega) \sim -Q''\gamma\,\hat{E}_D^{\,s''}\omega_n\int_0^1 dx \frac{x^{s''}}{x^2+\omega_n^2} \qquad (19)$$
$$Q'' = K''\beta n_0(1-c)(ea)^2$$

where $\hat{E}_D = E_D/\gamma$ and $\omega_n = \omega / E_D$. The integral in eq.(19) is given by the Hypergeometric function as

$$I = \int_0^1 dx \frac{x^\theta}{x^2+\omega_n^2} = \frac{{}_2F_1[1, \tfrac{1}{2}(1+\theta), \tfrac{1}{2}(3+\theta), -\tfrac{1}{\omega_n^2}]}{(1+\theta)\omega_n^2}. \quad (20)$$

Since we are interested in the region of $\omega\tau_D < 1$, $\sigma'_1(\omega)$ is approximately given by

$$\sigma_1'(\omega) \sim A'\omega^{s'} \quad (21)$$

$$A' = Q'\gamma^{1-s'}\frac{\pi}{2}\operatorname{cosec}\frac{s'\pi}{2}, \quad s' = 2-n. \quad (22)$$

Now the exponent is given as $n = 0.52$ in Fig.3, and so we have the frequency exponent $s' = 1.48$, which reproduces well the value 1.5 in Fig.2. Furthermore, from eq.(22) and using $\gamma \sim \exp(-\beta U_{\max})$ [2,8], we have

$$A' \sim \exp[-\beta(1-s')U_{\max}] \quad (23)$$

where the activation energies of A' leads to $(1-s')U_{\max} < 0$.

By further numerical analyses of eq.(11) or (13), it is also confirmed that the exponent s' depends on temperature such that $s' \to 2$ with increasing temperature. Illustratively at respective temperatures of $T = 125,\ 150,\ 200[\mathrm{K}]$, we have $s' = 1.54,\ 1.66,\ 1.82$ for $(U_{min}, U_{\max}) = (0.1,\ 0.4)[\mathrm{eV}]$, and $s' = 1.43,\ 1.50,\ 1.62$ for $(0.1,\ 0.6)[\mathrm{eV}]$, respectively. This temperature dependence is quite consistent with the MC results [13] where $s' \sim 1.5 \to 2$ for $\beta V = 60 \to 40$ with V being a typical Coulomb energy. Also this behavior is consistent with the experimental results reported in a variety of materials [9].

5. Quasi-Debye Law of Dielectric Function

From the conductivity, we straightforwardly obtain the dielectric function in the quasi-Debye regime by

$$\varepsilon'(\omega) - \varepsilon_\infty = \frac{Q'}{\varepsilon_0}\hat{E}_{\mathrm{D}}^{s'-1}\int_0^1 dx\frac{x^{s'}}{x^2+\omega_n^2}$$
$$\varepsilon''(\omega) = \frac{1}{\varepsilon_0}\frac{\sigma_0}{\omega} + \frac{Q'}{\varepsilon_0}\hat{E}_{\mathrm{D}}^{s''-1}\omega_n\int_0^1 dx\frac{x^{s''-1}}{x^2+\omega_n^2} \quad (24)$$

The first term $\chi' = \varepsilon'(\omega) - \varepsilon_\infty$ is a decreasing function of the frequency, and is numerically confirmed to give approximately in the form

$$\int_0^1 dx\frac{x^{s''}}{x^2+\omega_n^2} \sim a\left(1-b\omega_n^{s'-1}\right), \quad \omega\tau_{\mathrm{D}} < 1 \quad (25)$$

where $a = 1/(s'-1)$ and $b = 2/(s'+1)$. The above approximation is more complete if we take $a = 1.99$ and $b = 1.08$ for the present case $s' = 1.5$. It is also possible to approximate the above integration by the form:

$$\int_0^1 dx\frac{x^{s''}}{x^2+\omega_n^2} \sim \frac{a}{\left(1+b\omega_n^{c}\right)^d}, \quad 0 \le \omega_n \le 1. \quad (26)$$

6. Diffusion Coefficient

From the mean square displacement for the position operator R, the diffusion coefficient tensor D at time t has been obtained in hopping systems [14]. This leads to the xx-component of $D(t)$ as

$$\left\langle (\Delta R_x(t))^2 \right\rangle = 2D(t)\,t \quad (27)$$

$$D(t) = D_0 + D_1(t) \quad (28)$$

$$D_0 t = \lim_{q\to 0}\frac{E_{q,\varepsilon=0}}{q^2}t \quad (29)$$

$$D_1(t)t = \ell^{-1}\int_{\varepsilon\neq 0} dE_\varepsilon\, G(E_\varepsilon)\left(1-\exp(-E_\varepsilon t)\right). \quad (30)$$

It gives anomalous behavior which is surely observed in materials. Especially, in the quasi-Debye regime, we approximately have

$$D_1(t)\,t = K'a^2\int_0^{E_D} d\hat{E}\frac{1-\exp(-\hat{E}\tau)}{\hat{E}^{n}}, \quad \tau = \gamma t. \quad (31)$$

For $E_{\mathrm{D}}\, t >> 1$, we have

$$D_1(t)t \sim K'a^2\left\{\frac{1}{s'-1}\hat{E}_{\mathrm{D}}^{s''-1} - \left(\frac{1}{\gamma t}\right)^{s''-1}\Gamma(s'-1)\right\}$$
$$\sim \frac{K'a^2}{s'-1}\hat{E}_{\mathrm{D}}^{s''-1}, \quad E_{\mathrm{D}}t >> 1 \quad (32)$$
$$\Gamma(\alpha) = \int_0^\infty x^{\alpha-1}e^{-x}\mathrm{d}x, \quad \tau_{\mathrm{D}} < t < \tau_1$$

for $\tau_1 = 1/E_{\varepsilon=1}$.

7. Incoherent Scattering Function

Let us discuss the dynamic structure factor in this section which is directly related to the spin-lattice relaxation time [15]. It is formally given in d-dimensional system by [16]

$$\frac{1}{T_1} \sim S(\omega) = \left(\frac{L}{2\pi}\right)^d \int d\,\mathbf{q}\, S(\mathbf{q},\omega)$$
$$= \left(\frac{L}{2\pi}\right)^d \frac{1-c}{\pi}\sum_\varepsilon \int d\,\mathbf{q}\left|\langle\phi\,|\exp(-i\mathbf{q}\bullet\mathbf{R})|\,\psi_{q\varepsilon}\rangle\right|^2\frac{E_{q,\varepsilon}}{E_{q,\varepsilon}^2+\omega^2} \quad (33)$$

which is approximately evaluated in 3D-system as

$$S(\omega) \sim D_0^{-1}q_D + \frac{\pi}{10}K'q_D^5a^2\operatorname{cosec}\frac{s'\pi}{2}\gamma^{1-s'}\omega^{s''-2} \quad (34)$$

where the first term is originated in the diffusive branch and the second term holds for $\omega\tau_{\mathrm{D}} < 1$ in the quasi-Debye regime. Other contributions are not written. The temperature dependence in the quasi-Debye regime is the same as that of the conductivity. Extracting the temperature and frequency dependence from eq.(33), we have

$$S(\omega) \sim S_0 e^{\beta U_{\max}} + S' e^{\beta(s'-1)U_{\max}}\omega^{s''-2} \quad (35)$$

where S_0 and S' are constants independent of temperature and frequency. Thus two slopes $U_{\max}$ and $(s'-1)U_{\max}$

$\leq U_{max}$ exist in respective region divided at $E_{\varepsilon-1}$. The peak value is reached at $\omega \sim E_D$. If solids concerned are not vitreous, the quasi-Debye response is replaced by a near-Debye or Debye response.

8. Summary

We have proposed a new universality in the lowest frequency regime I in ionic conductors such that

$$\sigma'(\omega) - \sigma(0) = A''\omega^{s''}\Big|_{\mathrm{I}} + A\omega^{s}\Big|_{\mathrm{II}} + A'\omega^{s'}\Big|_{\mathrm{III}}$$

whose power in the regime I is $s'' \to 2$ with temperature. The mechanism is also due to a distribution of the ε-MSD per R-eigenvalue. The coefficient $\ln A''$ is an increasing function with respect to the inverse temperature β.

References

[1] T. Ishii: Prog. Theor. Phys. **77** (1987) 1364.
[2] T. Ishii: *Recent Res. Devel. Physics*, (Transworld, Res. Network. India) **3** (2002) 613.
[3] T. Ishii: J. Phys. Soc. Jpn. **69** (2000) 139.
[4] N. G. van Kampen: *Stochastic Processes in Physics and Chemistry* (North-Holland, Amsterdom, 1981).
[5] R. Kikuchi: Prog. Theor. Phys. Suppl. **35** (1966) 1.
[6] T. Ishii: Prog. Theor. Phys. Suppl. **115** (1994) 243.
[7] T. Ishii: J. Phys. Soc. Jpn. **65** (1996) 351.
[8] T. Abe and T. Ishii: J. Phys. Soc. Jpn. **70** (2001) 2622.
[9] A. K. Jonscher: *Dielectric Relaxation in Solids*, (Chelsea Dielectric Press, London, 1983).
[10] A. S. Nowick, B. S. Lim and A. V. Vaysleyb: J. Non-Cryst. Solids **172-174** (1994) 1243.
[11] T. Ishii and T. Abe: J. Phys. Soc. Jpn. **69** (2000) 2549.
[12] T. Ishii, T. Abe and H. Shirai: Solid State Commun. **127** (2003) 734.
[13] P. Pendzig and W. Dieterich: Solid State Ionics **105** (1998) 209.
[14] T. Ishii: Solid State Commun. **116** (2000) 327.
[15] P. M. Richards: *Physics of Superionic Conductors* ed. M. B. Salamon (Springer-Verlag, New York, 1979) p.141.
[16] T. Ishii: J. Phys. Soc. Jpn. **72** (2003) 335.

EXCITONS IN AgI-BASED-GLASSES AND -COMPOSITES

FUMITO FUJISHIRO
Department of Physics, College of Humanities and Sciences, Nihon University
3-25-40 Sakurajosui, Setagaya-ku, Tokyo 156-8550, Japan

SHOSUKE MOCHIZUKI
Department of Physics, College of Humanities and Sciences, Nihon University
3-25-40 Sakurajosui, Setagaya-ku, Tokyo 156-8550, Japan

We summarize our recent optical studies on different pristine AgI films, different AgI-based glasses and different AgI–oxide fine particle composites. The exciton spectra of these specimens give useful information about the ionic and electronic structures at the AgI/glass and AgI/oxide particle interfaces.

1. Introduction

Recently, there has been growing interest in fast ion conduction in glasses and composites, because they may yield new solid electrolytes. They are also useful specimens for studying the cooperative effects of mobile ions and electrons in ionic solids. Moreover, it is a hope that we can find new transparent electrical conductor, which is useful for some opto-electric device. Hitherto, a considerable work has been made by many workers, to clarify the mechanism of high ionic conductivity in AgI-based glasses [1] and AgI–oxide fine particle composites [2, 3].

AgI has three phases designated as α, β and γ at normal pressure in the order of decreasing temperature with the following properties. At the superionic transition temperature T_c (420 K), the superionic α phase transforms into the semiconductor β phase with a wurtzite lattice. The α phase has a body-centred cubic arrangement of I^- ions with highly mobile Ag^+ ions randomly distributed through the equivalent interstices, which has been known as the averaged structure. At room temperature, the semiconductor γ phase with a zincblende lattice appears as a metastable state. It is said that the stability of the γ phase is affected by slight nonstoichiometries and other defects.

Many models to explain the high ionic conductivity have been proposed by noting Ag^+ ion pathways near AgI/glass matrix interfaces and AgI/oxide particle interfaces. However, the following basic problems arise. What are the atomic and electronic structures of the AgI/glass matrix interfaces and AgI/oxide particle interfaces? How do Ag^+ ions delocalize in the glasses and composites, exhibiting fast ion conduction? AgI shows clear optical spectra due to excitons in the bulk [4], films [5 8] and microcrystals [9], even at room temperature. The exciton behavior is sensitive to the environment in which excitons are moving, and, therefore, the optical spectra due to excitons may provide useful information about the interfaces.

In the present paper, we summarize the recent morphological, structural and optical studies which we have conducted on different AgI-based glasses (AgI–$AgPO_3$ [10], AgI–Ag_2WO_4 [11]) and AgI-based fine particle composites (AgI–γAl_2O_3 [12], AgI–anatase [13], AgI–ZnO [14]), together with the studies of pristine AgI films.

2. Experiments

Pristine β- and γAgI films were prepared by different methods: (1) gradual vacuum evaporation on a room-temperature silica glass substrate, (2) flash vacuum evaporation on a room-temperature silica glass substrate. Besides the above fabrication methods, *in situ* optical studies of films grown on substrates by vacuum evaporation of AgI powder were carried out. These fabricated AgI films were characterized by x-ray diffraction (XRD) analysis with a Cu $K\alpha_1$ radiation. It is found that the βAgI films contain a small amount of γAgI, while the γAgI films contain a small amount of βAgI.

AgI-based glasses were prepared by fully mixing AgI powder and glass-forming oxides, melting and then cooling down rapidly to room temperature with a twin-roller to form glasses. AgI–oxide fine particle composites were prepared by fully mixing AgI powder and oxide fine particles, heating and then annealing at 373 K for 12 h to remove the stress induced at the phase transition point of AgI, 420 K. All specimens were characterized by the XRD at room temperature.

The morphology of the glass and composite specimens was studied with a scanning electron microscope (SEM) and an optical microscope with a charge-coupled device camera system. Distribution maps of the constituent elements in the specimens were obtained by using a SEM with an energy dispersive x-ray (EDX) fluorescence spectrophotometer (Shimadzu SSX-550).

The optical density (OD) spectra were measured with an optical multichannel analyzer (OMA) consisting of a spectrograph (focal length = 19 cm) and a photodiode array detector, or another OMA consisting of a spectrograph (focal length = 32 cm) and an image-intensified photodiode array detector. The photoluminescence (PL) spectra were measured with an OMA consisting of a spectrograph (focal length = 32 cm) and an image-intensified photodiode array detector. A Nd^{3+}: YAG laser, a Ti^{3+}: sapphire laser, an optical parametric oscillator, a N_2 laser and a monochromatic light source consisting of a 150 W xenon lamp and a grating monochromator (focal length = 20 cm) were used as the excitation source. The photoluminescence excitation (PLE) spectra were recorded by varying the photon energy E_{ex} with an apparatus consisting of the same monochromatic light source described above and by detecting the emission light intensity at the desired photon energy E_{ob} as a function of E_{ex} with a grating monochromator (focal length = 20 cm) and a synchronous light detection system. The time-resolved PL (TRPL) spectra were taken by using the same OMA to which two delay pulse generators were attached. The delay time and the gate time were set with these delay pulse generators which were controlled by a personal computer. A closed-cycle He refrigerator equipped with a temperature controller was used to change the specimen temperature between 7 K and 298 K.

3. Results and discussion

3.1. *Pristine AgI*

Figure 1 shows the OD and PL spectra of βAgI film at 7 K. The film was fabricated by the method (1). The W_1 (2.95 eV), W_2 (2.99 eV) and W_3 (3.79 eV) absorption peaks are assigned to the exciton transitions in the wurtzite structure. The W_1 peak is accompanied by faint the $Z_{1,2}$ peak which belongs to the zincblende structure, as a low energy shoulder. A PL peak appears at 2.90 eV, which is red-shifted from the W_1 peak. This indicates that the PL arises from radiative decay of shallowly trapped excitons by exciton-phonon interactions, or some defects, or impurities.

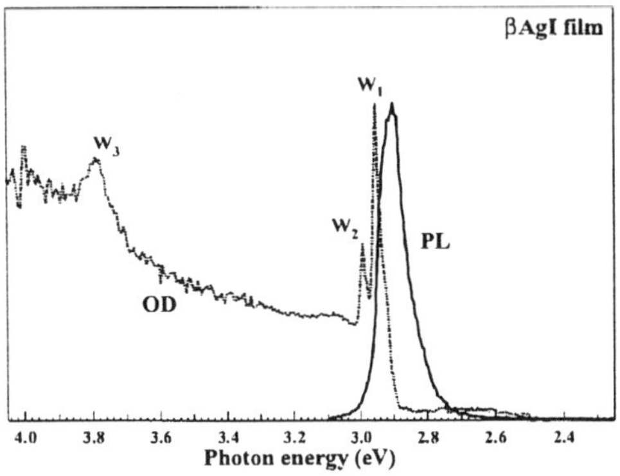

Figure 1. The OD and PL spectra of βAgI film at 7 K. The film was fabricated by the gradual vacuum-evaporation method.

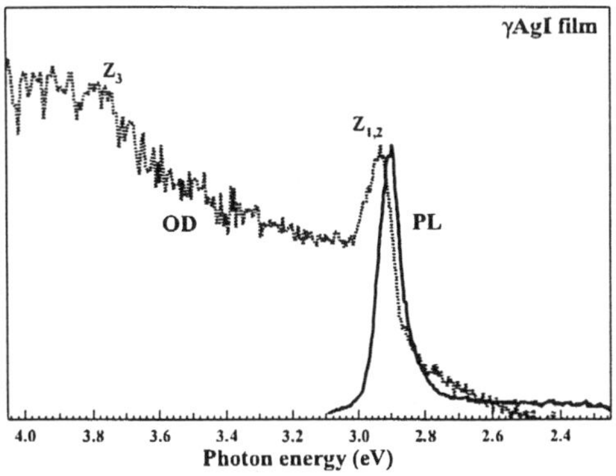

Figure 2. The OD and PL spectra of γAgI film at 7 K.

Figure 2 shows the OD and PL spectra of γAgI film at 7 K. The film was fabricated by the method (2). The $Z_{1,2}$ (2.93 eV) and Z_3 (3.77 eV) peaks are assigned to the exciton transitions in the zincblende structure. However, the $Z_{1,2}$ and Z_3 peaks are broader in comparison with the W_1, W_2 and W_3 peaks. The γ phase is thought to contain some stacking faults in real βAgI specimen. The results shown in Figures 1 and 2 suggest that real AgI films are affected by the wurtzite-zincblende bistability.

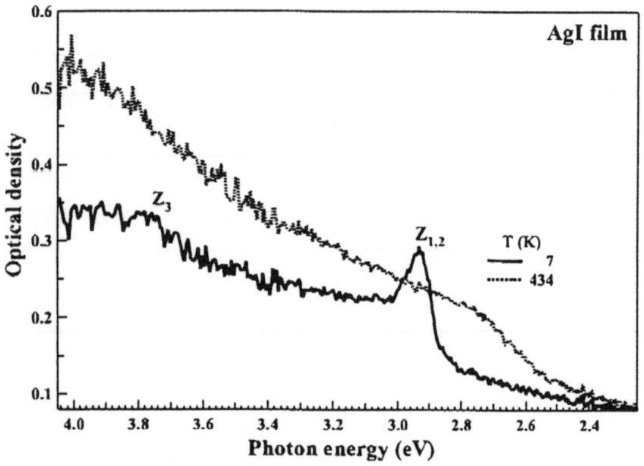

Figure 3. The OD spectra of AgI film at 7 K and 434 K.

We have also measured the OD spectrum of the γAgI film at different temperatures from 7 K to 453 K. As shown in Figure 3, the spectra at 7 K (γ phase) and 434 K (α phase) are compared. The spectrum at 434 K indicates that the optical absorption edge of αAgI is about 2.5 eV. The absorption tail of αAgI may be connected with the averaged structure.

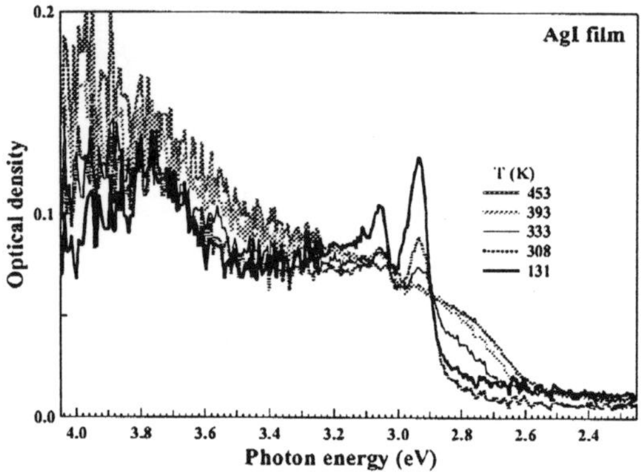

Figure 4. Temperature dependence of the OD spectrum of AgI film.

We have measured *in situ* the OD spectrum after evaporation without exposing specimen to air. In the experiment, very small amount of AgI powder was evaporated. After stopping the evaporation, the specimen was gradually cooled from 453 K to 131 K with an average cooling rate of 0.03 K/s. During the cooling process, the OD spectrum was measured as a function of temperature, at temperature intervals of 5 K. Of the spectra, five spectra are shown in Figure 4.

During cooling from 453 K to 398 K, the OD spectrum is almost independent of temperature. It is noted that the Urbach rule [15] does not hold in αAgI. This indicates that the contribution from the electron–phonon interaction counterbalances that from the thermal change of the bandgap energy with increasing temperature. No remarkable spectral change was observed near T_c. This is due to the hysteresis of the superionic conduction transition. At 393 K, several weak exciton absorption peaks and a shoulder appear faintly at 2.93 eV (peak), 2.98eV (shoulder), 3.06 eV (peak), 3.11 eV (peak), 3.77 eV (peak) and 3.98 eV (peak). On further cooling, these absorption peaks become prominent. In order to show the detail of spectral structure, the spectrum measured at 131 K is shown in Figure 5, comparing with that measured at 453 K.

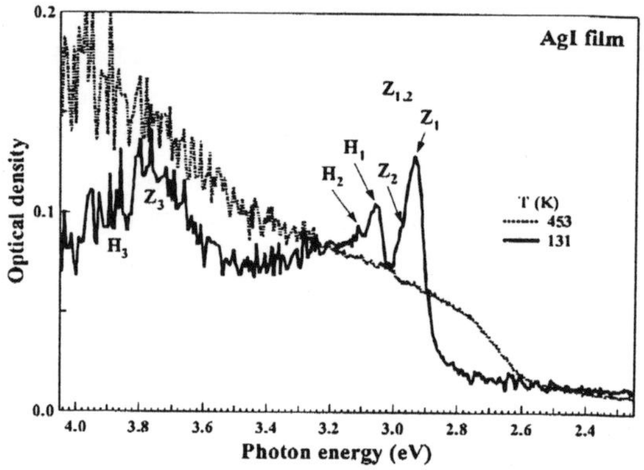

Figure 5. The OD spectra of AgI film at 131 and 453 K.

After the optical measurements, the crystal structure of the specimen was examined by the XRD analysis. It shows a considerable amount of γAgI and the broadening of the *(h0l)* diffraction lines of βAgI. The broadening arises from the stacking disorder of a hexagonal stacking sequence (ababab...) in βAgI, as pointed out by Lee *et al.* [16]. Taking into account this XRD data, as shown in Figure 5, absorption peaks and a shoulder observed at 2.93 eV (peak), 2.98 eV (shoulder) and 3.77 eV (peak) are assigned to the Z_1, Z_2 and Z_3 exciton absorptions of γAgI, while the peaks observed at 3.06 eV, 3.11 eV and 3.98 eV are assigned to the H_1, H_2 and H_3 exciton absorptions of some stacking-disorder-induced polytype structure of AgI. Similar exciton absorption spectrum due to the polytype structure was observed in CuI film evaporated onto a room temperature substrate [5].

Very recently, we have found another method for fabricating the AgI film with higher spectral quality, through different evaporation experiments. When AgI powder is thermally evaporated, I_2 molecules, Ag_mI_n clusters (m, n ≥ 1) and Ag_n clusters (n ≥ 1) appear in the vapour zone. If the temperature of the substrate is lower than 100 K, all the vapour species are adsorbed on the substrate and condense, forming a composite film. This composite film displays an extraordinary broad absorption spectrum due to I_2 molecules. During heating the film, re-evaporation of I_2 molecules in the film occurs suddenly at 220 K, showing clear exciton spectrum due to AgI.

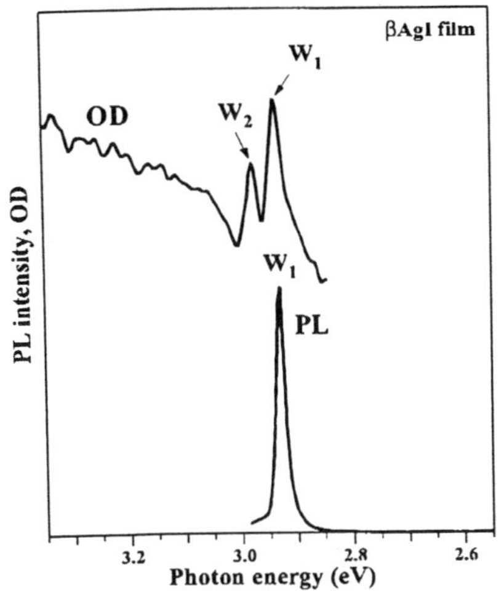

Figure 6. The OD and PL spectra of βAgI film at 13 K. This film was deposited at 200 K.

On the other hand, when a film is fabricated by deposition on the substrate whose temperature of 200 K and then annealing at room temperature, the film shows the OD spectrum resolved well into two peaks due to the W_1 and W_2 excitons of βAgI, as shown in Figure 6. Moreover, this film exhibits a sharp PL band at 13 K. The PL band has a very narrow full width at half-maximum, approximately 10 meV, with very small Stokes shift (< 7 meV) from the W_1 peak in the OD spectrum. More detail discussions are reported in [8].

3.2. *AgI–AgPO₃ glass*

The XRD patterns of different (x)AgI–(1–x)$AgPO_3$ glasses at room temperature are shown in Figure 7. The glass scattering of the $AgPO_3$ host glass at about 31 degrees is not greatly changed by doping AgI. This indicates that the AgI dopant hardly alters the short-range covalent bonding of phosphorous–oxygen network but is primarily introduced into some voids of the host phosphate glass structure. This glass structure also gives the prepeak at about 14 degrees. With increasing x, this prepeak shifts toward smaller angle side. This indicates that the structure of the intermediate range order varies with AgI doping. The present experiments show that the glass consists of interpenetrating silver phosphate and silver iodide regions with length scales of order 1 nm. For $x > 0.4$, the diffraction pattern shows a superposition of the glass scattering and Bragg powder diffraction of stacking-fault-contained βAgI, which precipitate in $AgPO_3$ glass matrix.

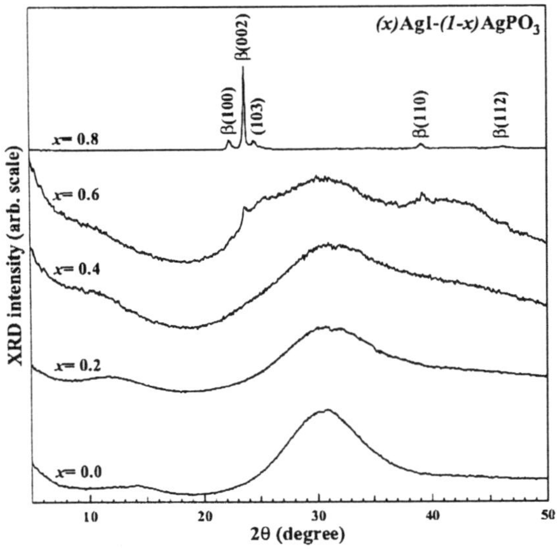

Figure 7. The XRD patterns of different (x)AgI–(1–x)$AgPO_3$ glasses at room temperature.

Figure 8 shows the PL spectra of (x)AgI–(1–x)$AgPO_3$ glasses at 10 K. The PL was excited by the 3.68 eV laser line of the N_2 laser. Each spectrum is normalized in intensity at each intensity maximum. Pure $AgPO_3$ glass shows a very broad PL band centred at 2.40 eV. Doping a small amount of AgI to this glass weakens considerably the 2.40 eV band, and new broad PL bands appear at about 1.77 eV and 2.36 eV. The attenuation of the 2.40 eV band is mainly due to laser light energy absorption of doped AgI. The 1.77 eV band may indicate that there are some deeply trapped exciton states at the AgI/glass matrix interfaces. At present, the morphology and the atomic structure of AgI in $AgPO_3$ glass are not clear.

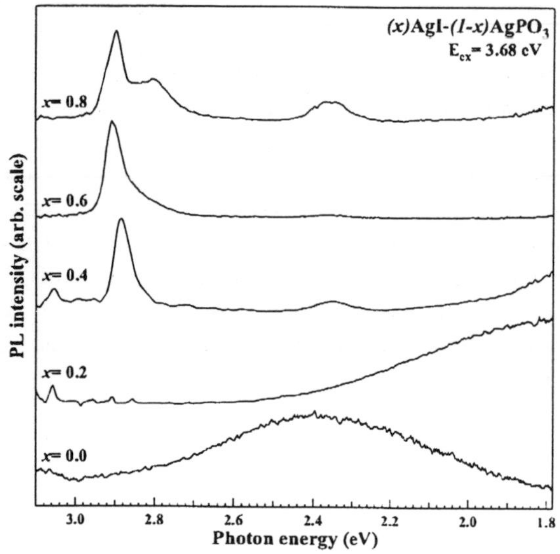

Figure 8. The PL spectra of different (x)AgI–(1−x)$AgPO_3$ glasses at 10 K.

With further AgI doping (x > 0.4), the intensity of the 1.77 eV band decreases, while a sharp PL band appears at about 2.89 eV. This 2.89 eV band arises from the radiative decay of excitons trapped by shallow level in AgI. It indicates that some precipitation of AgI crystallites occurs in AgI–$AgPO_3$ glasses, which is consistent with the XRD data. This PL band is accompanied by a shoulder band at about 2.80 eV. The radiative decay of exciton trapped by defects gives rise to the shoulder band. The 2.36 eV band has a long decay time connected to the radiative decay of deeply trapped excitons at the AgI/glass matrix interfaces. Since the precipitation of AgI in $AgPO_3$ matrix glass accompanies enhancement of ionic conductivity, the exchange of Ag^+–Ag^+ ion through such AgI crystallite/ matrix glass interfaces is thought to be easier at higher x region. Thus, such mobile Ag^+ ions at the interfaces may attract electrically the electrons of the free excitons in AgI crystallites, creating trapped excitons, as seen in the PL spectra.

3.3. *(0.85)AgI–(0.15)Ag_2WO_4 glass*

Depending on the cooling speed from the melt, two different types of AgI–Ag_2WO_4 glass specimen were obtained.

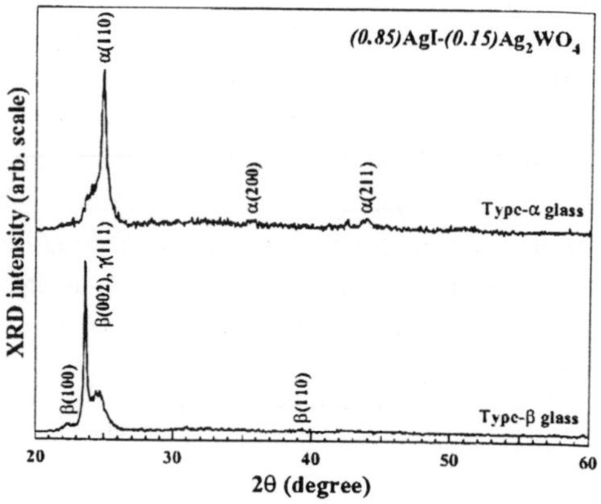

Figure 9. The XRD patterns of two kinds of (0.85)AgI–(0.15)Ag_2WO_4 glasses at room temperature.

As shown in Figure 9, a glass obtained under rapidly cooling condition predominantly shows the XRD pattern of αAgI phase (Type-α glass), while under less rapidly cooling condition, the XRD pattern of low-temperature AgI phases (Type-β glass) appears mainly. It has been elsewhere reported that AgI precipitated in glass are particles whose sizes are several ten nanometers [17].

Figure 10 shows the PL and PLE spectra of the type-α glass at 9 K. Each spectrum is normalized in intensity at each intensity maximum. We have found that the PL for the type-α glass cannot be excited with any ultraviolet light, but it can be excited with green lights. The PL spectrum has very broad bandwidth, and the intensity peaks of the PLE spectra appear near the absorption edge energy (about 2.5 eV) of αAgI. This indicates that the PL spectrum observed is due to radiative decay of excitons of αAgI particles precipitated in glass matrix. The PL spectrum is shifted by about 0.6 eV from the intensity averaged peak of the PLE spectra. This redshift may arise from radiative decay of deeply trapped exciton due to strong electron–phonon interaction in αAgI particle or deeply trapping potential at αAgI/glass matrix interfaces. The E_{ob} dependence of PLE spectra may arise from different trapped-states for excitons. The cause may arise from different shapes and different sizes of the αAgI particles.

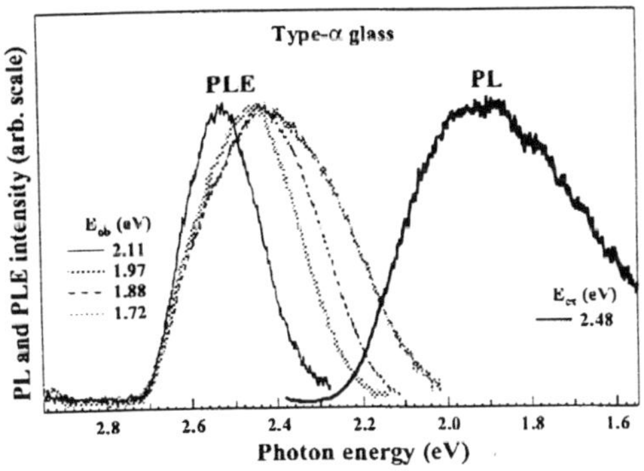

Figure 10. The PL and PLE spectra of type-α glass at 9 K.

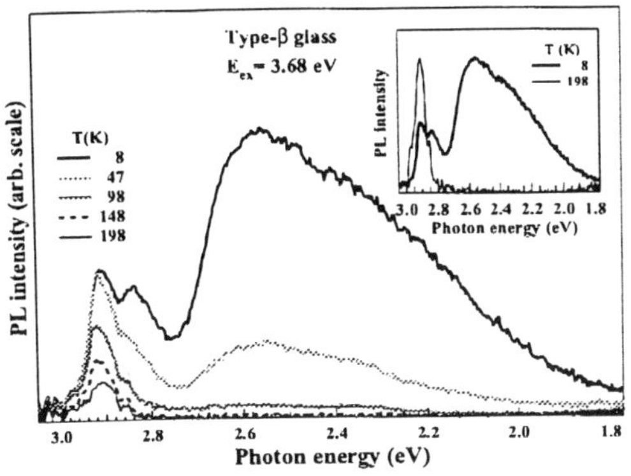

Figure 11. The temperature dependence of PL spectrum of (0.85)AgI–(0.15)Ag_2WO_4 glass. The PL at 8 and 198 K are compared in the inset.

Figure 11 shows the PL spectrum of the type-β glass at different temperatures. The PL was excited with the 3.68 eV laser line of the N_2 laser. The intensity peak at about 2.92 eV is assigned to radiative decay of free excitons in AgI particles, while the sideband at about 2.84 eV is assigned to radiative decay of excitons trapped by lattice defects and impurities. The TRPL measurements show that the decay time of PL due to the free exciton and the decay time of board PL band at about 2.56 eV are shorter than 10 ns and about 25 ns, respectively. The broadband peaked at about 2.56 eV is, therefore, assigned to radiative decay of deeply trapped excitons at β or αAgI/glass matrix interfaces. As shown in the inset, the PL bands except for radiative decay of free excitons are weaken, and the only PL band due to radiative decay of the free exciton becomes prominent with increasing temperature. This may be due to thermally activated reverse processes from trapped exciton states to free exciton state.

We apply the results of the present optical study to the ion conduction. Since the chemical potentials of Ag^+ ion in AgI and Ag_2WO_4 glass are difference, a transition layer is formed at the interfaces between AgI particle and Ag_2WO_4 glass matrix. Many crystal defects may be also contained at the interfaces, which gives rise to deep trap for free exciton and enhances ionic conductivity. Since the deeply trapped exciton states are able to connect to the transition layer, through this layer, the mobile Ag^+–Ag^+ ion exchange responsible for high ionic conduction occurs in the type-α and type-β glasses.

3.4. *AgI–γAl_2O_3 composite*

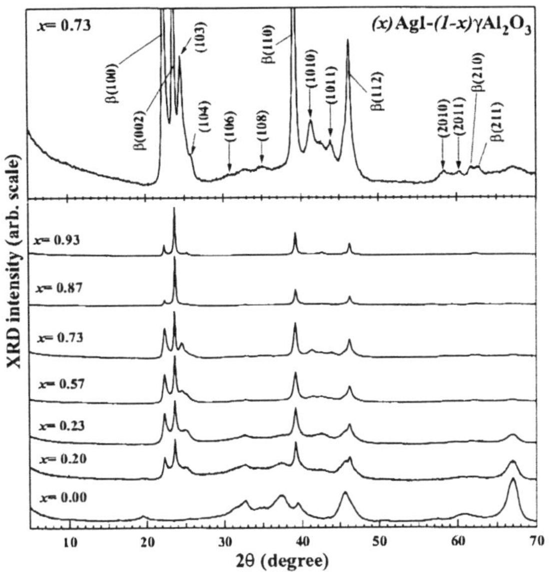

Figure 12. The XRD patterns of different of (x)AgI–(1–x)γAl_2O_3 composites at room temperature.

The XRD patterns of different (x)AgI–(1–x)γAl_2O_3 composites were measured at room temperature and the results for x = 0, 0.20, 0.23, 0.57, 0.73, 0.87 and 0.93 are shown in Figure 12. With increasing x, some weakening of the *(h0l)* peaks of βAgI become pronounced at 2θ = 25.4, 32.8, 42.6, 47.2, 52.0 and 66.5 degrees. On the other hand, new diffraction lines can be observed nearby reduced *(h0l)* peaks, as reported by Lee *et al* [3]. They can be referred to a new stacking sequence of the close packed planes in the wurtzite structure of AgI. They also explain the XRD features of AgI–γAl_2O_3 composite by taking account of a seven-hexagonal-layer polytype AgI (7H-AgI) with the stacking sequence ABCBCAC. The new diffraction lines at 24.5, 25.8, 30.7, 34.8, 41.5, 43.9, 58.2 and 60.2 degrees seem to correspond to the (103), (104), (106), (108), (1010), (1011), (2010) and (2011) peaks of the polytype structure, respectively. On further increasing x, the XRD pattern becomes close to that of the Type II AgI specimen, which has the 7H structure [16].

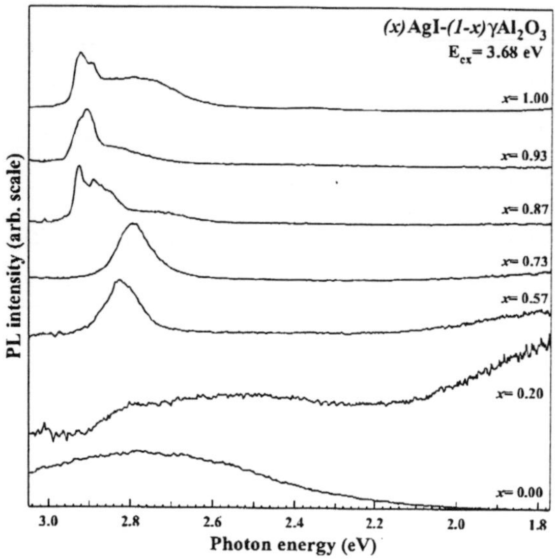

Figure 13. The PL spectra of different (x)AgI–(1−x) γAl_2O_3 composites at 10 K.

The PL spectra obtained for (x)AgI–(1−x)γAl_2O_3 composites with different compositions at 10 K are compared in Figure 13. The PL was excited with the 3.68 eV laser line of the N_2 laser. Each spectrum is normalized in intensity at each intensity maximum. Pure γAl_2O_3 shows a broad PL band centred at about 2.78 eV. Addition of small amount of AgI, x = 0.2, leads to a decrease of the PL intensity at higher energies than about 2.92 eV which corresponds to the $Z_{1,2}$ exciton energy of pristine AgI. The light absorption by surrounding AgI may decrease the intensity of light emitted from γAl_2O_3 particles. On the other hand, a very broad PL band centred at about 1.77 eV appears, which may be radiative decay of deeply trapped excitons at the AgI/γAl_2O_3 interfaces. On further increasing x, the PL intensity of the very broad band decreases and several PL bands around 2.82 eV become prominent. Since the $Z_{1,2}$ exciton energy of AgI is 2.92 eV, the bands around 2.82 eV may be assigned to radiative decay of the excitons trapped by a shallow potential inside AgI domains in the AgI–γAl_2O_3 composites. At the higher x region (x ≥ 0.87), the PL band at 2.92 eV due to radiative decay of the free excitons in AgI appears clearly.

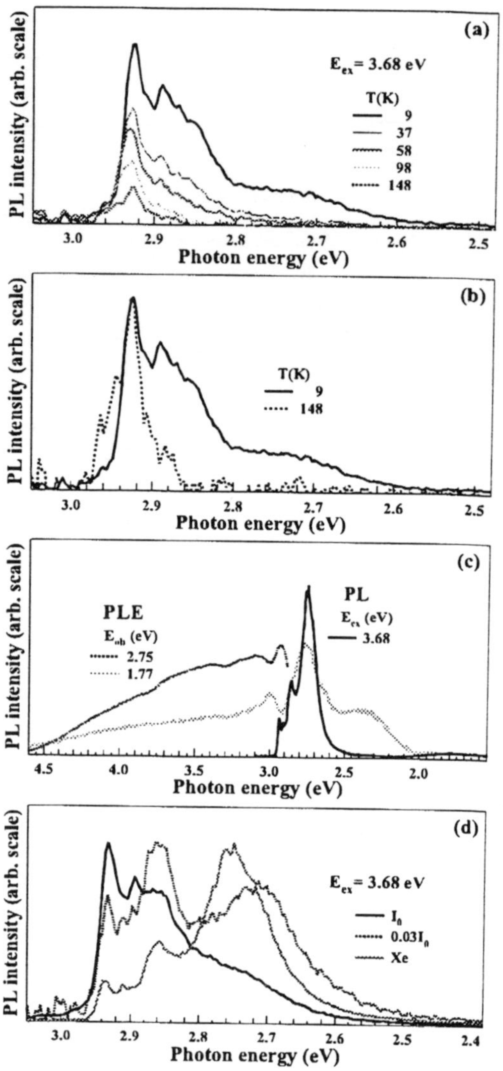

Figure 14. The PL spectra for (0.87)AgI–(0.13) γAl_2O_3 composites; (a), (b) temperature dependence, (c) PL and PLE spectra at 9 K, (d) excitation light intensity dependence at 9 K.

Figure 14(a) shows the temperature dependence of the PL spectrum of (0.87)AgI–(0.13)γAl_2O_3 composite. With increasing temperature, the PL intensity due to radiative decay of the free exciton prevails against that of the excitons trapped at shallow and deep levels. In Figure 14(b) the PL spectrum observed at 9 K is compared with that observed at 148 K, which may be due to thermal activation process from the shallowly and the deeply trapped exciton states to the free exciton state. Figure 14(c) shows the PL and PLE spectra at 9 K. This indicates that the shallow and the deep levels are below 0.16 and 0.56 eV from the bottom of the free exciton energy, respectively. The components of PLE spectra between about 3.00 eV and about 3.20 eV are assigned to the absorption of the free exciton in polytype-structured AgI. The excitation light intensity (I_{ex}) dependence of the PL spectrum was also studied. In Figure 14(d) the PL spectra under intense I_0 and weak $0.03I_0$ (peak power I_0 = 200kW) excitations at 9 K are compared typically. The results are summarized as follows. The PL band due to radiative decay of the free exciton increases with I_{ex}, while the other PL bands that have higher yields than that due to of the free exciton come to saturate under intense excitation. These saturations arise from finite number of the shallow and deep levels for PL centres.

On the basic of the results described above, we propose a structural model of AgI–γAl_2O_3 composites as follows, tentatively. The γAl_2O_3 particles are enveloped with the polytype AgI with a hexagonal-layer, which has high ionic conductivity. Both the structure of AgI/γAl_2O_3 interfaces and the polytype structure of AgI may play important roles in fast ion conduction. Through these structures, the mobile Ag^+–Ag^+ ion exchange responsible for fast ion conduction may occur. The free excitons are trapped by shallow potentials inside AgI domains and by deep potentials at the AgI/γAl_2O_3 interfaces. Since the mean free path of the free excitons is finite, the effects of the interfaces on the PL become obscure by the contribution from shallowly trapped excitons inside AgI domains with increasing AgI contents (*i.e.* AgI domain size), as shown in Figure 13.

3.5. *AgI–anatase composite*

We have also studied AgI–anatase composites. The surface morphology of these composites was observed by a SEM both on the raw surfaces and the fractured surfaces. It has found that, in the large x region, anatase fine particles are packed densely in AgI particle of several micrometres in size and small anatase particles are not so different in size from the initial average particle size of 50 nm, while other anatase particles coagulate to form irregularly shaped particles of several hundred nanometres in size. The EDX fluorescence analyses indicate that iodine-rich parts tend to appear inside pore walls and that the surroundings of anatase particles are silver-rich. The latter suggests that Ag^+ ions are adsorbed onto the anatase particle surfaces.

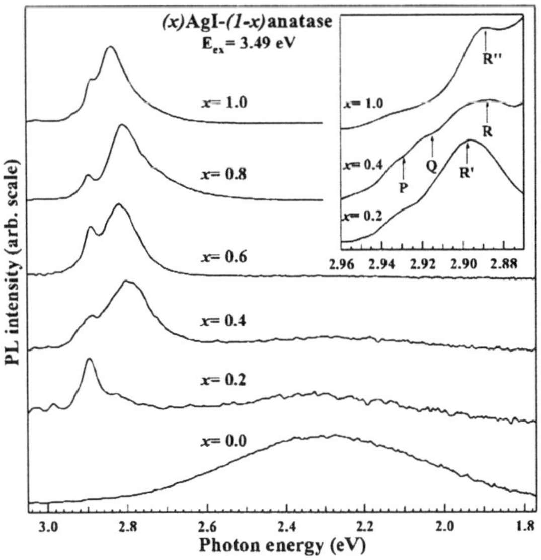

Figure 15. The PL spectra of different (x)AgI–(1–x)anatase composites at 8 K. In the inset, details of the PL band related to excitons in AgI–anatase composites.

The PL spectra obtained for (x)AgI–(1–x)anatase composites with different compositions at 8 K are compared in Figure 15. The PL was excited with the 3.49 eV laser line of the Nd^{3+}: YAG laser. Each spectrum is normalized in intensity at each intensity maximum. Pristine anatase shows a broad PL band centred at about 2.31 eV with a large Stokes shift of 0.89 eV. The present PLE spectrum measurement of this broad PL band at 8 K shows a sharp PLE edge, corresponding to the optical absorption edge, at about 3.2 eV which is close to the bandgap energy of pristine anatase. This PL band has been tentatively assigned to the radiative decay of the self-trapped excitons in pristine anatase by many authors [18–20]. The addition of a small amount of AgI, x = 0.2, leads to a considerable decrease of this PL due to anatase and gives a sharp PL band at 2.89 eV accompanied by shoulders at about 2.93 eV and about 2.83 eV. The spectrum is affected by noise at photon energies larger than 2.99 eV. The decreased PL intensity of the 2.31 eV band is due to the absorption of excitation laser light by the AgI domain in an anatase-packed AgI particle. The bands at about 2.92 eV and about 2.89 eV are assigned to the radiative decay of free exciton and shallowly trapped exciton, respectively. No measurable spectral change with x is observed in the 2.31 eV band. Further addition of AgI, x = 0.4, the sideband at lower photon energy increases and generate a broad band centred at 2.80 eV, and the 2.31 eV band reduces accompanied by a slight bandwidth narrowing at the lower photon energy side (E < 2.31 eV). On further increasing x from 0.4 to 0.8, this broad PL band becomes narrowed and blue-shifted to 2.81 eV, which centres at 2.84 eV on pristine AgI (x = 1.0). Therefore it should be interpreted that the broad PL band consists of the 2.80 and 2.84 eV bands at least.

We interpret the observed spectral variation with x, as follows. In the specimens with smaller x, since the free excitons are confined in smaller AgI domains, the free excitons reach the surfaces of AgI domain and the AgI/anatase interfaces, and then generate radiative decay. Since the bandgap energy (about 3.1 eV) of AgI is close to that (3.2 eV) of anatase, the potential for trapping the free exciton in AgI is thought to be shallow. In such a case, unlike the AgI–γAl_2O_3 composite, any PL band due to radiative decay of excitons trapped deeply at the AgI/anatase interfaces may not be observed as shown in Figure 15. However, a small number of excitons which are deeply trapped at the AgI/anatase interfaces are found through the excitation light intensity dependence of the PL spectrum.

In the inset, the spectral structures of higher energy region for x = 0.2 and 0.4 of (x)AgI–(1 x)anatase composites are compared with that for the pristine AgI specimen which was prepared in the same way (x = 1.0). The shoulders at about 2.929 eV and about 2.915 eV, which are respectively labelled P and Q, are seen for the (0.4)AgI–(0.6)anatase composite. As shown in the inset of Figure 16, both the P and Q shoulders remain clearly even at 129 K (= 11.1 meV) in spite of the small energy separation (14 meV) between P and Q. With increasing x, the P and Q shoulders tend to move toward about 2.92 eV. The $Z_{1,2}$ exciton originates from the doubly degenerate valence band Γ_8 with light and heavy hole masses. The Γ_8 band tends to split by some crystal-field change caused by mixing with the wurtzite structure or some internal stress caused by different thermal expansion coefficients between the AgI and anatase. Hence, the Γ_8 splitting should be taken into account for rigorous assignment of the 2.93 eV and 2.92 eV shoulder bands. Such an exciton spectra closely resemble that of the microcrystal specimen [9]. The spectral structure of the 2.89 eV peak (labelled R) assigned to shallowly trapped exciton changes with x, as shown in the inset. The changed bands are labelled R' and R". This indicates that there are several shallowly trapped exciton states. In such cases, the trapped excitons in lower energy states tend to populate in higher energy states with increasing temperature.

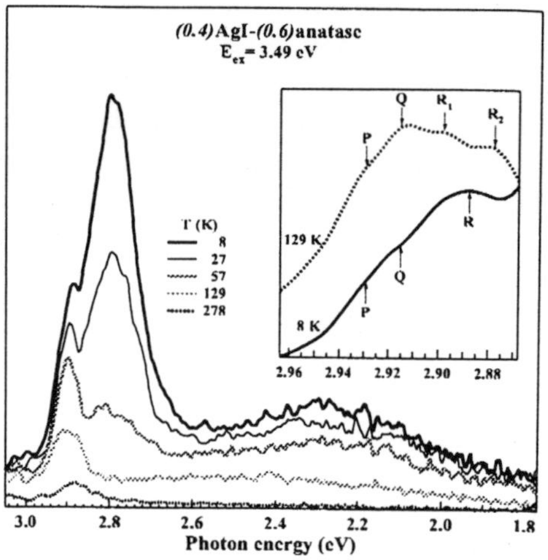

Figure 16. The temperature dependence of PL spectrum of (0.4)AgI–(0.6)anatase composite.

The PL spectra of AgI–anatase composites were studied at different temperatures between 8 K and 278 K. The results obtained for the (0.4)AgI–(0.6)anatase composite are shown in Figure 16, as a typical example. In comparison with the PL intensities of the P- and Q-shoulders and R-peak, the PL intensity of about 2.80 eV band decreases prominently with increasing temperature. This PL band at about 2.80 eV was assigned to radiative decay of excitons trapped by lattice defects and impurities in AgI domain. On further increasing temperature above 78 K, the 2.80 eV band weakens considerably and disappears absolutely at 278 K. Above 188 K, the PL bands of the intrinsic and the extrinsic exciton are slightly red-shifted, which is ascribed to the decrease of the bandgap energy of AgI with increasing temperature. Such a redshift was also observed at pristine AgI [9]. The disappearance of the 2.80 eV band may be due to a thermally activated reverse process from trapped states to the free exciton one. The 2.31 eV band becomes weaker and shifted to larger energy with increasing temperature. Such a blueshift was also observed in the pristine anatase specimen (x = 0). This indicates that the energy of exciton–lattice relaxation decreases with increasing temperature. Through the excitation light intensity dependence of the PL spectra, we have found that the broad 2.31 eV band has two bands at least. If these bands have different temperature dependences, this blueshift is also possible. The decrease of PL intensity of the 2.31 eV band with increasing temperature from 8 K to 98 K is about two times larger than that observed in pristine anatase. This may ascribe to the exciton trapping to generate the nonradiative decay of excitons at the AgI/ anatase interfaces.

3.6. *AgI–ZnO composite*

We have also studied the AgI–ZnO composites. The PL spectra obtained for (x)AgI–(1 x)ZnO composites with different compositions at 9 K are compared in Figure 17. The PL was excited with the 3.49 eV laser line of the Nd^{3+}: YAG laser. Each spectrum is normalized in intensity at each intensity maximum.

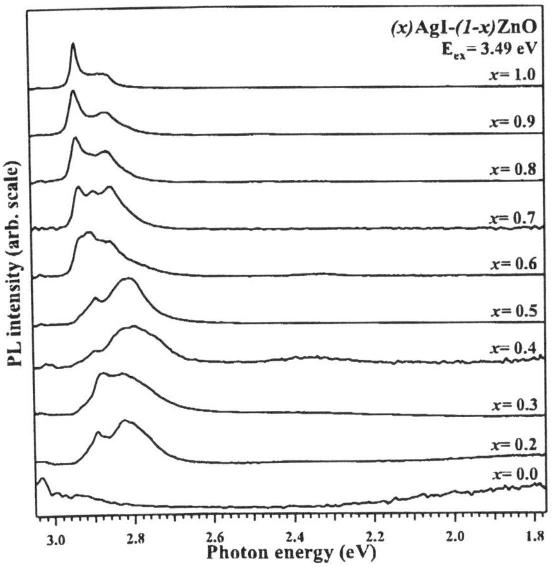

Figure 17. The PL spectra of different (x)AgI–(1 x)ZnO composites at 9 K.

The PL spectra of this composite resemble the AgI–anatase composite in features which is not observed the PL band due to radiative decay of deeply trapped exciton. However, the Γ_8 splitting like the P- and Q-peaks in the inset of Figures 15 and 16 is not observed in AgI–ZnO composite.

3.7. *Exciton nature in AgI–based glasses and AgI–oxide fine particle composites*

In summary above optical data, free, shallowly trapped and deeply trapped excitons have been observed in AgI–based glasses and AgI–oxide fine particle composites, depending AgI content. When band gap energy difference between AgI and glass (or oxide) material is large, the free excitons tend to be trapped deeply at the AgI/glass (or AgI/oxide particle) interfaces, while the difference is small, the free excitons tend to de trapped shallowly at the interfaces. The phonon-assisted reverse processes from the shallowly and deeply trapped exciton states to the free exciton states become prominent with increasing temperature. Thus, only PL band due to the radiative decay of the free excitons occurs as observed.

The optical spectra due to radiative decay of the shallowly trapped excitons are observed in most of the specimens. Such shallowly trapped exciton states may be related to halogen impurities (for example, bromine and chlorine) and crystal defects of AgI may also create such states. Since the exciton energies of the H_1, H_2 and H_3 excitons are close to those of the $Z_{1,2}$ and Z_3 excitons as shown in Figure 5, the depth of the exciton trapping potential at the interfaces between γAgI domain and polytype structured AgI is thought to be small (less than 0.2 eV) too.

Acknowledgments

This work was partially supported by a Grant-in-Aid for Scientific Research from the Ministry of Education, Science, Sports, Culture and Technology, Japan. This work is also partially supported by Project Research Grant from The Institute of Information Sciences of College of Humanities and Sciences (Nihon University) and by Cooperative Research Grant from The Institute of Natural Sciences (Nihon University).

References

1. B. Roling and M. D. Ingram, *Phys. Rev. B* **56**, 13619 (1997).
2. N. F. Uvarov, P. Vanek, M. Savinov, V. Zelezny, V. Studnicka and J. Petzelt, *Solid State Ionics* **127**, 253 (2000).

3. J. -S. Lee, St. Adams and J. Maier, *Solid State Ionics* **136-137**, 1261 (2000).
4. S. Mochizuki and F. Fujishiro, *Proc 6th Forum on Superionic Conductor Physics* (Kyoto: The Japanese Society of Ion Transport) 41-46 (2002).
5. M. Cardona, *Phys. Rev.* **129**, 69 (1963).
6. S. Mochizuki and Y. Ohta, *J. Lumin.* **87-89**, 299 (2000).
7. S. Mochizuki, *Physica B* **308-310**, 1042 (2001).
8. S. Mochizuki and F. Fujishiro, *J. Phys.: Condens. Matter* **16**, 3239 (2004).
9. S. Mochizuki and K. Umezawa, *Phys. Lett. A* **228**, 111 (1997).
10. F. Fujishiro and S. Mochizuki, *Nonlinear Optics* **29** , 443 (2002).
11. S. Mochizuki and F. Fujishiro, *phys. stat. sol. (c)* **0** , 767 (2003).
12. S. Mochizuki and F. Fujishiro, *phys. stat. sol. (c)* **0** , 763 (2003).
13. S. Mochizuki and F. Fujishiro, *J. Phys.: Condens. Matter* **15**, 5057 (2003).
14. F. Fujishiro and S. Mochizuki, Physica B **340-342**, 216 (2003).
15. F. Urbach, *Phys. Rev.* **92**, 1324 (1953).
16. J. -S. Lee, St. Adams and J. Maier, *J. Phys. Chem. Solids* **61**, 1607 (2000).
17. M. Tatsumisago, N. Itakura and T. Minami, *J. Non-Cryst. Solids* **232-234**, 267 (1998).
18. H. Tang, H. Berger, P. E. Schmid and F. Levy, *Solid State Commun.* **92**, 267 (1994).
19. N. Hosaka, T. Sekiya and S. Kurita, *J. Lumin.* **72-74**, 874 (1997).
20. A. Suisalu, J. Aarik, H. Mandar and I. Sildos, *Thin Solid Films* **336**, 295 (1998).

HOPPING MODELS FOR ION CONDUCTION IN NONCRYSTALS

Jeppe C. Dyre and Thomas B. Schrøder

Department of Mathematics and Physics (IMFUFA), Roskilde University, Postbox 260, DK-4000 Roskilde, Denmark.
Fax number: (+45) 46 74 30 20; E-mail address: dyre@ruc.dk

Abstract

Ion conduction in noncrystals (glasses, polymers, etc) has a number of properties in common. In fact, from a purely phenomenological point of view, these properties are even more widely observed: ion conduction behaves much like electronic conduction in disordered materials (e.g., amorphous semiconductors). These universalities are subject of much current interest, for instance interpreted in the context of simple hopping models. In the present paper we first discuss the temperature dependence of the dc conductivity in hopping models and the importance of the percolation phenomenon. Next, the experimental (quasi)universality of the ac conductivity is discussed. It is shown that hopping models are able to reproduce the experimental finding that the response obeys time-temperature superposition, while at the same time a broad range of activation energies is involved in the conduction process. Again, percolation is the key to understanding what is going on. Finally, some open problems in the field are listed.

Keywords: Dc conduction, ac conduction, hopping, glasses, polymers, scaling, universality, percolation.

1. Introduction

Ion conduction in crystals like NaCl is relatively well understood [1]. It proceeds via well-defined defects (vacancies or interstitials in pure crystals, or impurities). Ion conduction in noncrystals poses a much greater challenge to theorists [2-4]. In this paper we briefly review the random barrier model for ion conduction in disordered solids and discuss how it compares to experiment.

2. The random barrier model (RBM)

It should be emphasized from the outset that this model is highly simplified. The purpose of the model is not to account for differing details of ion conduction in differing solids, but the opposite: To give the simplest possible realistic model covering the overall features of experimental dc and ac ion conduction in disordered solids. The objective is to arrive at the analogue of the ***ideal gas model*** for ion conduction in noncrystals. How do we know that such a model exists? We don't know for certain *a priori*, but the surprisingly universal features of ion conduction in quite different disordered solids, and the fact that these features extend to electronic and polaronic conduction in an even larger class of solids, makes one optimistic in the search for an ideal gas model of ion conduction.

Let us briefly summarize the universal features one would like to understand within a simple model [5-10]:

1) The dc conductivity is Arrhenius temperature dependent.
2) The ac conductivity follows an approximate power law with an exponent smaller than 1 which, in a fixed frequency range, goes to 1 as temperature is lowered towards absolute zero.
3) The ac conductivity obeys time-temperature superposition, i.e., the same ac response is observed at different temperatures, just displaced in the log-log plot usually used. This is often referred to as ***scaling.***
4) Different solids show roughly the same ac response: (quasi)universality.

The random barrier model concerns the motion of completely non-interacting particles on a cubic lattice. Thus not only is Coulomb repulsion ignored, but so is the self-exclusion which is present in all ion conductors (i.e., the fact that there is only room for one ion in each potential energy minimum in the solid structure). This may seem completely unrealistic, but it is not difficult to arrive at the equation describing non-interacting particles by linearizing a more general master equation (see, e.g., Ref. 10 and its references).

The random barrier model (RBM) has one further simplification, namely that the potential felt by the non-interacting charge carriers has all equal minima (Fig. 1). Again this may be justified from more general principles [10] which we shall not discuss further here. Finally, the RBM assumes the ion sites are situated on a simple cubic lattice.

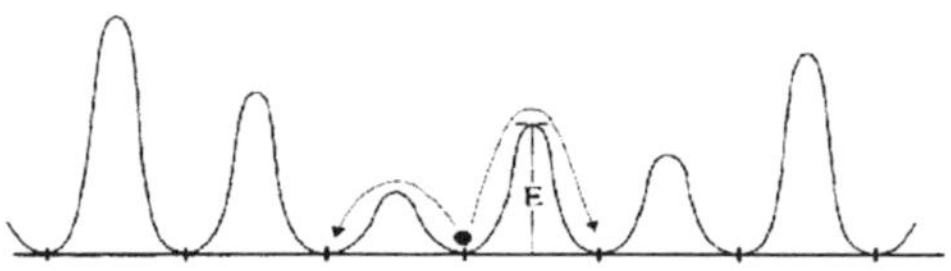

Fig. 1: *Typical potential for a system described by the RBM, shown here in one dimension. The barriers are assumed to vary randomly according to some probability distribution. The arrows indicate the two possible jumps for the charge carrier shown. In an external electric field the potential is tilted in one direction and a current flows.*

2. Three 'classical' arguments against barrier distributions

In disordered solids like glasses or polymers it seems eminently reasonable to assume that not all ion jumps involve barriers of exactly the same height. This was discussed in the literature already back in the 1950's. However, this idea was rejected because it was thought to be inconsistent with well-established experimental facts. The following three facts were used as arguments against barrier distributions:

1) The dc conductivity is Arrhenius temperature dependent.
2) The frequency marking onset of ac conduction is Arrhenius temperature dependent with the same activation energy as the dc conductivity.
3) The ac conductivity obeys time-temperature superposition, i.e., it is possible to scale data at different temperatures to one single master curve.

The reasoning [5,11-13] based on these 3 points which apparently rules out any but an extremely narrow barrier distribution goes as follows:

1) Whenever there are barriers of differing sizes involved in the conduction process one would not expect a simple Arrhenius temperature-dependent dc conductivity. After all, which one of the many barriers involved should be chosen as the overall dc conductivity activation energy? More precisely, one should expect a non-Arrhenius behavior with smaller activation energies dominating at low temperatures where it is important to take advantage of these (perhaps relatively few), and larger activation energies gradually coming into play as temperature is increased.
2) Ac conduction must be due to ion motion over limited distances while dc conduction involves motion over extended distances (this, incidently, is absolutely correct). Consequently, one expects ac conduction to involve smaller barriers than dc conduction. And since the frequency which marks onset of ac conduction is a characteristic of ac conduction, the activation energy of this frequency must be smaller than that of the dc conductivity.
3) Any process involving a distribution of barriers must violate time-temperature superposition (unless the barrier distribution itself has a peculiar temperature dependence), because in a log-log plot the response must broaden as

temperature is lowered. This is because the relevant quantity entering the transition rate is barrier divided by temperature, implying that at low temperatures more and more decades of jump frequencies are involved in the conduction process.

We now argue that these classical objections, which seem at first sight quite convincing, are in fact not valid. A more detailed discussion of this point may be found in a recent review [10].

3. Dc conductivity in the RBM

The first challenge to the RBM is to explain the fact that the dc conductivity is Arrhenius and to identify the activation energy. The answer is provided by percolation theory, a mathematical theory which was not invented when people in the early 1950's ruled out barrier distributions (and not known to them after 1957 when the percolation phenomenon was first discussed in the scientific literature [14]).

Suppose the barrier heights denoted by E vary randomly and uncorrelated from lattice link to lattice link of the RBM according to a probability distribution p(E). Consider the situation at low temperatures, i.e., where the barrier distribution is much broader than the thermal energy k_BT. We shall discuss ion motion in zero external field only, because according to the fluctuation-dissipation theorem the dc conductivity is proportional to the mean-square displacement per unit time in zero external field. Now, small barriers give rise to large jump rates, and the ions definitely prefer these. So most ion jumps proceed across small barriers. To extend the motion to infinity, however, some larger barriers obviously have to be overcome. The largest barrier which **must** be overcome to move to infinity becomes the dc conductivity activation energy. How to identify this? It is to answer this question [15,16] that percolation theory is necessary:

Consider the case of a two-dimensional simple square lattice. Suppose the links linking neighboring lattice sites are marked according to increasing barrier height. At some point in this process, referred to as the percolation threshold, an infinite cluster of marked links appear. This situation is illustrated in Fig. 2.

Fig. 2: *Percolation on a two-dimensional square lattice. The upper picture shows the situation below the percolation threshold where there is no infinite cluster of marked links, the lower picture shows the situation above the threshold. The largest barrier on the infinite cluster is the dc conductivity activation energy.*

When an infinite cluster of marked links appears (termed the percolation cluster) no larger barriers need to be overcome in order for the ion to move to infinity, corresponding to carrying a dc current. Thus, all barriers on the percolation cluster are in principle relevant for determining the dc conductivity activation energy, but no larger barriers. However, because temperature is by assumption low, the jump rates on the percolation cluster cover several decades. Consequently, the largest barrier on this cluster presents a bottleneck to the ion motion, and this largest barrier completely dominates the overall rate of motion. It is this largest barrier which becomes the dc conductivity activation energy. And it is because of the percolation phenomenon – in particular the identification of a definite bottleneck barrier – that the dc conductivity is Arrhenius temperature dependent despite the fact that a range of barriers are involved. Curiously enough, percolation theory is only relevant when a broad range of barriers is involved. So the classical argument that if there is any barrier distribution it must be quite narrow is as wrong as it could be: Only when the distribution is wide does one get an Arrhenius dc conductivity.

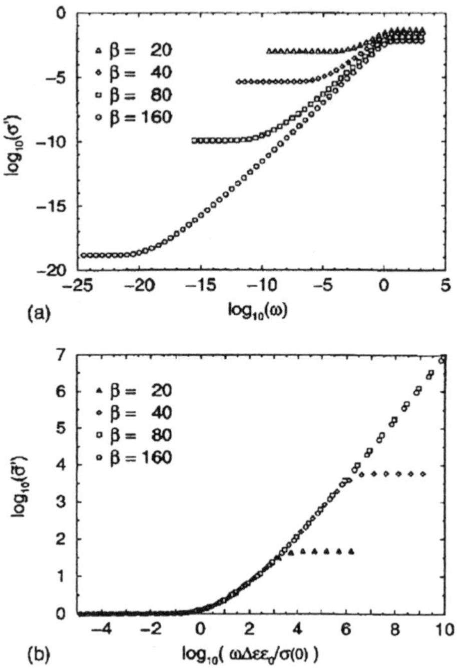

Fig. 3: *Computer simulations of the random barrier model in three dimensions with the uniform barrier distributions [10]. (a) shows the real part of the conductivity as function of angular frequency (in suitable units [10]) at different inverse temperatures β, (b) shows how these data may be scaled to one single master curve which is valid in a wider and wider frequency range as temperature is lowered.*

4. Ac conductivity in the RBM

What are the RBM-predictions for the ac conductivity? Figure 3 shows results of extensive computer simulations of the model. Clearly, a master curve is arrived at as the data of Fig. 3a are scaled by suitable displacements in the log-log plot (Fig. 3b), a master curve which as temperature is lowered applies in a wider and wider frequency range around the ac onset frequency. We conclude that the RBM obeys time-temperature superposition. This disproves the above classical argument 3) against barrier distributions. Our simulations [10] also show that the onset frequency has the same temperature dependence as the dc conductivity, a point we shall not dwell on here although it is an important manifestation of the celebrated Barton-Nakajima-Namikawa (BNN) relation [17-19].

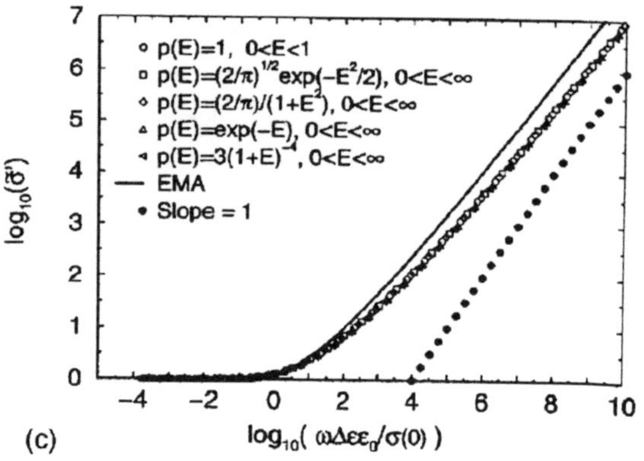

Fig. 4: *Computer simulations of the random barrier model for several barrier distributions [10]. The figure shows the ac conductivity relative to the dc conductivity as function of scaled frequency at low temperatures. The dots are shown as a guide to the eye giving a line of slope one, while the two other curves give two different analytical approximations to the universal ac conductivity. The full line is the classical effective medium approximation (EMA) (today more advanced analytical approximations are available [10]).*

There is one further important point where the RBM reproduces experiment, namely ac universality. Figure 4 shows results of computer simulations of the RBM for several different probability distributions. This figure shows that there is universality of the ac response in the RBM. We shall not discuss the correct recipe for scaling the frequency (which was recently a subject of some debate in the scientific literature [20-23]), but just note that the onset frequency is roughly proportional to the dc conductivity in the simulations. As already mentioned, this is also seen in experiment – apparently without exception.

Close inspection of Figs. 3 or 4 reveals that the universal ac conductivity is not an exact power law as function of frequency. Rather, the ac conductivity is an approximate power law with an exponent below one, which goes slowly to one as (scaled) frequency goes to infinity. This is in agreement with experiment. In particular, we note that as temperature is lowered, by measuring in a fixed frequency range, one effectively measures further and further out on the master curve and thus finds an approximate frequency power law exponent which goes to one. This is also observed.

The frequency exponent close to one referred to above found at low temperatures or very high frequencies is usually referred to as the "nearly constant loss" (NCL). This is another universal feature which the model reproduces, but as shown in the recent literature (see, e.g., [24,25] and their references) there are other possible explanations. Nevertheless it is encouraging, we feel, that the simple RBM is able to reproduce even this feature.

To summarize the computer simulations of the RBM, the model predicts a) time-temperature superposition at low temperatures, b) universality of the ac response, and finally c) that the frequency marking onset of ac conduction has the same activation energy as the dc conductivity. How can this be understood physically? The answer is, it turns out is again, that the percolation phenomenon is responsible. It is not possible here to argue for this in detail (see [10]), but we can very briefly sketch the reasoning: Once one has established that percolation explains the Arrhenius temperature dependence of the dc conductivity as due to the fact that the dc current mainly runs on the percolation cluster, it is tempting to guess that even the ac current runs mainly on the percolation cluster. This is not quite correct, but the universal part of the ac current does indeed run on the percolation cluster [work to be published], and the universal behavior is dominated by barriers close to the dc conductivity activation energy. This fact explains not only why the onset frequency has this activation energy but also universality: The only relevant number is the value of p(E) at the dc conductivity activation energy, and even this number is "scaled away" when simulation data are scaled to arrive at the master curve. Another way of expressing this is to say that at low temperatures (in the so-called "extreme disorder limit") any barrier distribution is effectively flat. Note also that this explains time-temperature superposition: Lowering the temperature really does not change anything but the values of the jump rates (not their probabilities relative to one another).

5. The RBM versus experiment

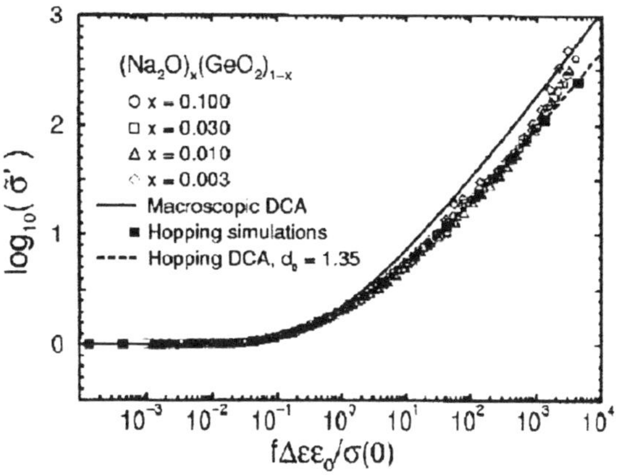

Fig. 5: *RBM-prediction for the ac conductivity versus data on Sodium Germanate glasses [21] with varying Sodium concentration. The open symbols are the experimental data while the full symbols give the RBM-universality prediction. "Hopping DCA" is the diffusion cluster approximation [10] (please ignore the "Macroscopic DCA").*

We have argued that thc random barrier model reproduces a number of characteristic features of experimental ion conduction. But how does the model compare to experiment quantitatively? We show in Fig. 5 typical ac data compared to the universal ac conductivity of the RBM. Not all data are identical as regards the ac response, of course, so instead of referring to universality perhaps the term "quasi-universality" is more appropriate. Nevertheless, most data are reasonably well fitted by the RBM, and certainly much better than one would expect *a priori*, given the fact that the model has no fitting parameters for its prediction of the ac conductivity once model predictions are written in terms of dimensionless variables.

6. Conclusion and open questions

We conclude that the RBM is a simple model which captures the essential physics of dc and ac ion conduction, at least from the theoretical physicists point of view. Thus the model does indeed, in our opinion, deserve the honor of being referred to as the analogue of the ideal gas model. Nevertheless, a number of important open questions relating to the model remains to be answered:

1) How are the predictions affected when one wants to be more realistic by modifying the model to take into account Coulomb interactions and self-exclusion [3,22]?
2) How is the universal ac conductivity modified when the model is generalized to deal with sites of differing energy? Is there still ac universality?
3) How is the physical insight that the current mainly runs on the percolation cluster utilized to arrive at precise quantitative predictions of the universal ac conductivity?

References

[1] C. P. Flynn, "Point Defects and Diffusion" (Oxford University Press, London, 1972).

[2] C. A. Angell, Chem. Rev., **90** (1990) 523.

[3] W. Dieterich and P. Maass, Chem. Phys., **284** (2002) 439.

[4] J. C. Dyre, J. Non-Cryst. Solids, **324** (2003) 192.

[5] J. O. Isard, J. Non-Cryst. Solids, **4** (1970) 357.

[6] A. K. Jonscher, Nature, **267** (1977) 673.

[7] A. E. Owen, J. Non-Cryst. Solids, **25** (1977) 370.

[8] S. R. Elliott, Solid State Ionics, **70/71** (1994) 27.

[9] J. R. Macdonald, J. Non-Cryst. Solids, **210** (1997) 70.

[10] J. C. Dyre and T. B. Schrøder, Rev. Mod. Phys., **72** (2000) 873.

[11] J. M. Stevels, "Handbuch der Physik", Vol. 20, edited by S. Flügge (Springer, Berlin, 1957), 350.

[12] H. E. Taylor, J. Soc. Glass Technol., **43** (1957) 350T.

[13] A. E. Owen, "Progress in Ceramic Science," Vol. 3, edited by J. E. Burke (Macmillan, New York, 1963), 77.

[14] S. R. Broadbent and J. M. Hammersley, Proc. Cambridge Philos. Soc., **53** (1957) 629.

[15] B. I. Shklovskii and A. L. Efros, Sov. Phys. JETP, **33** (1971) 468.

[16] V. Ambegaokar, B. I. Halperin, and J. S. Langer, Phys. Rev. B, **4** (1971) 2612.

[17] J. L. Barton, Verres et Refr., **29** (1966) 328.

[18] T. Nakajima, "Conference on Electric Insulation and Dielectric Phenomena," (National Academy of Sciences, Washington, DC, 1972), 176.

[19] H. Namikawa, J. Non-Cryst. Solids, **18** (1975) 173.

[20] B. Roling, A. Happe, K. Funke, M. D. Ingram, Phys. Rev. Lett., **78** (1997) 2160.

[21] D. L. Sidebottom, Phys. Rev. Lett., **82** (1999) 3653.

[22] M. Porto, P. Maass, M. Meyer, A. Bunde, and W. Dieterich, Phys. Rev. B, **61** (2000) 6057.

[23] T. B. Schrøder and J. C. Dyre, Phys. Rev. Lett., **84** (2000) 310.

[24] B. Roling, C. Martiny, and S. Murugavel, Phys. Rev. Lett., **87** (2001) 085901.

[25] K. L. Ngai and R. Casalini, Phys. Rev. B, **66** (2002) 132205.

COORDINATION ENVIRONMENT AND NETWORK STRUCTURE IN AgI DOPED As-CHALCOGENIDE GLASSES

TAKESHI USUKI, KOHEI NAKAJIMA, YASUO KAMEDA
Faculty of Science, Yamagata University, Yamagata 990-8560, Japan

MASAKI SAKURAI
Institute for Materials Research, Tohoku University, Sendai 980-8577, Japan

TOSHIO NASU
Faculty of Education, Yamagata University, Yamagata 990-8560, Japan

Diffraction measurements and extended X-ray absorption fine structure (EAXFS) studies together with their electrical and thermal properties have been carried out for $(AgI)_x(As_2Se_3)_{1-x}$ glasses in order to investigate the ionic conduction mechanism in AgI-doped chalcogenide glasses. The addition of AgI into As_2Se_3 glass is responsible for a pronounced increase in the electrical conductivity. EXAFS and diffraction measurements result in the concentration independence of all structural parameters, where As has three Se nearest neighbours at 2.41 Å and Ag has four I nearest neighbours at 2.78 Å, respectively. This suggests that the network matrix in the glass is made by covalent $As(Se_{1/2})_3$ pyramidal units and that a significant number of Ag ions have the tetrahedral coordination with I atoms, similar to the case in crystalline α-AgI. It is also suggested that one of the main effects of AgI doping is to expand the interlayer distance between $As(Se_{1/2})_3$ helical chains. Therefore, the structure model for the present $(AgI)_x(As_2Se_3)_{1-x}$ glasses can be proposed as the pseudo-binary mixture of the $As(Se_{1/2})_3$ network matrix and AgI-related conduction pathways. Structural characteristics of the host As_2Se_3 glass with flexible and strong networks play an important role in the ionic conduction of Ag^+ in the present glasses.

1. Introduction

Superionic conducting glasses containing Ag ions have received much attention because of scientific interests in their conduction mechanism as well as their application in solid-state electrochemical devices [1]. Several models for the conduction mechanism have been proposed. In one kind, the cluster model, it is proposed that the Ag halide is introduced into the glass network in clusters or micro domains [2,3]. In this model, the micro domains are assumed to have an internal structure similar to that of the dopant salt. In another kind of model, the diffusion pathways model, the ionic conduction takes place within connected conduction pathways [4,5]. This idea was first proposed to explain the conductivity of metal-oxide glasses and was later extended to the salt-doped glasses. The pathways are assumed to be inter-network channels built up by the anions of dopant salt. There are also various models which rest on the assumption that the ions of the dopant salt are homogeneously distributed in the glass, e.g., the modified random network model [6] and the free-volume model [7]. For example, in the free-volume model, the main effect of dopant salt is to expand the glass network, leading to an increase in the accessible free volume for the random distribution of ions of the dopant salt, where the cluster models are therefore inappropriate.

Although many different kinds of amorphous ionic conductors have been structurally investigated, with the aim of understanding the diffusion mechanism, especially in silver oxysalt systems, such as $AgI\text{-}Ag_2O\text{-}M_xO_y$, the mechanism is not yet fully understood. In addition, there has been a lack of information concerning physical properties and detailed microscopic structure for AgI doped 'non-oxide' glass systems. This paper will present results of diffraction and EAXFS studies for $AgI\text{-}As_2Se_3$ glasses and discuss the conduction mechanism in AgI doped chalcogenide glasses.

2. Experimental details

Appropriate amounts of AgI, As and Se, with those compositions expressed as $(AgI)_x(As_2Se_3)_{1-x}$ with $x \leq 0.6$, were sealed within an evacuated quartz ampoule and thoroughly mixed in a rocking furnace at 1000 K. Then, the melts were quenched rapidly in an ice-water mixture.

The total electrical conductivity was evaluated from the complex impedance method in the frequency range between 10 and 1.2×10^6 Hz. Each side of the disk-shaped samples was coated with silver paste as the electrodes. Transport numbers of Ag^+ were also determined by the Wagner's polarisation method [8]. Glass transition, crystallisation, and eutectic temperatures were determined by DSC measurements with a heating rate of 5 K/min.

Structural analysis was carried out using both X-ray (Mo-Kα radiation) and time-of-flight neutron diffraction measurements. X-ray diffraction measurements were performed in the reflection geometry using a θ-θ diffractometer (Rigaku Co.). Diffraction intensities were counted over the range of scattering vectors Q, $0.5 \leq Q \leq 17.1$ Å^{-1}, with a fixed counting time of 100 s. The whole range was scanned three times to minimize any long-term instrumental drift. The neutron diffraction measurements were performed on the HIT-II instrument installed at KENS-KEK (Tsukuba, Japan) using the time-of-flight technique. Scattered neutrons were detected by 104 ^{3}He counters covering scattering angles of $10 \leq 2\theta \leq 157°$. Glass samples were sealed in a cylindrical quartz cell. Measurements for background, standard vanadium rod and empty containers were also made to enable their corrections. Analyses for both diffraction measurements are identical to those described elsewhere [9,10].

Samples for EXAFS measurements were prepared by pelletising powdered samples together with BN fine powder. The EXAFS measurements at Ag, I, As and Se K-edges for present glasses were carried out on BL-12C and BL-10B XAFS stations in KEK-PF (Tsukuba, Japan). Storage ring energy was 3.0 GeV for Ag and I K-edge measurements and 2.5 GeV for As and Se K-edge ones, respectively. Measurement temperature was set to be 20 K. Data analyses for the EXAFS measurements are identical to those described elsewhere [9].

3. Results

3.1. *Electrical conductivity*

Figure 1 shows the temperature dependence of total conductivity σ for $(AgI)_x(As_2Se_3)_{1-x}$ glasses, together with those for the corresponding 'partially' crystallised samples for comparison. The crystallised samples were prepared by cooling slowly the melts. X-ray diffraction pattern for the crystallised samples has Bragg peaks only assigned to the β-AgI, that is, the 'partially' crystallised β-AgI is dispersed in host As_2Se_3 glasses. As seen in Fig. 1, a drastically increase of σ is observed at around 420 K for the crystallised samples (open marks), which is reasonably due to the β-α phase transition of crystalline AgI included in the samples. On the other hand, no step-like change is observed in the glass systems.

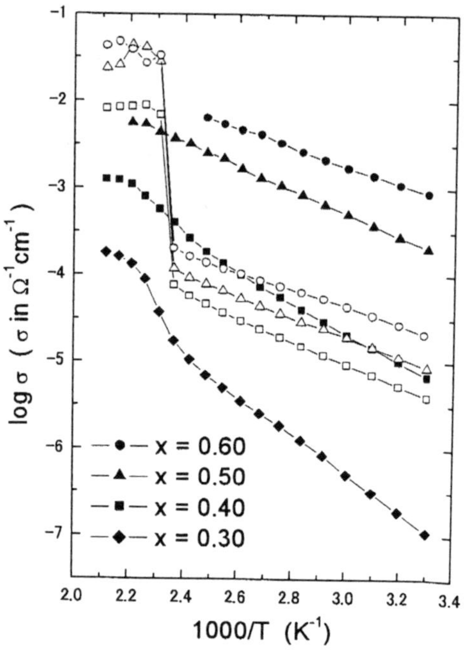

Figure 1. Temperature dependence of electrical conductivity for $(AgI)_x(As_2Se_3)_{1-x}$ systems (closed marks: glass samples, open marks: partially crystallised samples).

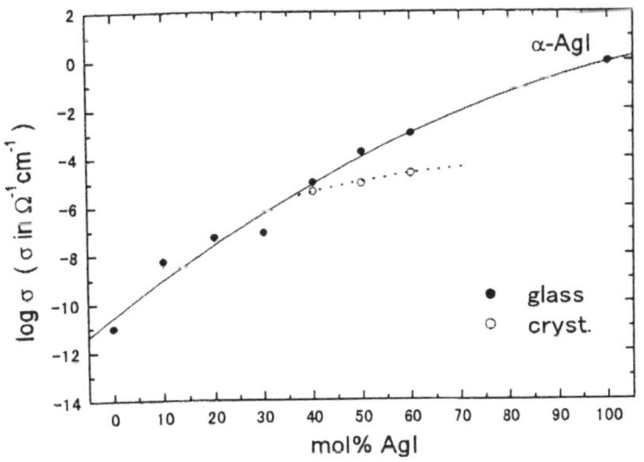

Figure 2. Conductivity isotherm at 300 K for $(AgI)_x(As_2Se_3)_{1-x}$ systems. The conductivity at 300 K for α-AgI was estimated by extrapola-tion of the linear σ values in α phase to the room temperature.

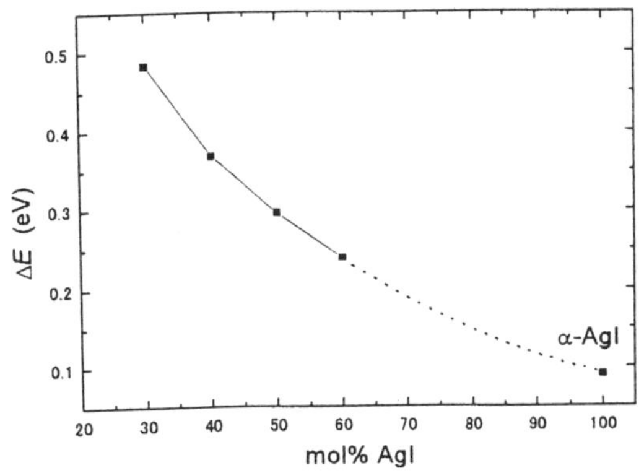

Figure 3. Activation energy for $(AgI)_x(As_2Se_3)_{1-x}$ systems.

The conductivity isotherm at 300 K is shown in Fig. 2. Incorporation of AgI into As_2Se_3 glass is responsible for a pronounced increase in the conductivity, which varies exponentially with increasing AgI content and approaches a high value of crystalline α-AgI (for the purpose of comparison, σ at 300 K for α-AgI was estimated by extrapolation of the linear σ values in α phase [11] to the room temperature). The room temperature conductivity of the glasses with $x \geq 0.4$ is greater than that of the corresponding crystallised samples, and is also greater than the value of β-AgI. In addition, electronic conductivity for the present glasses with $x \geq 0.4$ has a value of about 10^{-10} Scm^{-1}. The ion transport number is, then, almost unity, i.e., the contribution of electronic conduction is negligibly small in the present glasses. As shown in Fig. 1, an Arrhenius type temperature dependence of the conductivity is found for the present glasses. Therefore, one can calculate the activation energy for the conduction, results of which are illustrated in Fig. 3. The activation energy of the glass decreases monotonically with increasing AgI content, and a curve approaches smoothly a value of α-AgI.

These facts obtained by the conductivity measurements suggest that the superionic conducting state of α-AgI, which is thermodynamically stable only above 420 K, seems to be frozen at room temperature in the present AgI-doped As_2Se_3 glasses.

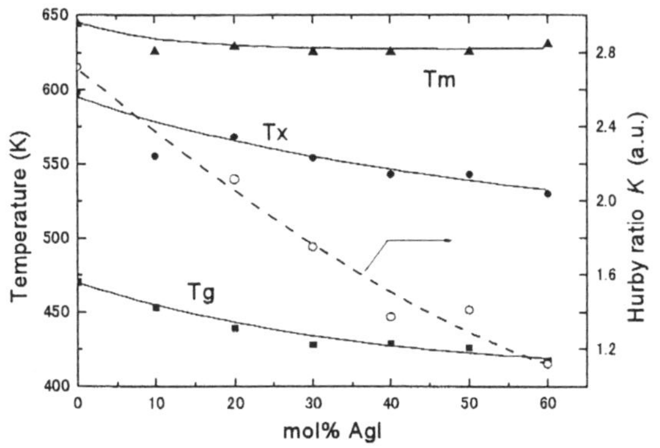

Figure 4. Characteristic temperature for $(AgI)_x(As_2Se_3)_{1-x}$ systems.

3.2. Thermal properties

Characteristic temperatures for $(AgI)_x(As_2Se_3)_{1-x}$ glasses obtained by DSC measurements are given in Fig. 4. Both glass transition T_g and crystallisation T_x temperature show a gradual decrease with increasing x, whereas eutectic temperature T_m is nearly constant. The Hurby ratio, defined as $K = (T_x - T_g)/(T_m - T_x)$ [12], decreases with increasing x, implying that the glass forming ability of the present glasses decreases with the addition of AgI. It was actually impossible to make glass samples with $x > 0.6$ by the usual rapid quenching method. It is worth of noting that the glass transition temperature at x = 0.6 is very close to the α-β phase transition temperature T_c of AgI (420 K). This implies that if T_g is greater than T_c, α-AgI like fragments may be stabilised in the host glasses before its α-β phase transition occurs. On the other hand, as the glass transition temperature is lowered below T_c, the β-α phase transition occurs before the amorphisation is reached. Therefore, when the AgI content is larger than 60 mol% for $(AgI)_x(As_2Se_3)_{1-x}$, the stabilisation of α-AgI like fragments cannot be accomplished, i.e., crystalline β-AgI particles precipitate in the host glass matrix.

Figure 5 gives the concentration dependence of molar volume V_M evaluated by density data. This figure allows two important points to be mentioned: (i) concentration dependences for glass samples are represented by a simple straight line without any minimum at intermediate concentrations, and (ii) experimental data points for glass samples appear to be very close to the calculated line which shows the averages of their values for the individual parts of crystalline α-AgI and As_2Se_3 glass. These observations suggest that the present glasses may be regarded as an ideal mixture of α-AgI like fragments and As_2Se_3 glass networks. This concept is also well supported by other experimental results of Raman spectroscopy and magnetic susceptibility measurements for the present glass systems [13].

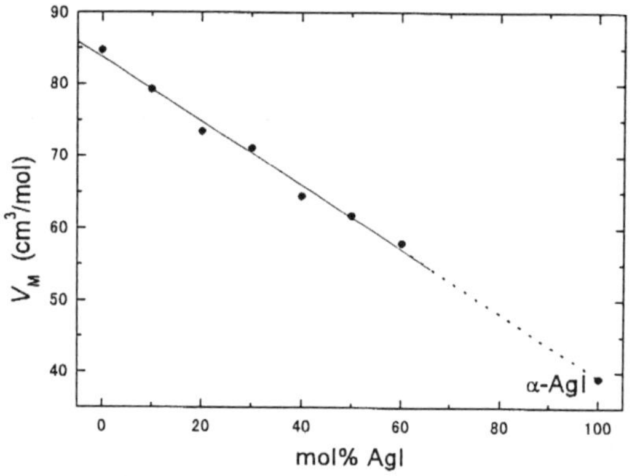

Figure 5. Molar volume for $(AgI)_x(As_2Se_3)_{1-x}$ systems.

3.3. *EXAFS data*

X-ray absorption fine structure technique is a powerful tool for investigating local structure around specified atoms. EXAFS oscillation functions $k^3\chi(k)$ and corresponding Fourier transforms $F(R)$ at As K-edge for the present glasses are shown in Fig. 6. A functional form of $k^3\chi(k)$ changes only slightly at any compositions. The As K-edge Fourier filtered experimental signals are well simulated by fitting structural parameters in a single shell model. Results indicate that, whatever the glass composition, covalent network units with As-Se interatomic distance of 2.41 Å do not change significantly with AgI content in the system (Table 1).

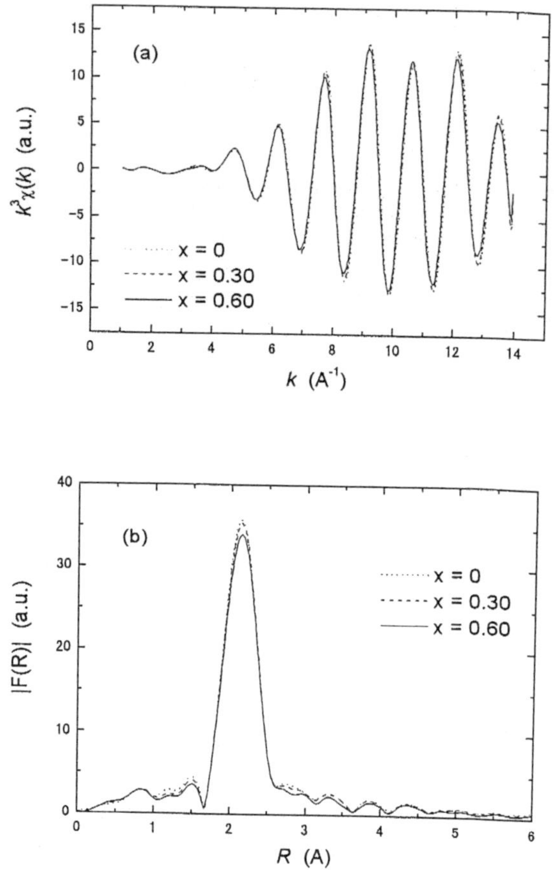

Figure 6. EXAFS oscillation functions (a) and corresponding Fourier transforms (b) at As K-edge for $(AgI)_x(As_2Se_3)_{1-x}$ systems.

Table 1. Interatomic distances r of As-Se, Ag-I and I-Ag correlations obtained by EXAFS analysis for $(AgI)_x(As_2Se_3)_{1-x}$ glasses.

x	As_2Se_3	0.3	0.6	β-AgI
r_{As-Se}(Å)	2.407(2)	2.411	2.414	-
r_{Ag-I}(Å)	-	2.769(3)	2.784	2.808
r_{I-Ag}(Å)	-	2.768(3)	2.787	2.807

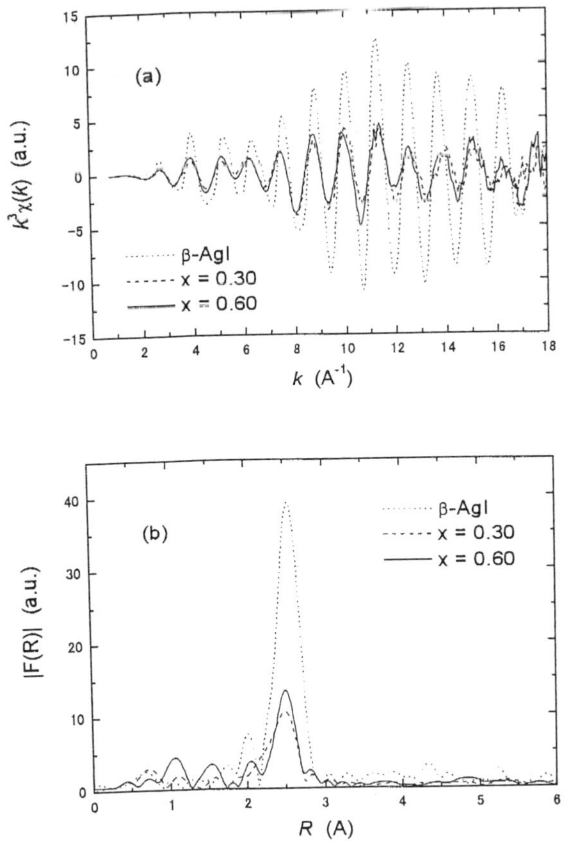

Figure 7. EXAFS oscillation functions (a) and corresponding Fourier transforms (b) at Ag K-edge for $(AgI)_x(As_2Se_3)_{1-x}$ systems.

Figure 7 shows $k^3\chi(k)$ and $F(R)$ at around Ag K-edge for the present glasses together with those for crystalline β-AgI. Because of the structural disorder of the glasses, signal intensities for the glasses were weaker compared with that for β-AgI. But reasonably good signals were obtained up to 18 $Å^{-1}$. A slight disagreement in the signal phase between glasses and β-AgI can be seen in the figure. The peak position of the Fourier transforms, similar for all of the glasses, becomes slightly lower than that for β-AgI. This feature may be directly related to the change in interatomic distance of Ag-I correlations. A least-squares curve fit with a single shell model with Ag-I correlation has been performed. Results are also summarised in Table 1. The interatomic distance of Ag-I decreases, although it changes very slightly, with decreasing AgI content. As listed in Table 1, this tendency is reasonably confirmed by the result for I K-edge EXAFS data. The possible structural aspect will be discussed in latter section.

3.4. *Diffraction data*

Figures 8 and 9 show respectively X-ray [13] and neutron structure factors $S(Q)$ for the present glasses. In terms of the atomic scattering amplitudes, X-ray data are more strongly weighted towards AgI component, and neutron data towards the network component of As_2Se_3. The neutron diffraction patterns at high Q region for AgI doped and undoped glasses change little, since this depends mostly on the short-range covalent bonding of host As_2Se_3 glass. However, there are large changes at low Q region. In particular, a first sharp diffraction peak (FSDP) at 1.34 $Å^{-1}$ in parent As_2Se_3 glass becomes weaker but remains clearly in the AgI doped glasses. The addition of AgI leads to a measurable shift in the position of FSDP from 1.34 in As_2Se_3 to 1.23 $Å^{-1}$ in 60 mol% AgI doped glass. This peak is less intense in the X-ray data, particularly for the higher AgI compositions.

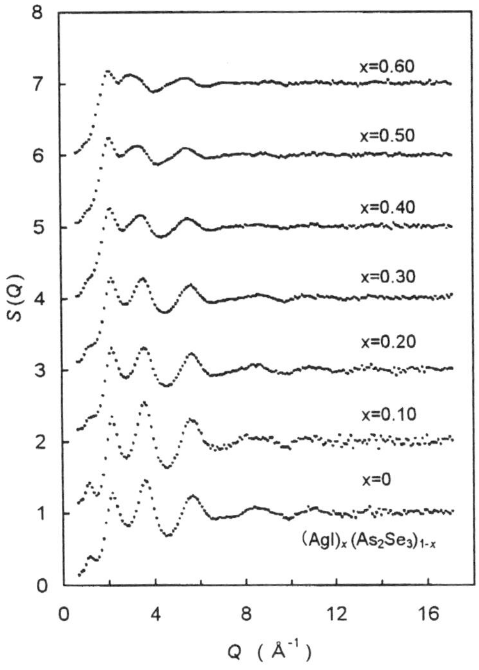

Figure 8. X-ray structure factors for $(AgI)_x(As_2Se_3)_{1-x}$ glasses.

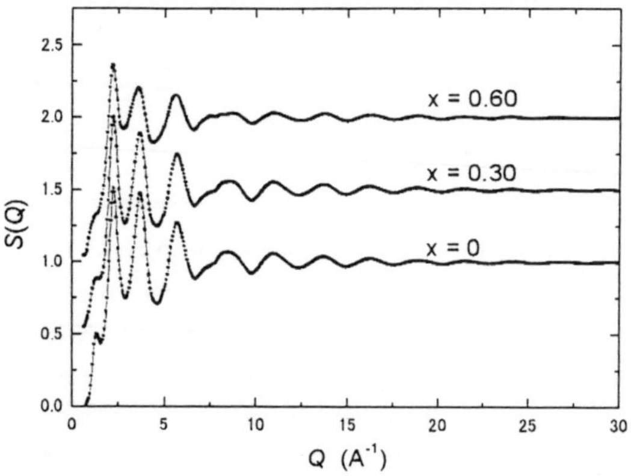

Figure 9. Neutron structure factors for $(AgI)_x(As_2Se_3)_{1-x}$ glasses.

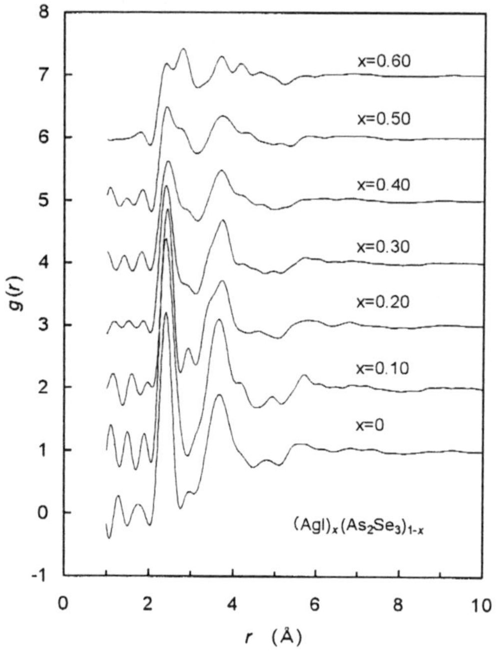

Figure 10. X-ray pair distribution functions for $(AgI)_x(As_2Se_3)_{1-x}$ glasses.

Figures 10 and 11 give pair distribution functions $g(r)$. A well-defined first peak is found at 2.41 Å in the host As_2Se_3 glass. With increasing x, this peak splits into two peaks. The intensity of an alternative peak at about 2.8 Å increases, and at the same time, a new component grows at 4-5 Å in the second shell. These changes correspond to the formation of the local atomic ordering constructed by Ag and I atoms. This feature can be seen clearly in the X-ray data, because X-ray data are more strongly weighted towards AgI component. Peak positions of these new components are in good agreement with nearest neighbour Ag-I and I-I distances in crystalline AgI ($r_{Ag\text{-}I} \cong 2.81$ Å and $r_{I\text{-}I} \cong 4.3 \sim 5.1$ Å in both α- and β-AgI [14]). These results allow us to predict that the environmental structure around Ag in the present glasses is similar to that in crystalline α- and/or β-AgI. In addition, the intensity at the second and third shell regions in $g(r)$ systematically decreases with increasing x, suggesting that the IRO in the present glasses changes gradually with x. This seems to be associated with the peak shift and disappearance of the FSDP in $S(Q)$ mentioned above.

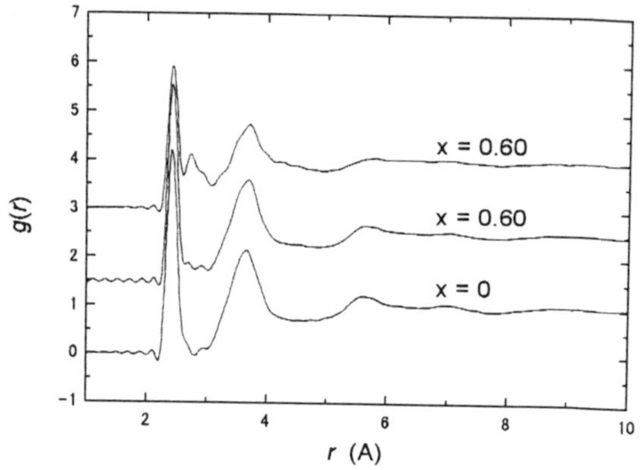

Figure 11. Neutron pair distribution functions for $(AgI)_x(As_2Se_3)_{1-x}$ glasses.

4. Discussion

4.1. *Coordination environments*

A basic conclusion from present experimental results can be pointed out that the structure of AgI doped As_2Se_3 glass can be considered as consisting of a host network component and an AgI component. Therefore, it is reasonable to assume that the first coordination shell of the present glasses is composed of two correlations, namely As-Se and Ag-I correlations. In order to discuss more quantitatively the local structure in the system, structural parameters for both correlations are determined by the least squares analysis for the total $S(Q)$ using a following model function,

$$S(Q) = 1 + \sum_{i,j} \frac{c_i n_{i-j} f_i f_j}{\left(\sum c_i f_i\right)^2} \exp\left(-\frac{1}{2} l_{i-j}^2 Q^2\right) \frac{\sin(r_{i-j} Q)}{r_{i-j} Q}, \quad (1)$$

where $r_{i\text{-}j}$, $l_{i\text{-}j}$ and $n_{i\text{-}j}$ denote the interatomic distance, root mean square displacement of *i-j* pairs and number of *j* atoms around a given atom *i*, respectively. The least squares fitting was performed at $Q \geq 9$ Å^{-1} independently for both X-ray and neutron data.

Table 2. Structural parameters of As-Se and Ag-I correlations for $(AgI)_x(As_2Se_3)_{1-x}$ glasses, obtained by least squares fitting for X-ray (XD) and neutron diffraction (ND) data.

	x	0	0.3	0.6
XD	$r_{As\text{-}Se}$(Å)	2.41(2)	2.41	2.41
	$l_{As\text{-}Se}$(Å)	0.099(5)	0.104	0.120
	$n_{As\text{-}Se}$	3.0(1)	3.0	2.7
	$r_{Ag\text{-}I}$(Å)	-	2.77(2)	2.79
	$l_{Ag\text{-}I}$(Å)	-	0.138(6)	0.140
	$n_{Ag\text{-}I}$	-	4.0(1)	3.8
ND	$r_{As\text{-}Se}$(Å)	2.41(2)	2.42	2.42
	$l_{As\text{-}Se}$(Å)	0.082(5)	0.082	0.087
	$n_{As\text{-}Se}$	3.0(1)	3.0	2.7
	$r_{Ag\text{-}I}$(Å)	-	2.77(2)	2.79
	$l_{Ag\text{-}I}$(Å)	-	0.107(6)	0.108
	$n_{Ag\text{-}I}$	-	4.0(1)	3.8

As summarised in Table 2, structural parameters derived from X-ray and neutron diffractions appear to be consistently close to each other within the data accuracy. The interatomic distance of As-Se pairs hardly changes on x. Coordination number of As is also nearly constant $n_{As\text{-}Se} \cong 3$. The addition of a large amount of AgI does not significantly affect the short-range ordering of the host network matrix, that is, $As(Se_{1/2})_3$ pyramidal units remain essentially intact on introducing AgI. In further detail, $n_{As\text{-}Se}$ seems to decrease slightly at $x = 0.6$, which may be related to the lower glass forming ability of heavily AgI-doped glasses.

Coordination number of Ag-I equals 4 at any compositions. Then, a significant number of Ag ions have a tetrahedral coordination environment with I atoms, similar to the case in crystalline AgI [14]. It should be pointed out that the interatomic distance of Ag-I in the present glasses seems to be somewhat shorter than that in the crystalline AgI (2.81 Å). In detail, $r_{Ag\text{-}I}$ increases slightly with increasing AgI content and reverts to the usual value of crystalline AgI. Similar results are found in the present EXAFS analysis (Table 1). Possible structural aspects can be pointed out as follows: (i) local compression effects where AgI components are confined by the host As_2Se_3 network, (ii) a volume contraction accompanied by a structure relaxation of AgI components in the amorphous state.

Anyway, the present analysis for the local structure around As and Ag atoms allows us to predict that the structure model for AgI-As_2Se_3 glasses is proposed to be a pseudo-binary mixture of the $As(Se_{1/2})_3$ network matrix and AgI-like components, where the addition of a large amount of AgI does not significantly affect the short-range ordering of both structural components.

4.2. *Network structures*

As shown in Fig. 9, the addition of AgI leads to a measurable shift in the position of FSDP on $S(Q)$, Q_P, from 1.34 Å^{-1} in As_2Se_3 to 1.23 Å^{-1} in 60 mol% AgI doped glass. The FSDP is normally considered to be an indication of intermediate-range ordering (IRO) on a correlation length scale of $L = 2\pi/Q_P$. For the host As_2Se_3 glass, L can be estimated to be ~4.7 Å, which probably represents a characteristic distance between As atoms located in the different network-chain layers constructed by $As(Se_{1/2})_3$ pyramidal units. This is strongly supported by the result that FSDP occurs only in the As-As partial structure factor recently derived from anomalous X-ray scattering measurements for amorphous As_2Se_3 system [15]. In other words, IRO in the distribution of As atoms brings about the appearance of FSDP in the structure factors. Then, the decrease of FSDP position with increasing AgI content in the present system suggests that one of the main effects of AgI doping is to expand the interlayer distance between $As(Se_{1/2})_3$ helical chains. The correlation length scale increases with AgI content from 4.67 Å in As_2Se_3 to 5.11 Å in 60 mol% AgI doped glass. That is to say, the glass network constructed by the $As(Se_{1/2})_3$ helical chains can be maintained even if the glass network expands with 10% increase in its interlayer distance when a large amount of AgI are doped.

These structural characteristics of As_2Se_3 glass with flexible and strong networks play an important role in the ionic conduction of Ag^+ in the present glasses. Firstly, it enables quasi-binary systems to amorphise and to stabilise the disordered structure over a wide range of composition. Secondly, it provides many conduction pathways for ionic migration. However, when the AgI content becomes larger than 60 mol%, not only network expansion but also destruction of As-Se bonds occurs. It reduces the strength of the glass network and weakens the glass structure. This leads to a decrease of glass forming ability, a decrease of T_g, and eventually a crystallisation of the system.

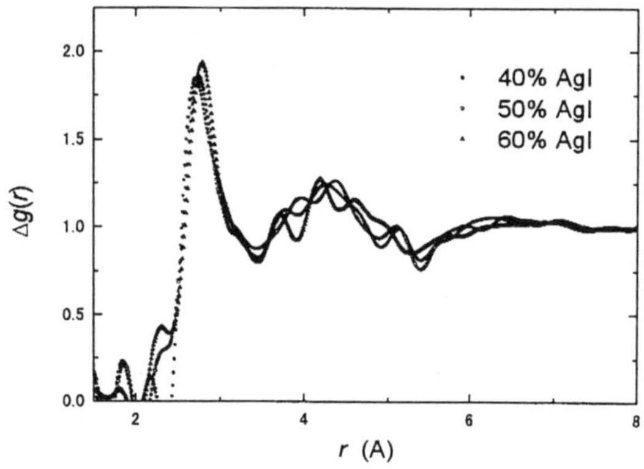

Figure 12. Glass network-eliminated quasi-partial pair distribution functions.

As mentioned in the former section, the tetrahedral coordination environment of Ag ions remains in the present glasses, similar to the case in crystalline AgI. In order to clarify the partial structure around Ag ions, we try to obtain quasi-partial structure factors, $\Delta S(Q)$, under an assumption that the partial structure factors are identical at all compositions. This assumption is reasonably supported by the result of molar volume data, in which the present glasses may be regarded as an ideal mixture of As_2Se_3 and AgI like fragments. Glass network-eliminated $\Delta S(Q)$, in which Ag-I, Ag-Ag and I-I correlations are emphasised, can be derived by the perfect cancellation of As-Se, As-As, and Se-Se correlations using X-ray $S(Q)$ data. Fig. 12 shows the corresponding quasi-partial pair distribution functions $\Delta g(r)$ calculated by the Fourier transformation of $\Delta S(Q)$ for the glasses with $x \geq 0.4$. The $\Delta g(r)$s have first and second peaks at around 2.78 and 4.3 Å, which are in agreement with the nearest neighbour Ag-I and I-I distances in AgI_4 tetrahedra of crystalline α-AgI. Moreover, there is a clear hump around 3 Å in the $\Delta g(r)$s. This hump is weak but would be one of the most important signals for the Ag^+ migration in the system.

In final step of the present study, the partial structure around Ag ions in the conduction pathways will be modeled using three $\Delta g(r)$s for different composition, with help of the reverse Monte Carlo (RMC) method [16]. The RMC is one of the powerful methods for structural modeling. It will produce a physically possible structure model, which is quantitatively consistent with the available data. For the present modeling, two additional constraints were applied: (1) a coordination constraint for Ag atoms to keep AgI_4 tetrahedra, which is strongly supported by the present analysis for diffraction data, and (2) a coordination constraint for I atoms to keep $n_{I\text{-}I} \cong 12$, which is consistent with the facts that $n_{I\text{-}I} = 12\sim14$ for crystalline AgI. Partial distribution functions derived by the RMC modeling are shown in Fig. 13. It is worthy of noting that the hump around 3 Å in the $\Delta g(r)$s corresponds to Ag-Ag neighbour contacts which do not exist in crystalline β- or λ-AgI but only in α-AgI. In the superionic phase of α-AgI, one can calculate Ag-Ag interatomic distances between several Ag ions located at 12d sites, for example, 1.78, 2.52, 3.08, 3.56 Å, etc. The position of Ag-Ag neighbour contacts around 3 Å is in wonderful agreement with the distance of third (3.08 Å) neighbour contacts in α-AgI. Similar contacts were observed in other Ag-rich chalcogenide and oxide glasses with ionic conduction, using neutron diffraction with isotopic substitution technique [5,17].

It is suggested that these relatively short Ag-Ag contacts indicate edge or face sharing of AgI_4 tetrahedral units, which form preferential conduction pathways for the Ag^+ ions migration. Coordination number distributions for Ag-Ag and Ag-I correlations determined from RMC modeling are shown in Fig. 14. The Ag-Ag coordination number reflects the connectivity of the Ag^+ related network: $n_{Ag\text{-}Ag} = 1$ for dimers, 2 for chains, 3 for 2D sheets, etc [5]. For the present glasses, $n_{Ag\text{-}Ag}$ is estimated to be 2.3 (68% of Ag have two-fold coordination and 23% three-fold coordination), indicating that edge or face-sharing AgI_4 tetrahedra form at least chains or tunnels as the Ag^+ related network. This connectivity seems to be responsible for the high Ag^+ conduction in the glass system.

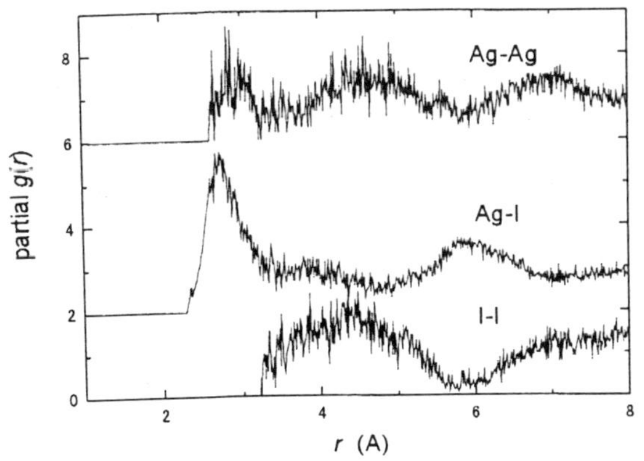

Figure 13. Partial pair distribution functions for Ag-Ag, Ag-I and I-I correlations determined from RMC modeling.

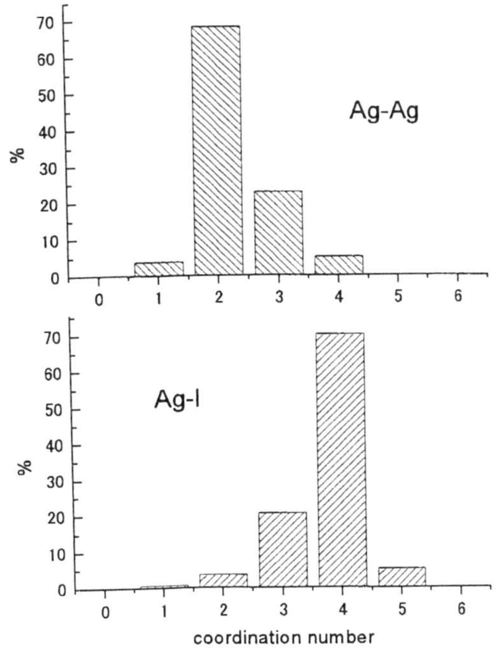

Figure 14. Coordination number distributions for Ag-Ag and Ag-I correlations determined from RMC modeling.

Therefore, the free volume model with random distribution of Ag and I ions in the glass network and/or the formation of oligomeric AgI clusters appear to be inappropriate to the case of the present AgI-As_2Se_3 glasses. There exists conduction pathways constructed by AgI_4 tetrahedral units, in which not only local coordination environment around Ag ions but also IRO of these units remains essentially similar to those in crystalline α-AgI, although a degree of disorder increases in the glasses. The correlation length, for example 5.11 Å for 60 mol% AgI glass estimated by FSDP position, is not a size of oligomeric AgI clusters but an average width of Ag^+ conduction pathways.

5. Conclusions

Superionic conducting glasses were obtained in the AgI doped As_2Se_3 system. Their ion transport properties are strongly associated with the environmental structure and glass networks. The present analysis for the short- and intermediate-range structure around As and Ag atoms allows us to predict that the structure model for AgI-As_2Se_3 glasses is proposed to be a pseudo-binary mixture of the $As(Se_{1/2})_3$ network matrix and AgI related conduction pathways, which would be responsible for the high mobility and diffusivity of Ag^+ in the present glass system.

Acknowledgements

The authors thank Prof. Uemura for the many fruitful discussions. This work was supported by a Grant-in-Aid for Young Scientists No.15760495 from the MEXT, Japan, and a grant from Nippon Sheet Glass Foundation. This work was performed under the inter-university cooperative research program of the IMR, Tohoku Univ.

References

1. see for example, E. Bychkov, V. Tsegelnik, Yu. Vlasov, A. Pradel, M. Ribes, J. Non-Cryst. Solids **208**, 1 (1996).
2. A. Fontana, F. Rocca, M.P. Fontana, Phys. Rev. Lett. **58**, 503 (1987).
3. C. Rousselot, M. Tachez, J.P. Malubani, R. Mercier, P. Chieux, Solid State Ionics **44**, 151 (1991).
4. T. Minami, J. Non-Cryst. Solids **73**, 273 (1985).
5. E. Bychkov, D.L.Price, C.J. Benmore, A.C.Hannon, Solid State Ionics **154-155**, 349 (2002).
6. G.N. Greaves, J. Non-Cryst Solids 71, 203 (1985).

7. J. Swenson, R.L. McGreevy, L. Borjesson, J.D. Wicks, Solid State Ionics **105**, 55 (1998).
8. C. Wagner, J. Chem. Phys. **26**, 1597 (1957).
9. T. Usuki, O. Uemura, S. Konno, Y. Kameda, M. Sakurai, J. Non-Cryst. Solids **293-295**, 799 (2001).
10. T. Usuki, Y. Murakami, K. Abe, O. Uemura, Y. Kameda, J. Phys. Soc. Jpn. **70**, suppl. A 259 (2001).
11. J.B. Boyce, B.A. Huberman, Phys. Rep. **51**, 189 (1979).
12. A. Hurby, Czech. J. Phys. B **22**, 1187 (1972).
13. T. Usuki, S. Saito, K. Nakajima, O. Uemura, Y. Kameda, T. Kamiyama, M. Sakurai, J. Non-Cryst. Solids **312-314**, 570 (2002).
14. L.W. Sorock, Z. Phys. Chem. B **25**, 441 (1934).
15. S. Hosokawa, Y. Wang, J. Berar, W. Pilgrim, S. Mamedov, P. Boolchand, Proc. 10th Conf. Phys. Non-Cryst. Solids (2003).
16. R.L. McGreevy, L. Pusztai, Mol. Sim. **1**, 359 (1988).
17. I.T. Penfold, P.S. Salmon, Phys. Rev. Lett. **64**, 2164 (1990).

SCALING PROPERTIES OF ION CONDUCTION AND WHAT THEY REVEAL ABOUT ION MOTION IN GLASSES

David L. Sidebottom

Department of Physics
Creighton University
2500 California Plaza
Omaha, NE 68178 (USA)
fax: 402.280.2140
email: sidebottom@creighton.edu

Abstract

The scaling properties of the ac conductivity of ion-conducting glasses have, in the last several years, led to considerable revision of our understanding of microscopic ion dynamics. Although the notion that frequency dependent dispersion of the ac conductivity is evidence for correlated (as opposed to purely random) ion motions remains intact, notions about the role of ion-ion interactions as a source for the correlated behavior are clearly inconsistent with the scaling properties of the ac conductivity. Instead, certain systematic variations in the dispersion highlight the possible importance of the cation's local structural environment to influence the correlated motion. Here, I review some of the recent scaling ideas and their consequences with the goal of providing the casual impedance spectroscopist a set of practical guidelines for how scaling analyses may help in understanding ion conductivity measurements.

Keywords: ionic relaxation, ion-conducting glasses, correlated motion, ac conductivity, scaling.

1. Introduction

Ion-conducting materials are finding niches in a variety of advanced technology devices including high energy density solid state batteries, chemical sensors, and electrochromic displays. Demand is being placed upon disordered ion-conducting materials to fill these duties since disordered materials offer compositional freedom to tailor mechanical as well as electrical properties. However, even as technological applications for ion-conducting materials continue to increase, a core understanding of the underlying motion of ions in these materials remains elusive. In fact, research in recent years has challenged many older notions regarding the nature of this ion motion and has led to some refreshing new insight on the process. This recent work focuses primarily upon features of the dielectric response of these ion-containing materials to an applied ac electric field.

The objective of this paper is to provide some suggestions as to how one might best dissect the information contained in the dielectric response. This task is broken into two parts. The first is a review of some fundamental statements relating ion motion to dielectric response and a review of past strategies for data dissection based upon the electric modulus emphasizing the strengths and limitations of this approach. The second portion is an examination of scaling properties observed in many ion-conducting glasses and a review of what information can be obtained using this approach.

2. Fundamentals

From linear response theory[1,2], the motions of ions in a material result in an ionic contribution to the complex conductivity of the form:

$$\sigma^{*}_{ion}(\omega) = -\omega^2 \frac{Nq^2}{6kT}\int_0^\infty \langle r^2(t)\rangle e^{-i\omega t}dt, \qquad (1)$$

where N is the ion density, q is the ion charge, and $\langle r^2(t)\rangle$ is the statistical mean-squared displacement (MSD) of the ions. This ionic conductivity is related to the complex permittivity as

$$\sigma^{*}_{ion} = \sigma'_{ion} - i\sigma''_{ion} = i\omega(\varepsilon' - \varepsilon_\infty + i\varepsilon'') \qquad (2)$$

where ε_∞ is a non-ionic contribution to ε' stemming from fast polarization (electronic, atomic) and is to be distinguished from the total complex conductivity

$$\sigma^*_{tot} = i\omega\varepsilon_o\varepsilon^*_{tot}, \quad (3)$$

which includes all forms of polarization processes (ionic, atomic, electronic).

Since the ac conductivity is merely a Fourier transform of the MSD, it follows that the former contains essentially the same information as the latter, or said differently, the ionic ac conductivity *is* the ions' MSD as it appears in a reciprocal space. The time dependent MSD is fundamental to the description of the ion motion and offers the most direct connection to theory and model predictions.

The primary features of ion motion are visible directly from examination of the real part of the ac conductivity. An example is provided in Fig. 1. In the absence of certain specimen dependent aberrations to be discussed below, the ac conductivity has an approximate frequency dependence of the form[3,4],

$$\sigma(f) \approx \sigma_o(1 + (f/f_o)^n) + Af. \quad (4)$$

One finds at low frequencies a frequency independent conduction, often referred to as the dc conduction

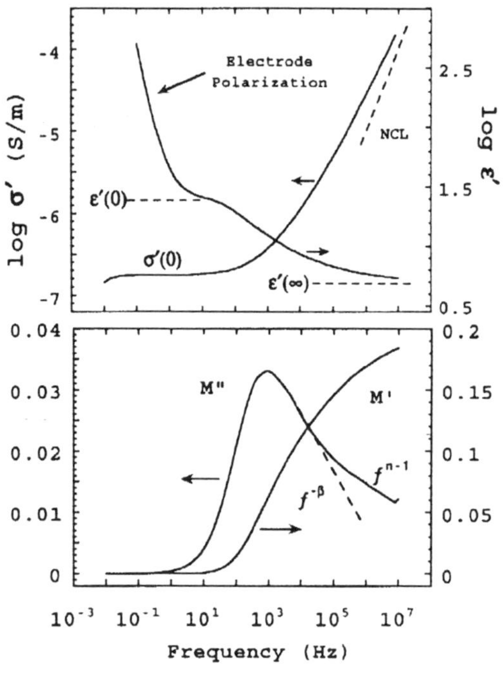

Fig. 1 A schematic representation of the ac conductivity and permittivity (top) found in ion-conducting glasses. The same data is displayed (bottom) in the electric modulus.

(σ_o), which from Eq. 1, is seen to be the reciprocal space representation of diffusive motion stemming from the random displacements of ions that occurs at long times and correspondingly long length scales. This is contrasted by the distinctly frequency dependent conduction observed at higher frequencies which appears to approach a linear dependence at very high frequencies. In Eq. 4 this "dispersion" has been rather arbitrarily approximated by two separate contributions: one a power law with exponent n and the other a linear contribution. This separation is only an empirical tool for discussing the overall frequency dependent behavior and is not intended to imply that two separate phenomena are present in the dynamics. Reflecting again upon Eq. 1, this dispersion indicates a "sub-diffusive" form of ion motion at short time and length scales, wherein the motion is not random, but rather correlated with a propensity for ions to return to their point of origin. The initial power law-like dispersion is often referred to as the Jonscher power law[3] (JPL) while the linear frequency dependent dispersion is commonly referred to as the nearly constant loss[4,5] (NCL). This latter dispersion, viewed through Eq. 1, would appear to reflect extremely short time ion dynamics wherein the MSD is not changing with time[5]. Such motion might conform to some form of quasi-oscillatory motion of the ion about a charge compensating site in the material. However, there is also speculation and support for the notion that part of this NCL stems from non-ionic sources[5,6].

In summary, it becomes clear that ion dynamics extends over considerable range of time scales and is distinguished by at least two regimes of motion: correlated motion at short times evolving into random motion at long time and length scales. While it is often the long range dynamics that pertain to technological devices, it is affirmed that only an unclouded understanding of the entire ion relaxation process at all relevant scales will aid in sustained advancement of engineering activity.

3. Electric Modulus Formalism

These same dispersive phenomena are also present in alternative representations of the dielectric response such as the electric modulus[7], M*. The electric modulus is formed by reciprocating the total complex permittivity, $M^* = 1/\varepsilon^*_{tot}$. Consequently, the modulus is a macroscopic quantity which does not discriminate between ionic and non-ionic polarizations. Introduced as an analog to the mechanical modulus which monitors stress relaxation in materials, the electric modulus monitors the bulk electric field decay associated with ion relaxation. The modulus has the added benefit of suppressing specimen dependent aberrations associated with electrode polarization (see Fig. 1). The frequency dependence of the modulus, $M^* = M' + iM''$, comprises a broad, asymmetric peaked function for the imaginary part with a corresponding step-like increase of the real part. An example can be found in Fig. 1. At high frequencies, M" varies as σ'/ω, again signaling the dispersion associated with the presence of correlated motion.

Following its introduction, there developed a prescribed strategy (formalism) for interpreting the modulus shape in terms of correlated ion motions[8]. In this formalism, the electric modulus is related to the electric field decay, $\phi(t)$ as:

$$M^*(\omega) = \frac{1}{\varepsilon_\infty}\left\{1 + \int_0^\infty \left(\frac{d\phi}{dt}\right) e^{-i\omega t} dt\right\}. \tag{5}$$

As an exponential decay of the electric field produces a symmetric modulus, the customary alternative is to assume a non-exponential decay function of the KWW form;

$$\phi(t) = A\exp\left[-(t/\tau)^\beta\right]. \tag{6}$$

The parameters (A, τ, and β) are then adjusted to provide the best fit to the experimental data. A and τ are simple scale parameters that establish the magnitude and position, respectively. The "stretching parameter", β, is effective in accounting for the variable broadness of the modulus; the width of M" increases with decreasing β. Typical fits are successful for low frequencies (corresponding to the dc regime) up to and including the peak in M" and out about an additional decade in frequency beyond the peak. However, it is well acknowledged that this success deteriorates at higher frequencies where the KWW fit (which varies as $\omega^{-\beta}$) commonly underestimates the actual high frequency behavior. Proponents[9] of this formalism argue that the deviation between fit and data at the high frequency regime is not failure of the KWW, but rather results from artificial "lifting" of the modulus by a non-ionic NCL contribution. While it is acknowledged that a non-ionic NCL contribution may be present, we will return to this issue later to show how this deviation is clearly a failure of the KWW.

One popular interpretation of the correlated motion comes from the so-called coupling model[10] which attributes the correlation to ion-ion interactions. These interactions are believed to retard the ionic relaxation process giving rise to a KWW decay of the electric field. Cast in this way, the stretching parameter (β) corresponds to a coupling parameter $1-\beta$, such that when coupling is strong, β is small and the asymmetry of M" is extreme. Conversely, when the coupling is absent, then the KWW approaches an exponential decay and M" becomes narrow and symmetric.

The strongest support for the coupling model comes from measurements by Patel and Martin[11] on sodium thioborate glasses of varying ion concentration. Patel and Martin observed that modulus narrowed and became more symmetric as the ion concentration was reduced. In other words, the KWW parameter used to characterize these data approached unity at extremely low ion concentrations. These findings fit nicely into the coupling model interpretation: lower ion concentration means decreased ion-ion interaction and thence a lower coupling (larger β).

This interpretation was challenged some years later. In a similar study of sodium germanate glasses of varying ion concentration[12,13], it was noted that despite the narrowing of the modulus with decreasing ion concentration, the ac conductivity displayed a concentration-independent dispersion. This concentration independence of the dispersion was demonstrated by scaling the ac conductivity so as to collapse onto a common curve:

$$\frac{\sigma'_{ion}}{\sigma_o} = F_1(f/f_o). \tag{7}$$

The ability to perform this scaling at all concentrations suggests that the correlated motion is not inherently influenced by ion-ion interactions. This finding prompted a review of the Patel and Martin study in which similar results were obtained[14]. There it was shown that both the ac conductivity and the ionic permittivity,

$$\frac{\varepsilon'_{ion}}{\Delta\varepsilon} = \frac{\varepsilon' - \varepsilon_\infty}{\Delta\varepsilon} = F_2(f/f_o), \tag{8}$$

could be scaled to the concentration independent mastercurves $F_1(x)$ and $F_2(x)$. In Eq. 8, De is the increase of permittivity resulting from the ionic relaxation process. Consequently, no evidence of ion-ion interaction is found for concentration variations of as much as three orders of magnitude.

At this juncture, let me briefly return to the issue raised earlier with regards to the failure of the KWW to describe the high frequency wing of M". Proponents of the modulus formalism argue this is due to a non-ionic NCL which artificially raises M" at high frequencies[9]. Here we see how this cannot be possible. With decreasing ion concentration, the KWW fit of M" requires an increasing β. For the ac conductivity, this KWW fit transforms into σ_{KWW} that resembles the JPL but with n replaced by $1-\beta$. Thus, with decreasing ion concentration, the KWW fit approaches a dispersion-free conductivity. It seems there can be no way to combine this changing σ_{KWW} with a linear frequency dependent (NCL) conductivity and still maintain a net ac conductivity with a concentration-independent shape.

The immense discrepancy in scaling properties of the ac conductivity and the modulus at first appears quite puzzling, since intuitively each would seem to be complimentary representations of identical data. The resolution to this puzzle lies in the fact that the two representations are not exactly complimentary[15]. For scaling to be performed on the ac permittivity (see Eq. 8), it was necessary to first remove the non-ionic polarization component, ε_∞. In other words, we find scaling only applies to the ionic contribution to the total complex conductivity, that part related to the MSD by Eq. 1. By comparison, the modulus retains ε_∞. The modulus thus contains contributions to the macroscopic electric field decay arising from all polarization process, ionic and non-ionic alike[16].

To better appreciate how inclusion of non-ionic ε_∞ modifies the scaling that is inherent in the solely ionic complex conductivity, let us use the scaling relations above (Eq. 7, Eq. 8) to form the imaginary part of the modulus,

$$M''(f) = \frac{1}{\Delta\varepsilon} \frac{F_1(x)/(2\pi x)}{(\lambda + F_2(x))^2 + (F_1(x)/(2\pi x))^2}, \quad (9)$$

where $\lambda = \varepsilon_\infty / \Delta\varepsilon$. Here it is obvious that inclusion of ε_∞ has forced a dependence of M*(*f*) upon λ that cannot be reduced. Thus, unless λ is constant, the modulus cannot be scaled as can σ_{ion} and ε_{ion}.

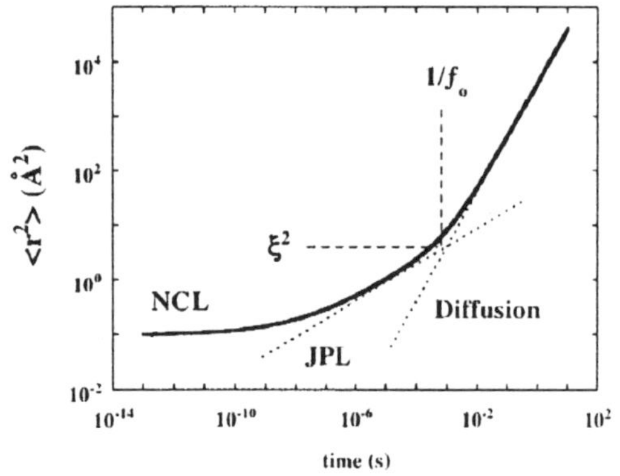

Fig. 2 A schematic representation of the mean-squared displacement of ions required to achieve Eq. 4.

Furthermore, this allows for resolution to why the modulus narrows with decreasing ion concentration. We have proposed (details below) that $\Delta\varepsilon \approx Nq^2\xi^2/6kT\varepsilon_o$, where ξ is some mean diffusion length scale which most likely scales as $\xi \approx N^{-1/3}$. Consequently, we expect $\lambda \approx N^{-1/3}$ and so increase with decreasing N. Analysis of Eq. 9, using $F_1(x)$ and $F_2(x)$ mastercurves obtained experimentally for the thioborate glasses with varying λ, does indeed reproduce the observed narrowing of the modulus with decreasing ion concentration. We are led to conclude then that the shape changes of the modulus for these systems is largely an artifact of the inclusion of ε_∞ into M*.

Another example is that of $0.4Ca(NO_3)_2$-KNO_3 (CKN). Measurements on this ionic salt[17] demonstrated an absence of shape changes for the modulus in the glass below T_g, but significant broadening of the modulus with increasing temperature in the liquid state above T_g. In the initial study, these shape changes were associated with the phenomena of "decoupling". Again the implication was that the correlated motion was being inherently altered on approach to the glass transition from above. However, in a later study[14,18] it was demonstrated that σ_{ion} and ε_{ion} in CKN could be scaled in accordance with Eq. 7 and Eq. 8 to produce a common mastercurve above and below T_g. This implies that the correlated motion is intrinsically undisturbed upon passage through the thermodynamic glass transition.

Again, we proposed that these changes in M" around the glass transition are chiefly the result of temperature dependent changes in $\Delta\varepsilon$. Indeed, $\Delta\varepsilon$ is observed to increase by about fourfold in going from the glass into the liquid. We speculate that the increase in $\Delta\varepsilon$ might arise from changes in the mobile ion population. While K^+ is regarded as the only mobile charge carrier at temperatures below T_g, we propose that above T_g, Ca^{2+} cations may become mobile and augment this K^+ population. Analysis[14] of our proposed relationship for $\Delta\varepsilon$ (presented above) accounts very nearly for the observed increase in $\Delta\varepsilon$.

4. Scaling Analyses

Having assessed the shape changes of M* as being prone to non-ionic artifacts, we now look with greater detail upon the shape of the conductivity mastercurve. A central concept in the scaling analysis is that for a given sample (fixed ion concentration and fixed composition) there will exist some invariant dynamic process ($\langle r^2(t) \rangle$) underlying the ion motion (see Fig. 2) which, owing to the temperature dependence of the characteristic frequency, is displaced relative to a given fixed frequency window of observation. We see from the figure that, provided no temperature dependence of the diffusion length scale, ξ, occurs, $\langle r^2(t)\rangle$ is displaced only horizontally (along t). However, since the dc conductivity depends upon f_o, the ac conductivity is displaced both horizontally (along f, by f_o) and vertically (along σ, by σ_o). The process of scaling the $\sigma(f)$ curves, obtained at different temperatures for a given sample, reconstitutes the dynamic process in the form of a mastercurve whose extent far exceeds the original window of observation. Again, these resulting mastercurves ($F_1(x)$ and $F_2(x)$) are the dynamic process as it appears in the reciprocal (i.e., frequency) space.

In the remainder of this section, it will be assumed that the dynamic process is invariant and that mastercurves can be obtained for a given sample for differing temperatures and for a homologous series of glasses varying only by ion concentration. In the following section, we will discuss the significance of cases in which the mastercurves evolve in response to changes in glass structure.

The first step in analysis would be to construct a mastercurve as defined in Eq. 7 from individual conductivity spectra ($\sigma(f)$) obtained at various temperatures for a given glass sample. There have appeared many recipes for performing this scaling, particularly with regard to the choice of frequency scale[12,13,19]. While the vertical scaling merely requires determination of the dc conductivity (as obtained from plateau of $\sigma(f)$ (provided electrode polarization effects are not severe), the horizontal scaling is more arbitrary. Some advocate Summerfield scaling[20] in which the frequency is scaled by $f_{scale} = \sigma_o T$, or modified Summerfield scaling with $f_{scale} = \sigma_o T/N$. However, the success of these rely upon their assumed validity, and we will see momentarily that Summerfield scaling is not always valid. A more direct approach to scaling the frequency is just to choose a scaling frequency which represents a well defined reference point on $\sigma(f)$. The common choice (and used in the following examples) is the frequency at which $\sigma(f_o) = 2\sigma_o$. From Fig. 2, this choice is seen to correspond with the approximate intersection of the long range diffusion with the shorter time correlated motion.

In addition to scaling the ac conductivity, it should be possible (in principle) to also scale the permittivity according to Eq. 8. However, in most instances determination of $\Delta\varepsilon$ is severely hindered by the early onset of electrode polarization. Nevertheless, the above scaling of σ_{ion} and ε_{ion} are connected by the Kramers-Kronig relations[14] with the consequence that

$$f_o \propto \frac{\sigma_o}{\varepsilon_o \Delta\varepsilon}. \qquad (10)$$

At this point, it is useful to comment upon the presence of experimental artifacts in the raw data. There are two primary artifacts common to ion-conducting materials. One is electrode polarization, which severely affects $\varepsilon'(f)$ (and to a lesser degree, $\sigma(f)$) at low frequencies below the onset of σ_o. These can be seen in Fig. 1. The other is due to finite lead impedance (resistive and inductive) which typically distort the data at high frequencies (above roughly 1 MHz). Being unrelated to the bulk ionic relaxation, data afflicted by these artifacts are not displaced by temperature in the same manner as the bulk ionic process. Consequently, scaling can aid in discerning the extent of these artifacts.

Having performed scaling of $\sigma(f)$ using $\sigma_o(T)$ and $f_o(T)$, the next obvious step would be to examine $f_o(T)$. From Fig. 2 it is apparent that this frequency corresponds to the random diffusion occurring on length scales of ξ, and so must, at the very least, be proportional to the ion hopping rate. Since ion hopping is traditionally thought to involve activated jumps of the ion over an energy barrier, an Arrhenius plot (f_o vs 1/T) is recommended. If the resulting plot is linear, the slope is a direct measure of the so-called activation energy.

The third step would be to examine the relationship between the scales (σ_o and f_o) required to complete the scaling. A quick gauge of this relation is achieved by plotting these two scales against another (actually $\sigma_o T$ vs f_o) on a double logarithmic scale. In order to interpret how σ_o and f_o may be related, let us return again to Fig. 2. The long range diffusion in this sketch can be described as $\langle r^2(t)\rangle = \xi^2(tf_o)$ and from Eq.1 leads to a dc conduction of the form:

$$\sigma_o = \frac{Nq^2\xi^2}{6kT} f_o. \qquad (11)$$

Thus we can express the expected relationship between σ_o and f_o in terms of the ratio

$$\frac{\sigma_o T}{f_o} = \frac{1}{6k}\left[N\left(q^2\xi^2\right)\right] = \frac{1}{6k}\left[N_o\left(pq^2\xi^2\right)\right] \qquad (12)$$

where we have replaced the mobile ion concentration, N, by $N = pN_o$, where N_o is the total ion concentration and p is the fraction thereof which are mobile. Let us now look at *three specific cases.*

4.1 Summerfield Scaling

In the first case, the quantity in square brackets in Eq. 12 is constant with temperature. In this case, $f_o \approx \sigma_o T$ and the dynamic process conforms to what is often referred to as Summerfield scaling. In this situation, a plot of $\sigma_o T$ versus f_o should exhibit linearity. An example of such a plot is provided for a series of monovalent metaphosphate glasses in Fig. 3. Plotted in this fashion one sees the striking appearance of not only a linear relation between $\sigma_o T$ and f_o, but a relation which is common for the variety of cation species considered. However, the double logarithmic plot can be deceptive. Although the plot, spanning some 25 (i.e., 5 by 5) square decades, appears linear, when we form the ratio of $\sigma_o T/f_o$, or rather consider the quantity

$$R(N_o,T) = p\xi^2 = \frac{\sigma_o T}{N_o f_o}\left(\frac{6k}{q^2}\right), \tag{13}$$

we find a residual weak temperature dependence is actually present. This is seen in the inset of Fig. 3 where R(T) shows a systematic increase with increasing temperature as well as compositional variations between the small (Li^+, Na^+) and large (Cs^+, Tl^+) cations. It is difficult to speculate on the origin of the temperature dependence, but we note that the structure of metaphosphate glasses is comprised of long PO_3 chains cross-linked by the cations[20]. It may be that the increase of R(T) seen

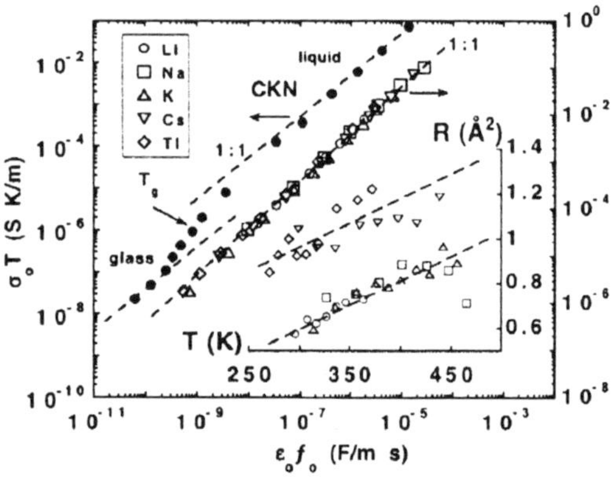

Fig. 3 Plots of $\sigma_o T$ vs. $\varepsilon_o f_o$ for CKN (solid circles) and a series of alkali metaphosphate glasses. Inset shows the temperature and composition dependence of the quantity R defined by Eq. 13 for the metaphosphate glasses.

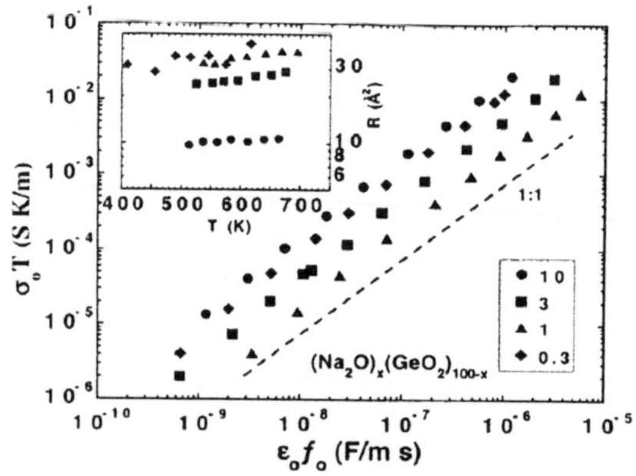

Fig. 4 Plots of $\sigma_o T$ vs. $\varepsilon_o f_o$ for a series of sodium germanate glasses. Inset shows the temperature and composition dependence of the quantity R defined by Eq. 13.

here is reflecting a corresponding increase in ξ^2 (see Eq. 13) that occurs as a direct result of thermal expansion.

4.2 Modified-Summerfield Scaling

A second case to consider is that for which the quantity in parenthesis in Eq. 12 remains constant for a series of glasses of varying ion concentration. In this case, $f_o \approx \sigma_o T/N_o$. This modified Summerfield scaling holds to a good approximation over limited ranges of N_o that correspond with considerable overlap of the coulomb wells associated with charge compensating sites in the glass[12]. However, at very low N_o the scaling fails[13]. As an example we consider the homologous series of sodium germanate glasses, $(Na_2O)_x(GeO_2)_{100-x}$, with x = 0.3, 1, 3, 10. Plots of $\sigma_o T$ versus f_o are provided in Fig. 4 and again show apparent linearity in the required double logarithmic representation. However, unlike the previous metaphosphate glasses, there is a significant compositional variation. Again we turn to the quantity R(N_o, T) defined in Eq. 13 which is reproduced in the inset of Fig. 4. Here we again see evidence of a weak temperature dependence, but also see a distinct concentration dependence: R increases with decreasing N_o. A simple interpretation of this trend follows from the association of the diffusion length scale ξ with the mean separation between charge compensating sites in the glassy network. For a homogeneous distribution of sites, this length scale increases with decreasing ion concentration as $\xi \approx N_o^{-1/3}$ Thus one anticipates $R \approx N_o^{-2/3}$. While there are an insufficient number of compositions to accurately test this prediction, the observed trend in not inconsistent and others[22] have reported similar findings.

4.3 Non-Summerfield Scaling

Lastly we consider a third case for which the quantity in parenthesis in Eq. 12 is neither constant with temperature nor with ion concentration. In such a case, we expect gross deviations from linearity in the plot of $\sigma_0 T$ versus f_0. Our first example of this is seen in CKN. The plot of $\sigma_0 T$ versus f_0 for CKN extends through the glass transition and is shown in Fig. 3. For CKN we see Summerfield scaling is approximately valid (i.e., linearity of $\sigma_0 T$ versus f_0 is apparent) both in the liquid and the glass (In reality, for both the glass and the liquid there is once more a weak temperature dependence to R), but fails in the vicinity (± 20 K) of the glass transition. As discussed earlier, we attribute this failure to changes in the population (and charge state) of the mobile ions. Below T_g, only monovalent K^+ ions are mobile. Above T_g this K^+ population is supplemented by a contingent of divalent mobile Ca^{2+} ions. From consideration of Eq. 12 ones sees that this change in mobile ion population will affect $\sigma_0 T/f_0$ near T_g in the fashion seen in Fig. 3.

Our second example where Summerfield scaling is violated appears in the sodium tellurite glasses. This failure was originally reported in recent work by Murugavel and Roling[20]. Here we examine (unpublished) measurements of our own on a similar series of sodium tellurite glasses where similar results are found. Glasses were prepared by melting appropriate amounts of Na_2CO_3 (99.995% by Aldrich) and TeO_2 (99.995% by Alfa) powders in an alumina crucible at temperatures of approximately 550°C. Samples (ca. 1 mm thick by 2 cm diameter) were cast and annealed, and concentric silver contacts were applied to opposite faces. Impedance measurements were obtained using a commercial impedance analyzer (Schlumberger 1260) for temperatures from ambient to approximately 1.1 T_g.

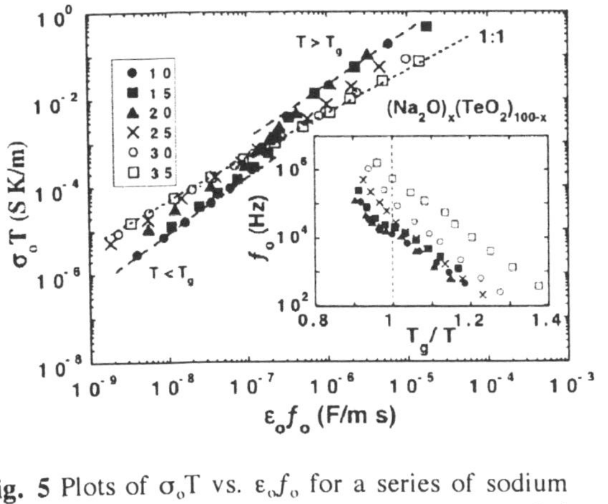

Fig. 5 Plots of $\sigma_0 T$ vs. $\varepsilon_0 f_0$ for a series of sodium tellurite glasses. Inset shows the temperature (scaled to T_g) and composition dependence of the scaling frequency.

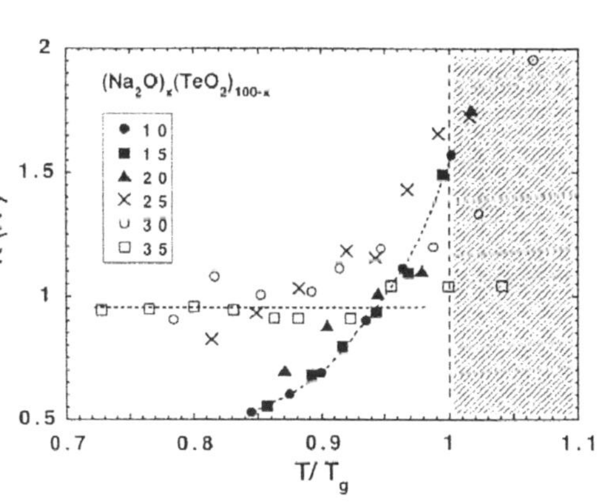

Fig. 6 Plots of the quantity R defined by Eq. 13 for the series of sodium tellurite glasses. In the glass state, R is constant for high sodium glasses but exhibits a strong temperature dependence for the low sodium glasses.

Plots of $\sigma_0 T$ versus f_0 are provided in Fig. 5 for glasses with x = 10, 15, 20, 25, 30, and 35. We find that the two high N_0 glasses (x = 30, 35) exhibit Summerfield scaling. In distinct contrast, the low N_0 glasses (x = 10, 15) show pronounced violation of Summerfield scaling; $\sigma_0 T$ versus f_0 is non-linear both above and below T_g with a step-like transition occurring in the vicinity of T_g. These peculiarities are also seen in the variation of f_0 with temperature, shown in the inset to Fig. 5. Here we find Arrhenius behavior for the high N_0 glasses but non-Arrhenius temperature dependence for the low N_0 glasses and a flattening of the curve on approach to T_g. In Fig. 6 we consider the compositional and temperature dependence of $R(N_0,T)$ below T_g. We observe two sorts of behavior for the high and low N_0 glasses respectively. The high N_0 glasses show $R(N_0, T)$ that is constant while the low N_0 glasses display significant temperature dependence of $R(N_0,T)$ which appears to scale with the approach to T_g.

These all suggest something akin to a transition between two dynamic phases; one at high N_0 ($x > 25$) where $R \approx$ constant and a second phase at low N_0 ($x < 20$) for which R varies with temperature. Such transitions are not unknown to the tellurite glass system. An NMR study[23] of the same series of sodium tellurite glasses reported a pronounced, discontinuous increase in the second moment of the

NMR spin echo decay occurring between $x = 15$ and $x = 20$. This increase was attributed to a transition from randomly distributed Na^+ ions at low N_o to Na^+ distribution at high N_o that have "extensive intermediate order" (i.e., clustering of the ions). Even more recently, an ultrasonic study[24] of divalent copper tellurite glasses also indicates evidence for some form of structural transition in the tellurite glasses at intermediate ion concentrations. While further investigation is needed, it is interesting that these transitions may also be showing themselves in the ac conductivity, once an appropriate analysis is conducted.

5. Variations in Mastercurve Shape

Although the underlying dynamic process is often invariant with regards to temperature and concentration, there are cases in which systematic changes in the mastercurve shape have been reported[25,26]. An important issue is how one might parameterize the shape of the dispersion so as to categorize the various shapes. Roling[25] has suggested that plots of the running slope of log σ/ log f can be compared to assess differences in shape. A simpler, although less precise parameter is the power law exponent obtained by fitting Eq. 2 to the mastercurve obtained for a given sample. A survey of literature data reveals some interesting features regarding this exponent[27]. First, many oxide glasses (SiO_2, GeO_2, etc.) exhibit a reasonably common $n \approx 2/3$, indicating a common dynamic process for the motion of ions in the oxide networks. Second, there appears to be a weak trend in which the exponent decreases with decreasing dimensionality of the ion conduction space. In hindsight, owing to the smallness of the range ($\xi \approx 2$ Å) over which the correlated motion extends, it is more accurate to say that the exponent responds to changes in the coordination of the ions local conduction space (CLCS). That is, changes in the average local availability of doorways leading to conduction paths within the structure.

In our own work we have found cases of systematic variation of the dispersion. A study of AgI-doped Ag metaphosphate glasses and the series of alkali metaphosphate glasses[26], alluded to earlier with regards to Fig. 3, revealed a peculiar "constriction effect" wherein the exponent responded to changes in the ratio of cation size to mean phosphate chain separation. For small values of this ratio, corresponding to an unconstricted CLCS, $n \approx 2/3$ was observed. However, when this ratio exceeded about 1/2, the exponent was significantly reduced. The current interpretation of these changes is that larger sized cations experience a loss of CLCS (i.e, doorways) due to the increased crowding of the structure (phosphate chains) about their position.

In summary, although the dispersion of many ion-conducting glasses shows remarkably common shape, this shape does appear to undergo modest modification as a result of changes in the CLCS. Interestingly, this finding is consistent with simulations of the random barrier model for ion conduction for which similar modest, but systematic, variations in the mastercurve result from changes in the dimension of the simulation lattice[28]. Further examination of these changes should aid in sorting out the proper origin for correlated motion in these and other materials.

6. Conclusion

"Ionic relaxation" has been studied for quite some time and the image invoked by the term has changed during that period. For some, ionic relaxation conjures up the image of the electric modulus, its asymmetric shape and the corresponding parameterization of that shape via the KWW relaxation function. For others, ionic relaxation is something of a misnomer, for the "relaxation" consists merely in the continuous temporal evolution of ions MSD which is exhibited in frequency space by dc conduction and ac conductivity dispersion. Here we have reviewed many of the pros and cons of these two perspectives. Focusing upon the more recent one, we have strived to demonstrate what useful insight can be obtained via scaling analysis of the ac conductivity and its corollaries. The challenge as we see it now is to compare mastercurve shapes (1) to each other for differing glasses, crystals, etc., and (2) to that of model and theory prediction. By so doing, we anticipate further success in understanding the origins of correlated motion.

References

[1] R. Kubo, J. Phys. Soc. Japan **12** (1957) 570.

[2] B. Roling, C. Martiny, and K. Funke, J. Non-Cryst. Sol. **249** (1999) 201.

[3] A. K. Jonscher, Nature **267** (1977) 673.

[4] A. S. Nowick, B. S. Lim, and A. V. Vaysleyb, J. Non-Cryst. Sol. **172-174** (1994) 1243.

[5] K. L. Ngai, J. Chem. Phys. **110** (1999) 10576.

[6] D. L. Sidebottom and C. M. Murray-Krezan, Phys. Rev. Lett. **89** (2002) 195901.

[7] V. Provenzano, L. P. Boesch, V. Volterra, C. T. Moynihan, and P. B. Macedo, J. Am. Ceram. Soc. **55** (1972) 492.

[8] C. T. Moynihan, L. P. Boesch, and N. L. Laberge, Phys. Chem. Glasses **14** (1973) 122.

[9] K. L. Ngai and C. T. Moynihan, Mat. Res. Soc. Bull. **23** (1998) 51.

[10] K. L. Ngai, Phys. Rev. B **48** (1993) 13481; J. Chem. Phys. **98** (1993) 6424; Solid State Ionics **105** (1998) 225.

[11] H. K. Patel and S. W. Martin, Phys. Rev. B **45** (1992) 10292.

[12] B. Roling, A. Happe, K. Funke, and M. D. Ingram, Phys. Rev. Lett. **78** (1997) 2160.

[13] D. L. Sidebottom, Phys. Rev. Lett. **82** (1999) 3653.

[14] D. L. Sidebottom and J. Zhang, Phys. Rev. B **62** (2000) 5503.

[15] D. L. Sidebottom, B. Roling, and K. Funke, Phys. Rev. B **63** (2000) 024301.

[16] K. L. Ngai and C. Leon, Phys. Rev. B **60** (1999) 9396.

[17] F. S. Howell, R. A. Bose, P. B. Macedo, and C. T. Moynihan, J. Phys. Chem. **78** (1974) 639.

[18] D. L. Sidebottom, P. F. Green, and R. K. Brow, Phys. Rev. B **56** (1997) 170

[19] T. B. Schroder and J. C. Dyre, Phys. Rev. Lett. **84** (2000) 310.

[20] S. Murugavel and B. Roling, Phys. Rev. Lett. **89** (2002) 195902.

[21] R. K. Brow, J. Non-Cryst. Sol. **263-264** (2000) 1.

[22] B. Roling, Solid State Ionics **105** (1998) 185.

[23] J. W. Zwanziger, J. C. McLaughlin, and S. L. Tagg, Phys. Rev. B **56** (1997) 5243.

[24] A. Paul, P. Roychoudhury, S. Mukherjee, and C. Basu, J. Non-Cryst. Sol. **275** (2000) 83.

[25] B. Roling and C. Martiny, Phys. Rev. Lett. **85** (2000) 1274.

[26] D. L. Sidebottom, Phys. Rev. B **61** (2000) 14507.

[27] D. L. Sidebottom, Phys. Rev. Lett. **83** (1999) 983.

[28] J. C. Dyre, Phys. Rev. B **48** (1993) 12511.

STUDY ON SUPERIONIC CONDUCTORS BY OPTICAL MEASUREMENTS*

TAKESHI HATTORI†
Institute of Multidisciplinary Research for Advanced Materials, Tohoku University
Katahira, Sendai 980-8577, Japan

Usefulness of optical measurements for studying ion dynamics in superionic conductors will be discussed using the results of β-aluminas as an example; namely, their Raman scattering, reflection and luminescence spectra and hole burning spectroscopy.

1. Introduction

In superionic conductor, generally, ionic conductivity is high as comparable to liquid electrolyte and activation energy is low, its order is about 0.1 eV. The typical materials are α-AgI, β-alumina, YSZ, perovskite-type proton conductors and so on [1].

Main problems of the investigations of superionic conductors will be as follows: (1) Syntheses of new materials, (2) applications to solid state batteries, fuel cells, sensors and so on, and (3) analyses of their conduction mechanisms.

If we will understand perfectly fundamental properties of superionic conductors, we will able to expect a breakthrough of development of applications and of synthesis of new materials. Of course, the properties of superionic conductor themselves bring us the interesting features in the fields of fundamental physics and chemistry.

Many attempts have been made experimentally and theoretically to analyze the conduction mechanisms in superionic conductors. In those cases, optical spectroscopy has strongly contributed to this analysis [2,3]. For example, the crystallographic structure, localized structure and/or disordered nature have been analyzed using Raman scattering spectra [4,5], luminescence spectra [6] and so on. The electronic structure has been also studied by optical absorption and reflection spectroscopy in vacuum-ultraviolet wavelength region [7]. With regard to ion dynamics, the attempt frequency for ionic conduction has been observed in Raman scattering spectra [4,8]. The activation energy for ionic conduction is also obtained from analysis of quasi-elastic light scattering spectra [9,10]. The analysis of bare potential barrier through the conduction path is important for the understanding the conduction mechanism. This is obtained from the analysis of hole burning spectra [11,12]. Moreover, relaxation frequency and time for ion motion will be obtained from the analyses of non-linear optical effects such as hyper Raman Scattering [13] and optical Kerr effect, respectively [14].

In this paper, the results of Raman scattering, luminescence and reflection spectra, and hole burning spectroscopy of β-aluminas are discussed. Why are only β-aluminas discussed? Because there are following reasons: (1) Good quality single crystal was easily got. (2) The crystal is stable optically. (3) Many investigations have been performed in order to analyze their fundamental properties and to apply in batteries, sensors and so on. (4) The conduction ions, usually Na ions, can be easily exchanged to other cations, and so on. (5) There are two types of crystals in β-alumina family, namely β-alumina and β"-alumina. Both are famous superionic conductors [15].

With regard to investigation of Ag-compound such as AgI, $RbAg_4I_5$, and so on, please see reference [3].

2. Light Scattering

2.1. *Introduction*

Three types of light scattering under the linear response of light, namely, Raman scattering, Brillouin scattering and quasi-elastic light scattering had been reported in order to analyze the properties of β-aluminas. The first interest investigation of Raman scattering of β-aluminas had been performed by Hao et al [4].

---

* This work is partially supported by CREST of JST (Japan Science and Technology.)

† Present Address: *Department of Applied Physics, Tokyo University of Science, Tokyo 162-8601, Japan*

We will get the information about structure and attempt frequency from their Raman scattering spectra. The lattice vibrations and the attempt frequencies had been discussed from the results of Brillouin scattering of β-aluminas [8, 16]. Suemoto and Ishigame had developed quasi-elastic light scattering spectroscopy of superionic conductors [2,9]. The activation energy had been obtained from optical method using this method [10].
Sub-headings should be typeset in boldface italic and capitalize the first letter of the first word only. Section number to be in boldface roman.

2.2. *Raman scattering*

Hao et al. had reported first the interesting results of Raman scattering spectra of β-aluminas with various conduction ions [4]. The same and different points among the spectra among different kinds of conduction ions are as follows: (1) The spectra above 100 cm^{-1} observed in all β-aluminas with different conduction ions are the same. These are lattice modes of β-alumina. (2) The characteristic features of Raman scattering spectra in β-aluminas appear in the spectra below 100 cm^{-1} [4].

Figure 1 shows the Raman scattering spectra of β-aluminas with various conduction ions, namely Tl, Ag, Cs, Rb, K and Na below 120 cm^{-1} [16]. These bands were measured under the condition of the observation of E_{2g} mode. Namely, in this mode, only the conduction ion is mainly vibrated parallel to the conduction plane. The bands below 100 cm^{-1} correspond to the attempt frequencies for ionic conduction of β-aluminas.

Other interesting results of the Raman scattering spectra of β-alumina had been reported by Colomban and Lucazeau [5]. They had prepared stoichiometric compound of Na β-alumina. They compared the differences between the Raman scattering spectra of stoichiometric and non-stoichiometric Na β-aluminas. They had found that disordered characters of β-alumina appeared in the broad scattering bands of non-stoichiometric Na β-alumina.

Sections, sub-sections and sub-subsections are numbered in Arabic. Flush left all paragraphs that follow after section headings.

3. Reflection Spectra

A characteristic feature of reflection spectra of β-alumina was studied by using a beam line with a synchrotron radiation [7,17]. This work had been performed in order

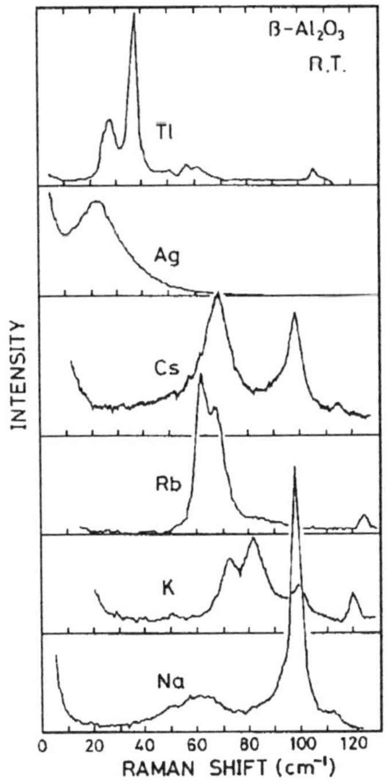

Figure. 1 The Raman scattering spectra (E_{2g} mode) below 120 cm^{-1} of β-aluminas with various conduction ions.

to analyze mixed-cation effects of β-alumina [17]. As shown Fig. 2, new extra bands in the reflection spectra of mixed-cation β-aluminas in addition to the exciton bands of the end components (x=0 and 1) were observed at the concentration regions (x) where the ionic conductivity shows a minimum in $Na_{1-x}K_x$ and $Na_{1-x}Ag_x$ β-aluminas. On the other hand, no additional band was observed in $Na_{1-x}Tl_x$ β-aluminas. New extra exciton bands in $Na_{1-x}K_x$ and $Na_{1-x}Ag_x$ β-aluminas will correlate to a special structure of the conduction plane. This will be related strongly to the origin of the mixed-cation effect in β-aluminas. This will be the experimental direct evidence of the origin of the mixed-cation effect in β-aluminas. These results were also suggested by luminescence measurements of mixed cation β-aluminas.

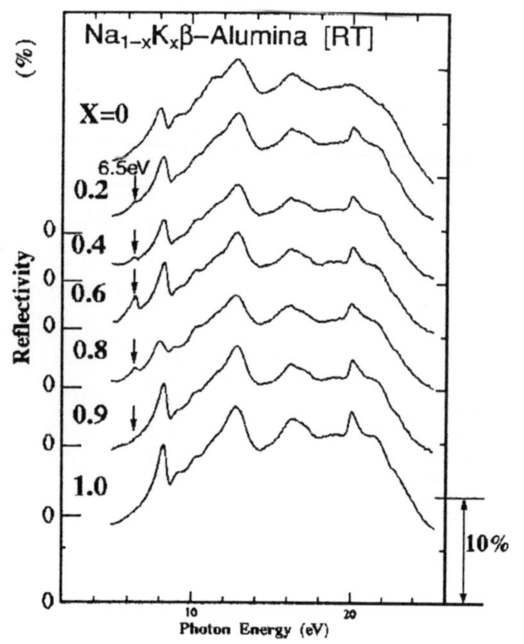

Figure 2(a)

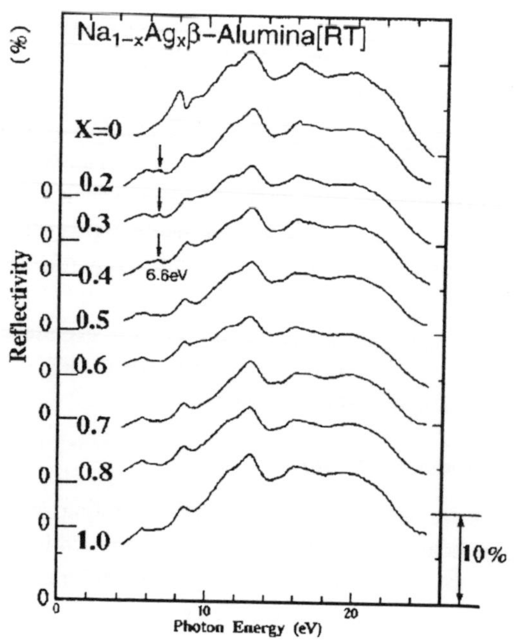

Figure 2(b)

Figure 2. The reflection spectra of mixed-cation β-aluminas ((a) $Na_{1-x}K_x$, (b) $Na_{1-x}Ag_x$ and (c) $Na_{1-x}Tl_x$ β-aluminas).

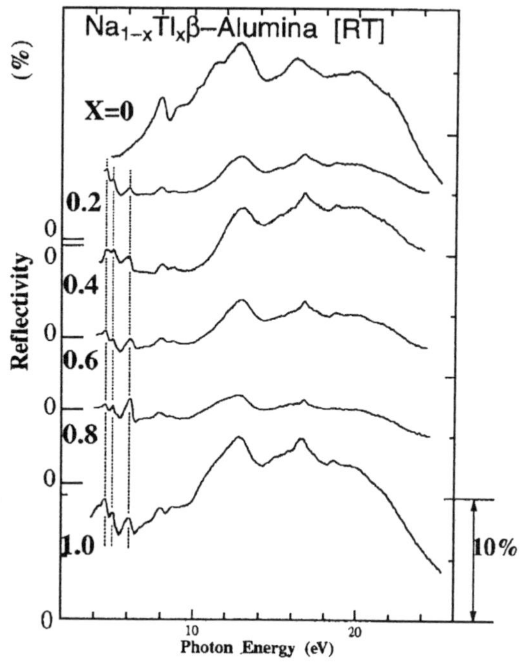

Figure 2(c)

4. Luminescence

Figure 3 shows one of the time resolved luminescence spectra of Cr ions as residual impurities in mixed-cation β-aluminas, $Na_{0.3}K_{0.7}$ β-alumina. From the analysis of the decay feature of luminescence intensity, it had been concluded that the structure of the conduction plane at the conductivity minimum of the mixed cation $Na_{1-x}K_x$ β-aluminas would be related to the fractal property [6,18].The final hard copy that you submit must be absolutely clean and unfolded. It will be printed directly without any further editing. Use a printer that has a good resolution printout (600 dpi or higher). There should not be any corrections on the printed pages, nor should adhesive tape cover any lettering. Photocopies are not acceptable.

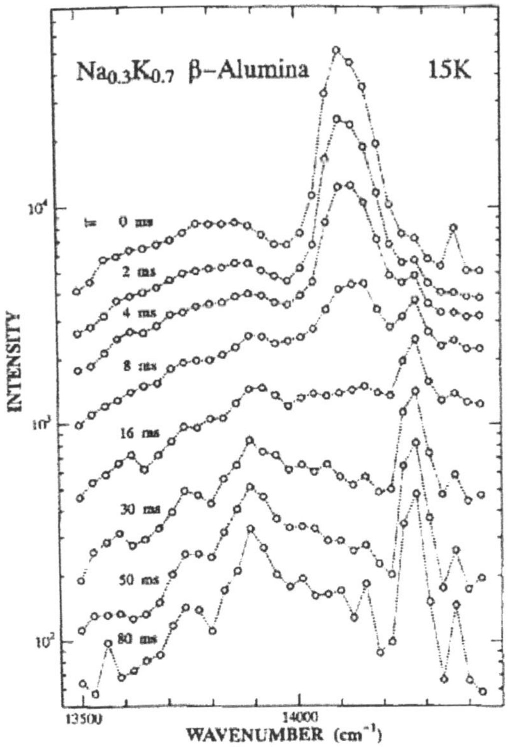

Figure 3. The time resolved luminescence spectra of Cr ions in $Na_{0.3}K_{0.7}$ β-alumina at 15 K.

5. Hole Burning

5.1 *Introduction*

When optical centers such as rare-earth metal ions are introduced in suitable disordered host materials, they often have an inhomogeneous absorption band, which is usually broad. Laser irradiation at a specific frequency within that absorption band makes a dip in the absorption spectrum. When irradiation terminated, this dip usually vanished within a lifetime of the electron in the excited state. This is a transient hole. However, the dip can be observed often for a long time in the special case. This is a persistent hole. Usually, this is called hole or spectral hole. Such phenomenon is call hole burning. Persistent hole is available to study on disordered materials.

Three origins of persistent holes have been known for the holes using a lanthanoid metal ion as optical center: The first is due to phtoionization of rare-earth metal ion as optical center. Although this is important for application to optical memory, it depends on the kind of rare-earth metal ion as optical center. The second is due to optical pumping between hyperfine splitting levels of lanthanoid metal ions. And, the third is due to local-structure change surrounding the optical center. This hole is important to obtain the potential energy along the conduction path in superionic conductors. Namely, in superionic conductors, the locally structural change is strongly related to the elementary migration of conductive ions. Needless to say, homogeneous bandwidth for the hole is important for discussion of disordered nature of the materials. The persistent hole is necessary for understanding of structure and electronic properties of disordered materials [19], especially, ion dynamics in superionic conductors [11,12].

In experimental aspect, hole burning experiments are normally carried out at low temperature, typically bellow 10K. This is because the relaxation to the ground state occurs fast at high temperatures, that the spectral hole vanishes quickly. Hole burning experiment consists of two processes. One is burning process, which indicates a creation of a spectral hole, that is, irradiation of intense laser light of fixed frequency to the sample. The other is a scanning process, that is, a measurement of an absorption spectrum around the hole. Thus hole burning spectroscopy is one of the pump-probe spectroscopy in broad sense.

5.2 *Hole spectra*

In this work, Na β”-alumina sample doped with 90-mol % Eu^{3+} ions were prepared. Hole spectra burned at 577.6 nm and 578.7 nm in Eu^{3+} exchanged Na β”-alumina sample at 10 K are shown in Fig. 4 [12]. The anti-holes and the side-holes were observed in addition to a main hole in the spectrum burned at 577.6 nm. These were mainly observed below 14 K and those lifetimes are about 120s. It is concluded from these data that this hole is attributed optical pumping of nuclear quadruple levels because the nuclear spin quantum number of Eu^{3+} ion is 5/2 and the electron spin quantum number is zero.

One the other hand, a hole burned at 578.7 nm was observed with a single Lorentzian shape as shown in Fig. 4. The hole with the same intensity and the same shape

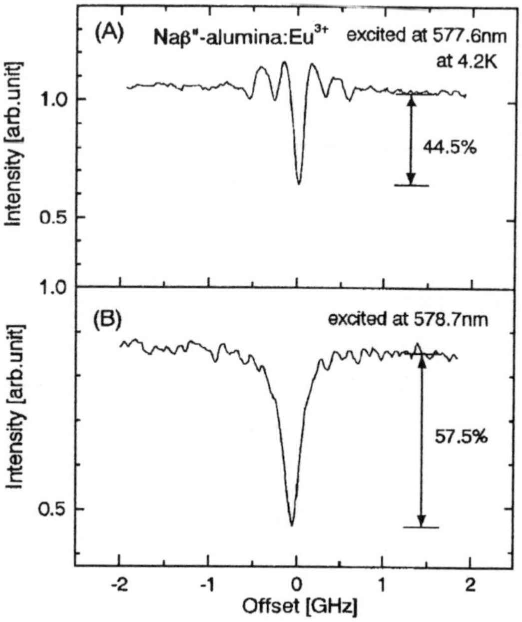

Figure 4. Hole spectra burned at 577.6 nm and 578.7nm in 90 % Eu^{3+} ions exchanged Na β"-alumina sample at 4.2K.

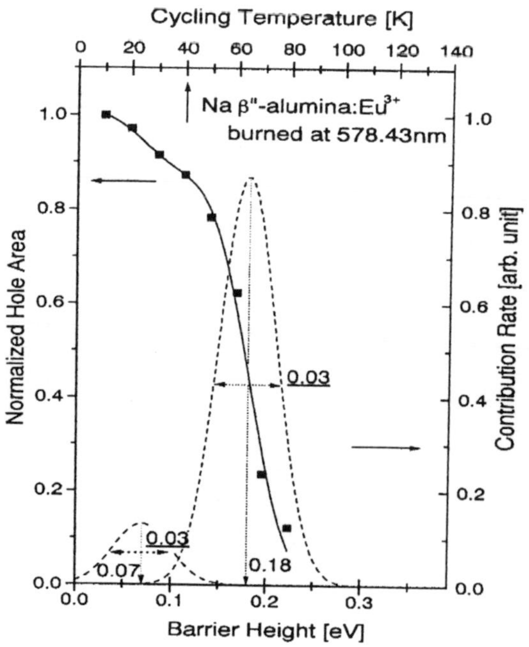

Figure 5. Cycling temperatures vs. normalized hole area in a hole burned at 578.7 nm of Eu^{3+} ions exchanged Na β"-alumina sample. Solid circles show the experimental points and solid line the calculated one fitted to the experimental points. About two broken lines please show the text.

was observed some hours after the burning laser was switched-off. Namely, lifetime will be infinity. The hole (B) was observed relatively high temperatures: 70 K in Na β"-alumina. It is concluded from these experimental result that the hole (B) is caused by a light-induced local-structural change by ionic motion of monovalent Na ions surrounding Eu^{3+} ions.

5.3 *Result of Temperature Cycling Experiments*

The barrier height or the potential energy for the light-induced local motion of ions or the local-structure change surrounding Eu^{3+} ions was determined from the analysis of the thermal decay-profile of the persistent hole, which was obtained from the so-called temperature cycling experiment. Figure 5 shows the result of the temperature cycling experiment for the persistent hole burned at 578.52 nm of Eu^{3+} ions exchanged Na β"-alumina [12]. Hole is burned and measured at 10 K and sample is kept at a cycling temperature for a period of 2 minutes. Solid circles show the experimental points and solid line the calculated curve fitted to the experimental points. The calculated curve was obtained under the assumption of two potential energies (V_1=0.07 eV, V_2=0.18 eV) and both energies have Gaussian-type distributions. The two broken lines show those two potential energies.

5.4 *Potential Curve*

From the results by Collin et al [20], Na ions in the conduction plane locate in two sites namely, the sites (1) and the site (2). Therefore, two potential energies obtained from the hole burning spectroscopy will corresponds to the barriers between the site (1) and the sate (2) and between the site (2) and the site (2), respectively. From the experimental results, we obtained a potential curve for monovalent conductive ion along the conduction path as shown in Fig. 6 , because the

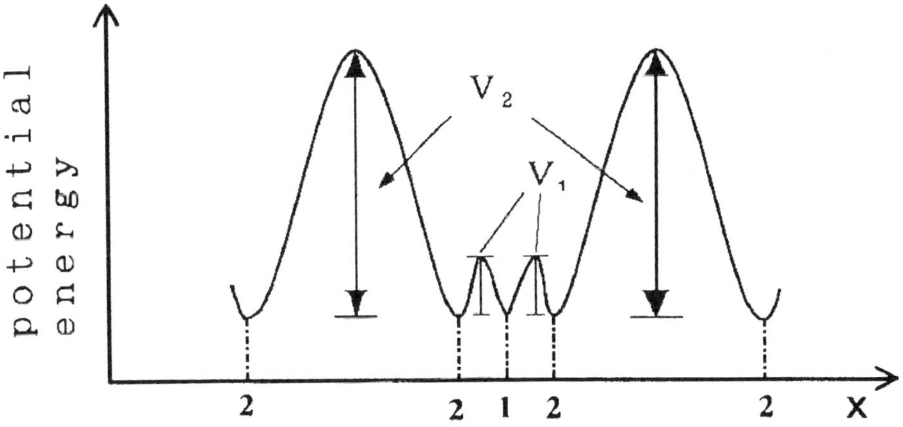

Figure 6. A potential curve for monovalent ion along the conduction path.

conduction ion moves as site (2), site (2), site (1), site (2) and so on. Since the concentration of Eu^{3+} ions is about 90-mol %, conducting Na ions of 10-mol % are isolated from each other.

The activation energy (V=0.09 eV for Na β"-alumina [21]) for ionic migration of the conductive monovalent Na ions is larger than V_1 (=0.07 eV) but smaller than V_2 (=0.18 eV). If we considered the ionic interaction between the conductive ions, because in normal β"-alumina, many conduction ions are located in the conduction plane, larger potential energies of V_2 than the activation energy for ionic conduction can be concluded. This conclusion will correspond to the suggestion in the vacancy model by Wang and Pickett Jr.. [22] It is considered that the ionic interaction plays an important role for the ion migration in Na β"-alumina.

6. Summary

Usefulness of optical measurements for studying ion-dynamics in superionic conductors was described using the results of β-aluminas as an example, namely their Raman scattering, reflection and luminescence spectra and Hole burning spectroscopy.

Acknowledgments

The author would like to thank many persons for kindly cooperation; Prof. A. Mitsuishi, Prof. M. Ishigame, Prof. S. Shin, Prof. H. Yugami, Prof. N. Sata and my students (Dr. H. Okamura, Dr. I. Kawaharda, Dr. R. Yagi, Dr. H. Kurokawa, Mr. H. Nkata, T. Imanishi, Mr. Y. Chiba, Mr. H. Kobayashi, Mr. K. Aso, Mr. H. Yashima, and so on).

References

1. S. Chandra, "Superionic Solids, Principle and *Applications*" (North-Holland Pub. Com., Amsterdam-New York-Oxford, 1981).
2. M. Ishigame, S. Shin, and T. Suemoto, *Solid Sate Ionics* **47**, 47 (1991).
3. I. Kh. Akopyan, A. A. Klochikhin, B. V. Novikov, M. Ya. Valakh, A. P. Litvinchuk, and I. Kosazkii, *phys. stat. sol.* (a)**119**, 363 (1990).
4. C. H. Hao, L. L. Chase, and G. D. Mahan, *Phys. Rev.* B**13**, 4306 (1976).
5. Ph. Colomban and G. Lucazeau, *J. Chem. Phys.* **72**, 1213 (1980).
6. T. Hattori, H. Kobayashi, H. Yugami and M. Ishigame, *Solid State Ionics* **79**, 21 (1995).
7. T. Hattori, and M. Ishigame, *Solid State Ionics* **109**, 197 (1998) & T. Hattori, Y. Chiba and M. Ishigame, *Solid State Ionics* **113-115**, 527 (1998).
8. T. Hattori, H. Nakata, T. Imanishi, H. Kurokawa, and A. Mitsuishi, *Solid State Ionics* **2**, 47 (1981).
9. T. Suemoto, and M. Ishigame, *Phys. Rev.* **B32,**

4126 (1985).
10. I. Kawaharada, T. Hattori, M. Ishigame, and S. Shin, *Solid State Ionics* **69**, 79 (1994).
11. S. Matsuo, H. Yugami, and M. Ishigame, *Phys. Rev.* B**64,** 024302 (2001).
12. T. Hattori, R. Yagi, K. Aso, and M. Ishigame, *Solid State Ionics* **136-137**, 409 (2000).
13. S. Shin, Y. Tezuka, A. Sugawara, and M. Ishigame, *Phys. Rev.* B**44**, 11724 (1991).
14. T. Hattori, unpublished data.
15. J. H. Kennedy, "*Topics in applied physics*, vol. 21 *Solid State Electrolytes*" edited by S. Geller (Springer-Verlag, Berlin, 1977) Chap. 5.
16. T. Hattori, H. Nakata, T. Imanishi, and A. Mitsuishi, *Solid Sate Ionics* **3/4**, 69 (1981).
17. T. Hattori, S. Yashima, I. Kawaharada, N. Sata, M. Ishigame, SSI-14, Monterey, California, USA, June, 2003.
18. T. Hattori, H. Yugami, and M. Ishigame, *Solid State Ioncis* **70/71**, 417 (1994).
19. "*Persistent Spectral Hole-Burning: Science and Applications*", edited by W. E. Morener (Springer-Verlag, Berlin, 1988).
20. G. Collin, J. P. Boilt, Ph. Colomban, and R. Comes, *Phys. Rev.* B**34**, 5838 (1986).
21. G. C. Farrington, B. Dunn, and J. O. Thomas, *Apply. Phys.* A**32**, 159 (1983).
22. J. C. Wong and D. F. Pickett Jr., *J. Chem. Phys.* **65**, 5378 (1976).

PHASE STABILIZATION AND HEAT CAPACITY OF ZIRCONIA

TAKEO TOJO, HITOSHI KAWAJI, TOORU ATAKE
Materials and Structures Laboratory, Tokyo Institute of Technology, 4259 Nagatsuta-cho, Midori-ku, Yokohama, 226-8503 Japan

TOSHIYUKI MORI
National Institute for Materials Science, 1-2-1 Sengen, Tsukuba, 305-0047 Japan

HIROSHI YAMAMURA
Department of Applied Chemistry, Faculty of Engineering, Kanagawa University, 3-27-1 Rokkakubashi, Kanagawa-ku, Yokohama, 221-8686 Japan

Heat capacity of 4 mol% yttria stabilized zirconia $\{(ZrO_2)_{1-x}(Y_2O_3)_x; x = 0.0395\}$ has been measured between 13 and 300 K using a high precision adiabatic calorimeter. Large excess heat capacity has been found in yttria stabilized zirconia compared with pure zirconia and pure yttria on the basis of simple additivity rule. From analyses of the excess heat capacity and the phonon density of states derived from neutron inelastic scattering experiments, a softening of the phonons near the Brillouin zone boundary is suggested in stabilized zirconia. The soft phonons are assigned to the vibrations of zirconium ion in ZrO_8 polyhedron. The eightfold coordination of oxide ions around the zirconium ion in stabilized zirconia is of over-coordination compared with that in the most stable monoclinic phase. The frequency of the soft phonon in tetragonal stabilized zirconia is higher than that in cubic stabilized zirconia, which should reflect the Zr-O bond strength due to the tetragonal distortion of ZrO_8 polyhedron. The number of the soft phonons decreases with increasing yttria content because of formation of oxygen vacancies, which leads to the decrease in the coordination number of Zr ion and to the relaxation in the anion sublattice.

1. Introduction

It is well known that zirconia (ZrO_2) is highly functionalized by introducing defects into the crystal lattice sites. By doping lower valent oxides, so-called stabilized zirconia exhibits high oxide ion conductivity and fracture toughness. The defect structure of stabilized zirconia has been studied extensively in order to elucidate the mechanism and to improve the functions. The crystal structure of pure zirconia at room temperature under atmospheric pressure is monoclinic (space group $P2_1/c$), which is a distortion of cubic fluorite type [1-3]. In the monoclinic phase, Zr atom has sevenfold coordination of oxygen atoms. With increasing temperature, zirconia undergoes two structural phase transitions; the monoclinic-to-tetragonal at 1440 K and the tetragonal-to-cubic at 2640 K, and the cubic melts at 2983 K [4]. In the cubic phase, zirconia has the fluorite type structure and Zr and O atoms occupy the special positions of 4(a) and 8(c) of space group *F*m3m. The tetragonal phase of zirconia has space group $P4_2/nmc$, and Zr and O atoms are at the special positions of 2(a) and 4(d) [5]. In the high temperature phases, Zr atom has eightfold coordination of oxygen atoms, and the Zr-O distance is the same for all the 8 pairs in cubic zirconia. In the case of tetragonal phase, however, the ZrO_8 polyhedron is distorted and the length of four Zr-O bonds is short and the other four bonds are long. It is well known that the incorporation of lower valent oxides such as CaO, Y_2O_3, etc. into zirconia stabilizes the cubic or tetragonal forms of zirconia even below room temperature. In stabilized zirconia, doped cations substitute for Zr atoms at random and O vacancies are formed in the anion sublattice to maintain the electrical neutrality. At elevated temperatures, stabilized zirconia exhibits high oxide ion conductivity via these vacancies. The ionic conductivity depends on the dopant concentration and the maximum conductivity is obtained at 8-10 mol%, in the case of yttria doping, which is the least amount of dopant required to stabilize the cubic phase and further doping leads to fuller stabilization of the cubic phase and also to the decrease of conductivity. Although several mechanisms, such as formation of defect associates [6], defect-defect interaction [7], blockage by the static aggregates [8] and so on, have been proposed for the decrease of conductivity in higher dopant concentrations, it is still an open question, why the dopant concentration of (cubic+tetragonal) and cubic phase boundary coincides with that of the maximum conductivity. The formation of O vacancies introduces distortions in anion sublattice [9-17]. While the cation sublattice maintains its almost perfect fcc structure, the O atoms beside vacancies are shifted significantly from the ideal fluorite 8(c) positions without significant crystallite strain. Such distortion in the atomic arrangement should alter the atomic interactions and the lattice vibrations. The vibrational properties of stabilized zirconia have been studied extensively by Raman [18-20], IR [21] and neutron inelastic scattering [22-24] experiments. The lattice vibration is influenced by the local environments of atoms, and the studies of lattice vibrations should provide information about the defect structure. The low temperature calorimetry is also a good way to investigate the lattice properties of materials because of the intimate relationship between the thermodynamic properties and phonon excitations. In the previous papers [25,26], low temperature heat capacities of monoclinic zirconia and cubic stabilized zirconia doped with 8-11 mol% yttria have been reported. Cubic stabilized zirconia has larger heat capacity value than that expected from an assumption of simple additivity rule with the heat capacities of pure zirconia and pure yttria.

The excess heat capacity suggests softening of the lattice vibrations in cubic stabilized zirconia. The excess heat capacity decreases with increasing yttria content, although it is caused by the doping of yttria. In the stabilized zirconia based on the fluorite-type cubic symmetry, the decrease of coordination number of Zr ion and the relaxation in the anion sublattice due to the O vacancies should alter the phonon density of states. In the present study, low temperature heat capacity of tetragonal stabilized zirconia doped with yttria has been measured, and the results have been compared with those of monoclinic zirconia and cubic stabilized zirconia. The difference in coordination environment of Zr atom in monoclinic, tetragonal and cubic polymorphs has been reflected in the excess heat capacity. The neutron inelastic scattering experiments have been made to verify the softening modes. The relationship between the phonon density of states and the local structure of stabilized zirconia have been discussed.

2. Experimental

The sample of 4 mol% yttria stabilized zirconia $(ZrO_2)_{1-x}(Y_2O_3)_x$; $x = 0.0395$} (4YSZ) was synthesized by a hydrolysis method [25,26]. The powder obtained was calcined at 1120 K, and pressed into a plate uniaxially under 50 MPa followed by isostatic cold-pressing under 200 MPa. The green compact was sintered at 1770 K in air for 4 h. By X-ray diffractometry, the product was identified as a mixture of tetragonal and cubic forms. The volume fractions of those phases were 70 % of the tetragonal and 30 % of the cubic estimated by the intensities of $(400)_c$, $(400)_t$ and $(004)_t$ diffraction peaks. The 4 mol% yttria stabilized zirconia is so-called partially stabilized zirconia. The samples of pure zirconia and 8,10,11 mol% yttria stabilized zirconia {$(ZrO_2)_{1-x}(Y_2O_3)_x$; $x = 0$, 0.0776, 0.0970, 0.1135} (8, 10, 11YSZ) have been confirmed to be a single phase of monoclinic and cubic, respectively, by X-ray diffractometry. The samples of yttria stabilized zirconia are abbreviated as 4-11YSZ, respectively, hereafter. Heat capacity measurements have been made using a homemade high precision adiabatic calorimeter at the temperatures between 13 and 300 K. The working thermometer was a platinum resistance thermometer calibrated between 13.8 and 303 K at National Physical Laboratory (England) in terms of the International Temperature Scale of 1990 (ITS-90). The accuracy of the calorimeter was confirmed to be within 0.1 % by measuring the standard reference material (SRM 720; synthetic sapphire) provided by NIST (U.S.A.)[27]. The amount of the 4YSZ sample used for the heat capacity measurement was 22.748 g (0.17872 mol). The heat capacity of the sample was obtained by subtracting the contribution of the empty calorimeter vessel from the total heat capacity. As the contribution of the sample to the total heat capacity including the calorimeter vessel was about 30, 37, and 50 %, at 50, 100, and 300 K, respectively, the accuracy of the measured heat capacity should be 0.1 %. The details for the heat capacity measurements for pure zirconia and 8-11YSZ were described elsewhere [25,26]. Neutron inelastic scattering experiments have been performed at 17 K by time-of-flight (TOF) method using LAM-D spectrometer [28] in KENS facility at the High Energy Accelerator Research Organization (Tsukuba, Japan). The TOF spectra were collected at the scattering angles of 40° and 85°. Sintered samples of 8-11YSZ of about 20 g were mounted in an aluminum container. The background scattering from the sample container was measured separately for further corrections.

3. Results and discussion

The heat capacity of the tetragonal stabilized zirconia (4YSZ) has been measured by adiabatic calorimetry between 13 and 300 K. No anomalous behavior such as supercooling or long thermal relaxation behavior was observed during the measurements. The molar heat capacity of 4YSZ in chemical formula $\{(ZrO_2)_{1-x}(Y_2O_3)_x;\ x = 0.0395\}$ is shown in Fig. 1. The heat capacity of pure zirconia and 8-11YSZ, which have been measured previously [25,26], are also plotted for comparison. Except for the very low temperatures, the heat capacity of YSZ increases with increasing yttria content, which is mainly because of the increase of the number of atoms in the chemical formula $(ZrO_2)_{1-x}(Y_2O_3)_x$ with increasing the value of x. At room temperature, the heat capacity value is approximately proportional to the number of atoms in the chemical formula. The low temperature heat capacity, however, is correlated to the phonon density of states (DOS) by quantum effects and thus the difference of DOS is reflected in the value of heat capacity. In the lowest temperature region, the heat capacity can be compared in terms of $C_p \cdot T^{-3}$ vs. T (see the inset of Fig. 1), which shows that the YSZ has much larger heat capacity than pure zirconia. In the group of YSZ, 8-10YSZ have largest heat capacity compared with 4 and 11YSZ. Interestingly, this corresponds to the fact that the maximum oxide ion conductivity is realized in 8-10YSZ. Standard thermodynamic functions at 298.15 K of pure zirconia and YSZ samples calculated from the heat capacity data are given in Table I, where the heat capacity below 14 K was estimated by graphical smooth extrapolation down to 0 K, and $S^o_0 = 0$ was assumed. Figure 2 shows the standard enthalpy ($H^o_{298.15}$) as a function of yttria content x. The $H^o_{298.15}$ of YSZ increases linearly with increasing yttria content, that is, the plots of 4, 8, 10 and 11YSZ are on a single line. On the other hand, the plot of pure zirconia deviates from the line of the YSZ. Clearly, the YSZ should have an excess enthalpy compared with pure

Table I. Standard thermodynamic functions at 298.15 K. entropy ($\Delta S^o_{298.15}$), enthalpy ($\Delta H^o_{298.15}$) and Gibbs energy ($\Delta G^o_{298.15}$) of yttria stabilized zirconia. $S^o_0 = 0$ and $H^o_0 = 0$ were assumed.

Y_2O_3 (mol%)	$\Delta S^o_{298.15}$ ($J \cdot K^{-1} \cdot mol^{-1}$)	$\Delta H^o_{298.15}$ ($J \cdot mol^{-1}$)	$\Delta G^o_{298.15}$ ($J \cdot mol^{-1}$)
0	49.79	8711	6133
3.95	55.11	9362	-7069
7.76	56.47	9595	-7242
9.70	57.16	9723	-7320
11.35	57.46	9805	-7327

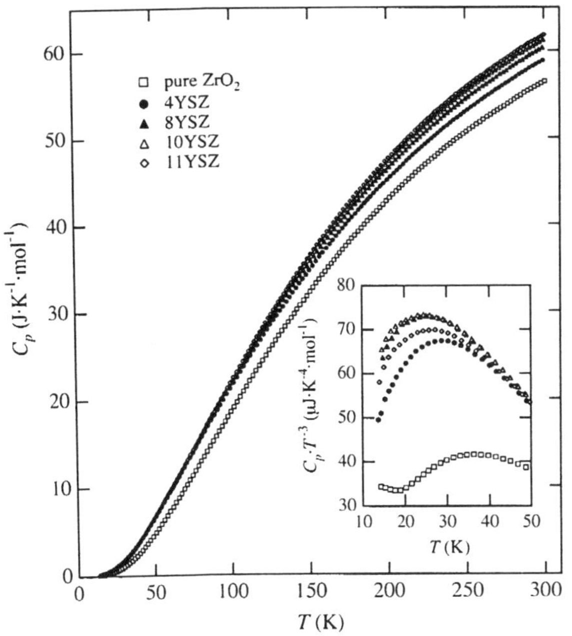

Figure 1. Molar heat capacity of pure zirconia and 4-11YSZ $\{(ZrO_2)_{1-x}(Y_2O_3)_x;\ x = 0, 0.0395, 0.0776, 0.0970, 0.1135\}$ [25,26].

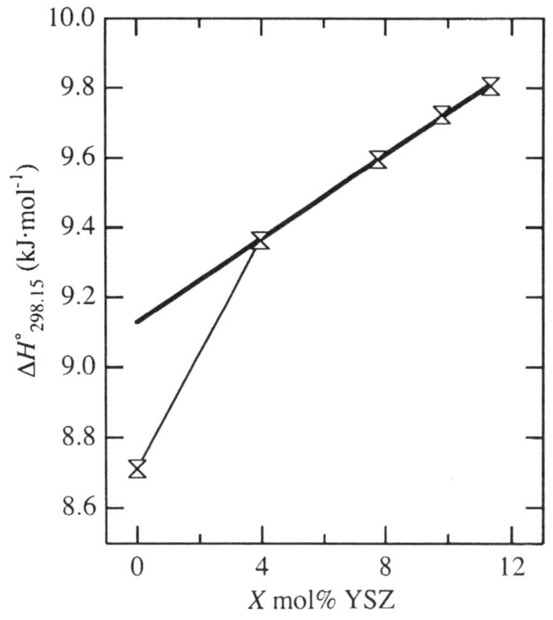

Figure 2. Standard enthalpy of the pure zirconia and 4-11YSZ $\{(ZrO_2)_{1-x}(Y_2O_3)_x;\ x = 0, 0.0395, 0.0776, 0.0970, 0.1135\}$.

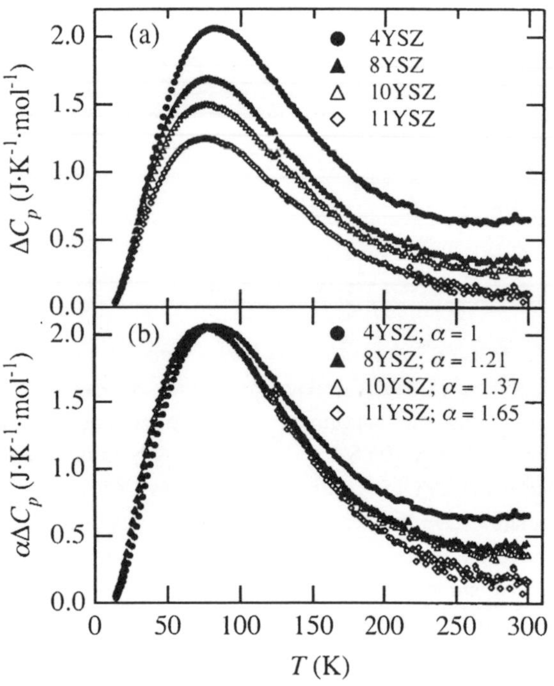

Figure 3. (a) Deviation of the heat capacity of 4-11YSZ from the assumption of the simple additivity of the heat capacities of pure zirconia and pure yttria. (b) Scaled plots of the deviations. α is a normalizing factor.

zirconia. Such excess enthalpy in YSZ should be attributed to the structural difference between the monoclinic zirconia and cubic/tetragonal YSZ because the phonon DOS strongly depends on the crystal structure. The heat capacity difference, ΔC_p, of YSZ is derived by subtracting the values of the heat capacity of pure zirconia and pure yttria from the measured heat capacity of YSZ on the basis of simple additivity rule as follows:

$$\Delta C_p = C_p^{\mathrm{YSZ}} - \{(1 - x)\cdot C_p^{\mathrm{zirconia}} + x\cdot C_p^{\mathrm{yttria}}\} \qquad (1)$$

where C_p^{YSZ}, C_p^{zirconia} and C_p^{yttria} stand for the molar heat capacities of YSZ, pure zirconia and pure yttria [29], respectively, and x is the yttria content that is the fraction ratio in YSZ. The calculated excess heat capacity ΔC_p is shown in Fig. 3(a). Apparently, the YSZ shows excess heat capacity with a broad hump at about 100 K, and the all the curves are very similar to Schottky type of anomaly. It is also found that the excess heat capacity decreases with increasing yttria content. A comparison is given in Fig. 3(b), where the curves are normalized to coincide the peak of the excess heat capacity; α is a factor for normalization. It is clearly seen that all of the curves of the excess heat capacity are very similar to each others. In Fig. 3(b), however, the curve of 4YSZ deviates from the cubic single phase samples of 8-11YSZ. The 4YSZ has higher peak temperature and larger excess heat capacity at high temperatures than the other YSZ. This should be caused by the fact that the 4YSZ is the mixture of cubic and tetragonal structure, while the others are the single phase of cubic structure. The Schottky anomaly can be represented by two Einstein functions with characteristic frequencies of ν and 2ν as follows [30,31];

$$C_{\mathrm{Schottky}}(\nu) = C_{\mathrm{E}}(\nu) - C_{\mathrm{E}}(2\nu) \qquad (2)$$

where $C_{\mathrm{E}}(\nu)$ is the Einstein heat capacity as a function of characteristic frequency. That is to say, the excess heat capacity is realized as a result of the simultaneous decrease and increase at the frequencies 2ν and ν, respectively, in the phonon DOS of zirconia. Thus the phonon softening should occur in YSZ. The peak temperature T_{p} of the Schottky heat capacity is given as; $T_{\mathrm{p}} = 0.42\ h\nu/k_{\mathrm{B}}$, where h is Planck's constant, and k_{B} is the Boltzmann's constant. Using this relation, the ν is estimated as to be 3.6 THz (15 meV) for 8-11YSZ and 4.1 THz (17 meV) for 4YSZ. The neutron inelastic scattering experiments have been performed to verify the phonon DOS of YSZ samples. The result of 10YSZ is shown in Fig. 4. The phonon DOS of

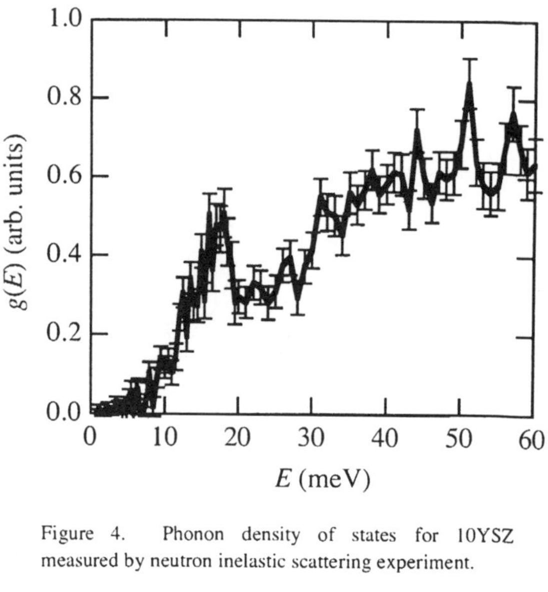

Figure 4. Phonon density of states for 10YSZ measured by neutron inelastic scattering experiment.

10YSZ exhibits a sharp band at about 16-18 meV, which corresponds to the softening mode frequency estimated from the excess heat capacity. It has been reported that the phonon dispersion of stabilized zirconia could not be defined well at higher energy region because of the rapid phonon dumping, and optical branches were not observed [22-24]. On the other hand, Liu et al. [22] have reported the phonon dispersion curves and phonon DOS of YSZ by lattice dynamics calculations. Ishigame and Yoshida [19] also performed lattice dynamics calculations of cubic stabilized zirconia and reported similar result. Considering the results of their calculations, the observed sharp band at about 16-18 meV in this study should be of a van Hove singularity [32] due to the TA phonons near the Brillouin zone boundary. Although the phonon DOS of the monoclinic zirconia has not been reported, a relationship is supposed to the soft vibration of the zone boundary TA phonons in stabilized zirconia. It should be noted that the excess heat capacity decreases with increasing yttria content without change of its temperature dependence (Fig. 3). This implies the hardening, the simultaneous decrease and increase at the frequencies ν and 2ν, respectively, in the phonon DOS of YSZ. Thus the soft phonon and the hard phonon, having a similar frequency to the inherent phonon in monoclinic zirconia, should coexist in stabilized zirconia. Strauß et al. [23] have reported that the phonon peak observed in neutron inelastic scattering splits into two peaks near the zone boundary due to the local breaking of symmetry. The two zone boundary TA phonon frequencies should correspond to ν and 2ν derived from the excess heat capacity using the Schottky model.

Apparently such alteration in the lattice vibration of stabilized zirconia correspond to changes in the local defect structure. The most important point is that the yttria-doping provides the zone boundary mode of monoclinic phase within the cubic or tetragonal frame of zirconia. A primary difference between monoclinic zirconia and cubic or tetragonal YSZ is in the coordination number of cation. In cubic and tetragonal stabilized zirconia, Zr ion has eightfold coordination of oxide ions, forming ZrO_8 polyhedron. On the other hand, in monoclinic zirconia, it has 7 coordination, which means a distorted ZrO_7 polyhedron in the crystal. The average Zr-O bond length is elongated in cubic/tetragonal YSZ compared with that in monoclinic zirconia; in the tetragonal YSZ, four Zr-O bonds are longer, the other four bonds are shorter than that of the ZrO_8 polyhedron in cubic YSZ [3]. Apparently the softening is enhanced by increasing the coordination number of zirconium cation, i.e. the formation of the ZrO_8 polyhedron, and the difference in Zr-O bond strength due to the change in Zr-O bond length is reflected in the zone boundary TA mode frequency. This may be caused by the local instability around the over-coordinated Zr ions. On the other hand, the hardening which leads to the decrease of the excess heat capacity is caused by the formation of oxygen vacancies and corresponding relaxation in anion sublattice. The decrease in average coordination number of Zr ion and significant displacement of the anions provide the resemblance of the structural environment of Zr ion to that in monoclinic zirconia. Similar conclusion has been given by Catlow et al. [33] from the EXAFS study.

The frequency of the TA phonons near zone boundary is decreased by cubic/tetragonal phase stabilization with yttria doping. On the other hand the softening is partially suppressed by the formation of oxygen vacancy. Such mechanisms compete and give a maximum in a physical property of stabilized zirconia at the phase boundary composition. The high temperature ionic conductivity of stabilized zirconia depends on the amount of doping, and exhibits a maximum at the phase boundary composition; 8-10 mol% in the case of YSZ. If the significant distortion of the anion sublattice around the vacancy is to reduce the local instability in the cubic/tetragonal form of zirconia at low temperatures, some change similar to the monoclinic-to-tetragonal phase transition in anion sublattice should be expected at elevated temperatures. It has been reported that the activation energy for ionic conductivity in stabilized zirconia decreases with increasing temperature and the change is most significant around 1000-1200 K; just a little below the monoclinic-to-tetragonal phase transition temperature of pure zirconia. This has been attributed to the break up of the vacancy clusters. Although it may be caused by a partial phase transition of anion sublattice. The diffuseness of such phase transition should be owed to the variation of local defect structures and the phase transition temperature is lowered by the stabilization of cation sublattice. Further theoretical and experimental studies are needed to clarify the mechanism.

4. Conclusion

Cubic/tetragonal stabilized zirconia doped with yttria has an excess heat capacity compared to pure monoclinic zirconia, which should be attributed to the softening of the phonons near the Brillouin zone boundary. The soft modes should be inherent vibrations of ZrO_8 polyhedron in tetragonal and cubic stabilized zirconia. The frequency of the soft phonon decreases with increasing the yttria content up to the point where the cubic structure of the cation sublattice is fully stabilized. On the other hand, the formation of oxygen vacancies and the subsequent relaxation of anion sublattice significantly reduce the number of the soft phonons and the excess heat capacity, which implies the enhancement of the local stability in stabilized zirconia by the oxygen vacancy.

References

1. J. D. McCullough and K. N. Trueblood, Acta Crystallogr. **12**, 507 (1959).
2. D. K. Smith and H. W. Newkirk, Acta Crystallogr. **18**, 983 (1965).
3. C. J. Howard, R.J. Hill and B. E. Reichert, Acta Crystallogr. Sect. B **44**, 116 (1988).
4. R. J. Ackermann, S. P. Garg and E. G. Rauh, J. Am. Ceram. Soc. **60**, 341 (1977).
5. G. Teufer, Acta Crystallogr. **15**, 1187 (1962).
6. R. E. W. Casselto, Phys. Status Solidi A **2**, 571 (1970).
7. H. Schmalzried, Z. Phys. Chem. Neue Folge **105**, 47 (1977).
8. J. P. Goff, W. Hayes, S. Hull, M. T. Hutchings and K. N. Clausen, Phys. Rev. B **59**, 14202 (1999).
9. D. Steele and B. E. F. Fender, J. Phys. C: Solid State Phys. **7**, 1 (1974).
10. J. Faber, Jr., M. H. Mueller and B. R. Cooper, Phys. Rev. B **17**, 4884 (1978).
11. M. Morinaga, J. B. Cohen and J. Faber, Jr., Acta Crystallogr. Sect. A **35**, 789 (1979).
12. M. Morinaga, J. B. Cohen and J. Faber, Jr., Acta Crystallogr. Sect. A **36**, 520 (1980).
13. H. Horiuchi, A. J. Schults, P. C. Leung and J. M. Williams, Acta Crystallogr. Sect. B **40**, 367 (1984).
14. C. J. Howard and R. J. Hill, J. Mater. Sci. **26**, 127 (1991).
15. K. J McClellan, S. -Q. Xiao, K. P. D. Lagerlof and A. H. Heuer, Philos. Mag. A **70**, 185 (1994).
16. M. Yashima, S. Sasaki, M. Kakihana, Y. Yamaguchi, H. Arashi and M. Yoshimura, Acta Crystallogr. Sect. B **50**, 663 (1994).
17. D. N. Argyriou, M. M. Elcombe and A. C. Larson, J. Phys. Chem. Solids **57**, 183 (1996).
18. A. Feinberg and C. H. Perry, J. Phys. Chem. Solids, **42**, 513 (1981).
19. M. Ishigame and E. Yoshida, Solid State Ionics **23**, 211 (1987).
20. M. Yashima, K. Ohtake, M. Kakihana, H. Arashi and M. Yoshimura, J. Phys. Chem. Solids **57**, 17 (1996).
21. D. W. Liu, C. H. Perry and R. P. Ingel, J. Appl. Phys. **64**, 1413 (1988).
22. D. W. Liu, C. H. Perry, A. A. Feinberg and R. Currat, Phys. Rev. B **36**, 9212 (1987).
23. B. Strauß, H. Boysen, F. Frey, U. Steigenberger, F. Güthoff, A. Krimmel, H.M. Mayer and D. Welz, J. Phys.: Condens. Matter **7**, 7823 (1995).
24. D. N. Argyriou and M. M. Elcombe, J. Phys. Chem. Solids **57**, 343 (1996).
25. T. Tojo, T. Atake, T. Mori, and H. Yamamura, J. Chem. Thermodyn. **31**, 831 (1999).
26. T. Tojo, T. Atake, T. Mori and H. Yamamura, J. Therm. Anal. Calorimetry **54**, 447 (1999).
27. T. Atake, H. Kawaji, A. Hamano and Y. Saito, Rep. Res. Lab. Eng. Mater., Tokyo Inst. Tech. **15**, 13 (1990).
28. K. Inoue, T. Kanaya, Y. Kiyanagi, K. Shibata, K. Kaji, S. Ikeda, H. Iwasa and Y. Izumi, Nucl. Instrum. Methods **A327**, 433 (1993).
29 K. S. Gavrichev, V. E. Gorbunov, L. N. Golushina, G. E. Nikiforova, G. A. Totrova and I. S. Shaplygin, Russ. J. Phys. Chem. **67**, 1554 (1993).
30. E. F. Westrum, Jr., J. Chem. Thermodyn. **15**, 305 (1983).
31. E. S. R. Gopal, "Specific Heats at Low Temperatures", (Plenum Press, New York, 1962) 158.
32. L. Van Hove, Phys. Rev. **89** (1953) 1189.
33. C. R. A. Catlow, A. V. Chadwick, G. N. Greaves and L. M. Moroney, J. Am. Ceram. Soc. **69**, 272 (1986).

COMPUTER SIMULATION STUDY OF ANOMALOUS DIFFUSION IN β-AGI

KAZUHIRO TAKAHASHI
Kobe Women's University Seto Junior College, Okayama 709-0863, Japan

TADAO ISHII
Faculty of Engineering, Okayama University, Okayama 700-8530, Japan

The dynamics of Ag^+ ions in β-AgI is studied at high temperatures by means of molecular dynamics simulation. It is observed that the mean square displacement has three different time-regions. Especially a hopping behavior of Ag^+ ion in each time region is investigated.

1. Introduction

Diffusion is one of the fundamental physical properties in ion transport, and especially, anomalous diffusion is of recent interest [1-3]. Ishii investigated the hopping diffusion of an ideal (noninteracting) lattice gas, on the basis of the relaxation mode theory [3,4], in one dimensional random lattice, and proposed that the anomalous diffusion in the intermediate time domain exists and originates in the localized and extended nondiffusive modes [4].

In this report, we study the mean square displacement (MSD) of Ag^+ ions in β−AgI using molecular dynamics simulation (MD), and investigate whether the anomalous hopping diffusion exists.

2. Model and Numerical Simulation

The model system of β-AgI consists of the soft-core potential. The interactions V_{ij} between ions i and j at distance r are given by

$$V_{ij}(r) = \varepsilon \frac{(\sigma_i + \sigma_j)^n}{r^n} + \frac{Z_i Z_j (fe)^2}{r} \tag{1}$$

where σ_i and Z_i are the effective core radius and the valence of ith ion, respectively, and f is the ionicity. In the present calculation, the parameters are taken as $\varepsilon = 0.085\,\text{eV}$, $\sigma_{\text{Ag}} = 0.63$ Å, $\sigma_{\text{I}} = 2.2$ Å, $f = 0.6$ and $n = 7$, which are the same values as a set of parameters used in the previous work for α - AgI [5]. The calculations are performed on the 288-ions system with the pressure constant (NPT) algorithm and time step of 2.0×10^{-15} s (2fs). The defect concentration is averaged over 4.8×10^4 time steps after equilibration. In case of the MSD calculation, the average is taken over more than 4×10^6 time steps. The simulation is done based on the program MXDORTO developed by K.Kawamura.

3. Results and Discussion

3.1. *Defects*

The primitive unit cell of β-AgI has three octahedral interstitial sites and one tetrahedral interstitial site. The

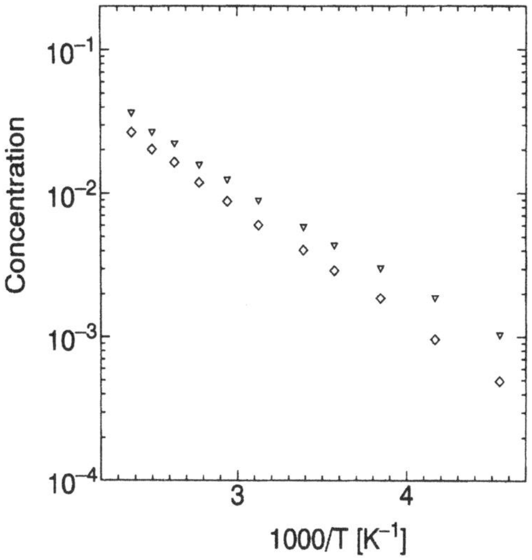

Fig.1 Temperature dependence of defect concentration: tetrahedral (▽) and octahedral (◇) defects.

defect concentration is obtained at different temperatures between 220K and 420K, and is shown in Fig.1. The concentration of tetrahedral defect (▽) is larger than that of octahedral one (◇) as investigated by Zimmer et al. [6]. As the temperature T is raised, the defect concentration c increases where $\log(c)$ approximately varies linearly with $1/T$.

3.2. *Mean square displacement*

The MSD of a particle of species α is given by

$$<\mathbf{r}_\alpha^2(t)>=\left\langle\left|\mathbf{r}_i^\alpha(t+t_0)-\mathbf{r}_i^\alpha(t_0)\right|^2\right\rangle \qquad (2)$$

where < > denotes an average over the positions r_i^α of all particles $\{i\}$ of α and over the starting times t_0.

Figure 2 gives a result of the time dependence of MSD at temperatures T=400, 410 and 420K. It can be seen that I^- ions are not diffusive while Ag^+ ions become diffusive. The MSD of Ag^+ ions shows a characteristic feature with the ballistic motion following $<\mathbf{r}_{Ag}^2(t)>\propto t^2$ in the shortest time region, with a plateau regime at intermediate time region, and with diffusive motion at longer times.

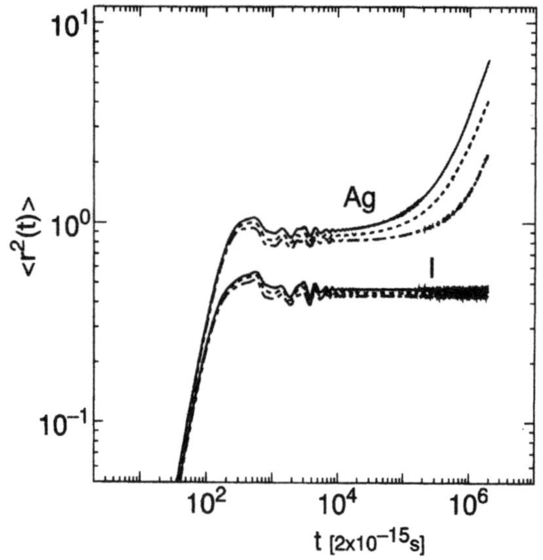

Fig.2 Mean square displacement as a function of time at T=400 (dashed-dotted), 410 (dashed) and 420K (solid).

It is unknown what sorts of motions contribute to the behaviors in respective time regions of an Ag^+ ion. In order to separate a purely hopping motion from the above curves for mobile Ag^+ ions, we define the coordinate of the ion $r(t)$ by the

$$\mathbf{r}(t)=\mathbf{R}(t)+\mathbf{u}(t) \qquad (3)$$

where $u(t)$ is the local displacement within a sitting cell, and $R(t)$ is the averaged position in the cell such that $<r(t)>=R(t)$. Then, the mean square displacement for $R(t)$ defines a hopping diffusion of Ag^+ ion. The hopping MSD $<R_{Ag}^2(t)>$ calculated at temperature 410K is shown in Fig.3, together with $<r_{Ag}^2(t)>$ and the difference of these quantities $<r_{Ag}^2(t)>$ - $<R_{Ag}^2(t)>$ being equal to $<u^2(t)>$. As is naturally expected, the time dependence of $<r_{Ag}^2(t)>$ - $<R_{Ag}^2(t)>$ is constant at intermediate and longest time region, which is quite characteristic of lattice vibrations as in the MSD of I^- ion $<r_I^{\,2}(t)>$ in Fig.2: it should be notable that both behaviors are much the same.

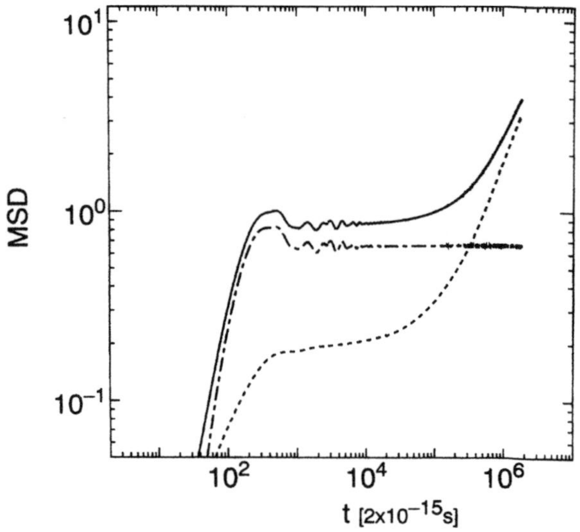

Fig.3 Mean square displacement as a function of time at T=410K: $<r_{Ag}^2(t)>$ (solid), $<R_{Ag}^2(t)>$ (dashed) and $<r_{Ag}^2(t)>$ - $<R_{Ag}^2(t)>$ (dashed-dotted).

Now let us investigate a characteristic feature of the MSD of hopping motion. As has been found in experiments [1,2] and theory [3], we may put the time dependence of MSD in terms of an exponent k as

$$<\mathbf{R}_{Ag}{}^2(t)>\sim t^k \qquad (4)$$

It is remarkable to see from Fig.4 that $<R_{Ag}{}^2(t)>$ shows a linear dependence on time at shortest times and then $k=1$, while it has an approximate exponent of $k\sim 0.06$ at plateau regime as theoretically predicted [3]. Further in the longer time region, the slope of the MSD increases with time to approach to $k=1$ again.

This indicates that the hopping mean square displacement is at least composed of two components, nondiffusive and diffusive components [3,7]. In a one dimensional random lattice system with a uniform distribution of activation energies, Ishii investigated the above two types of diffusions, due to nondiffusive and diffusive. In the former nondiffusive contribution, there exist two further components, which are due to localized and extended modes. However, it is quite difficult at present to discriminate into two separately. Thus we analyze the obtained MSD result by means of the diffusive and nondiffusive components, on

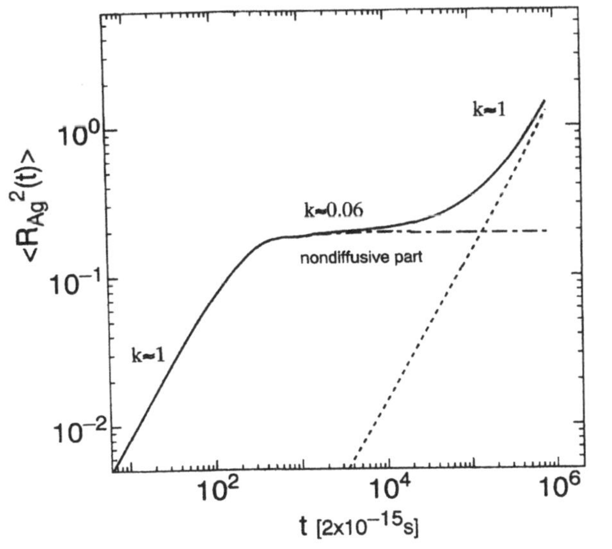

Fig.4 Mean square displacement as a function of time at T=410K (solid), nondiffusive (dashed-dotted) and diffusive part (dashed).

condition that the latter component saturates at a constant value at longer times. Figure 4 shows the result where there surely exist the contributions of nondiffuisve (dashed-dotted) with $k = 1 \rightarrow 0.06$ and diffusive (dashed) modes with $k = 1$.

Now, we examine a real microscopic behavior of hopping motion of Ag^+ ions. In the first place, we present a typical example of hopping motion of Ag^+ ion at T=410K. In Fig5(a), a coordinate of an Ag^+ ion, $\mathbf{R}_{Ag}(t) = (X_{Ag}(t), Y_{Ag}(t), Z_{Ag}(t))$ is plotted against the time. There appears a localized hopping among a regular and interstitial site, and a jump over a lattice spacing.

The localized hopping is shown in Fig.5(b) for $Z_{Ag}(t)$ between the time range 1×10^6 and 1.12×10^6.The Ag^+ ion sits in the regular site (R) for a long time and occasionally visits tetrahedral (T) or octahedral interstitials (O).

In order to see the mechanisms of diffusions in a real space, we define an average number of ions at distance R=$|R|$ reaching at time t by $n(R,t)$: $n(R,t)$ is the number of Ag^+ ions existing in the interval $(R\text{-}\ \delta\ R/2,\ R\text{+}\ \delta\ R/2)$, which is normalized by the total number of samplings. We calculate it with $\delta\ R$=0.04Å and show the result in Figs.6(a)-(c). When no transition has occurred within the time t, the event is not counted nor shown in the figures. Figure 6(a) shows the $n(R,t)$ at t=20, 40 and 60 in the shorter time region. The three

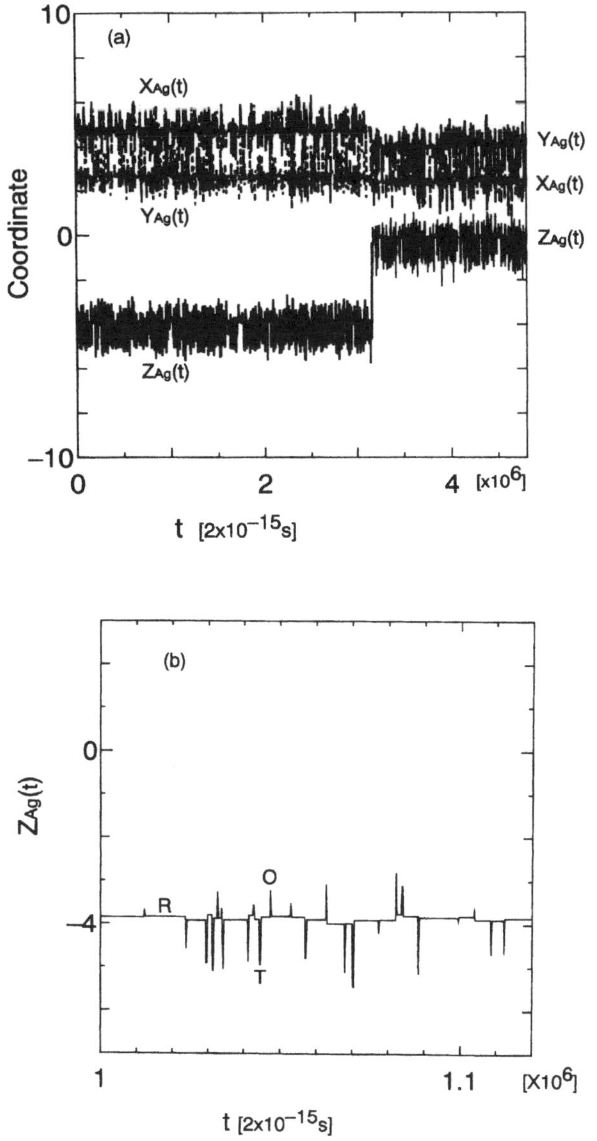

Fig.5(a). Coordinates of an Ag^+ ion plotted against the time at T=410K. (b) $Z_{Ag}(t)$ coordinate of the ion in an enlarged time scale: sitting in the regular site(R), tetrahedral interstitial (T) and octahedral interstitial sites (O).

curves essentially display the same peak at $R \sim 1.1$ Å which is the distance between a regular site and one of its nearest neighbor interstitials. The magnitude of the peak increases with t. In Fig.6 (b), the distributions at three different times t=1000, 4000 and 10000 are shown. A peak appears conspicuously near R=0 and the magnitude of the peak increases with t. This means that Ag^+ ions return to their initial sites. In Fig.6 (c), the distributions at t=1x10^5, 4x10^5 and 1x10^6 are shown. A large peak exists at $R \sim 4.6$ Å, being the distance of a lattice spacing. The magnitude of the peak also increases with t.

By the present investigation for Ag^+ ions, we can see that the localized hopping contributes to the nondiffusive part and the no-return jump over lattice spacing contributes to the diffusive part.

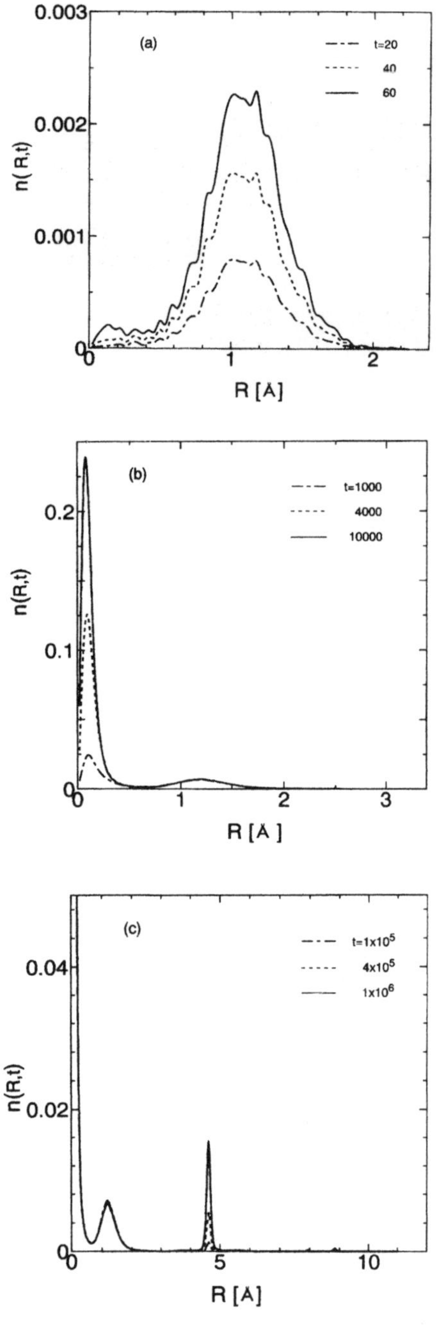

Fig.6(a)-(c). Distribution $n(R,t)$ of Ag^+ ion as a function of displacement R: (a) at t=20 (dashed-dotted), 40(dashed) and 60(solid) in the short time region, (b) at t=1000 (dashed-dotted), 4000(dashed) and 10000(solid) in the intermediate time region, and (c) at t=1x10^5 (dashed-dotted), 4x10^5(dashed) and 1x10^6 (solid) in the long time region.

4. Summary

We have investigated the dynamics of Ag^+ ion in β-AgI at high temperatures by means of MD simulation. It is confirmed that two kinds of defects are created at tetrahedral and octahedral sites. The defect concentration of tetrahedral site is larger than that of octahedral one as obtained by Zimmer et al.[6].

From the MSD study at temperatures T=400, 410 and 420K, it is surely recognized that the anomalous hopping diffusion exists. Especially the hopping MSD shows three different regimes including a plateau region, originated in the nondiffusive modes at earliest and intermediate time regions, and diffusive modes at longest time region.

References

1. W. Smith, G.N.Greaves and M.J.Gillan, cited in, K.L.Ngai, H.Jain and D.Gupta (Eds.), *Diffusion in Amorphous Materials*, Proc. Int. Symp., 1993, p17.
2. D. Knödler, P. Pendzig and W. Dieterich, *Solid State Ionics.* **86-88**, 29 (1996).
3. T. Ishii, *Solid State Commun.* **116**, 327 (2000).
4. T. Ishii, *Prog. Theor. Phys.* **77**, 1364 (1987).
5. A. Fukumoto, A. Ueda and Y. Hiwatari, *J. Phys. Soc. Jpn.* **51**, 3966 (1982).
6. F. Zimmer, P. Ballone, J. Maier and M. Parrinello, *J. Chem. Phys.* **112**, 641 (2000).
7. T. Ishii and T. Abe, *J. Phys. Soc. Jpn.* **69**, 2549 (2000).

NMR STUDY ON Li^+ IONIC DIFFUSION IN $Li_xV_2O_5$ PREPARED BY SOLID-STATE REACTION

K. NAKAMURA† AND T. KANASHIRO

Department of Physics, Faculty of Engineering, The University of Tokushima, 2-1 Minami-Josanjima-Cho, Tokushima 770-8506, Japan

† E-mail: koichi@pm.tokushima-u.ac.jp

M. VIJAYAKUMAR AND S. SELVASEKARAPANDIAN

Solid State and Radiation Physics Laboratory, Department of Physics, Bharathiar University, Coimbatore-641 046, Tamilnadu, India

^{7}Li -NMR measurements of $Li_xV_2O_5$ prepared by solid-state reaction were performed. The line width narrowing related to the Li^+ ionic diffusion was observed in the static NMR spectrum. Both the activation energy and the onset temperature of the narrowing are lower in the β-phase than in the other phases. A steep temperature dependence of line width was observed in the sample with Li content above $x = 0.6$. It produces the higher activation energy in the ε-phase over a wide range of Li content at high temperature. ^{7}Li MAS-NMR spectra show that the samples are in the multi-phase over the entire range of $0.4 < x < 1.4$. The Li^+ ionic diffusion is discussed in relation to a complicated change in the β-, δ-, ε-phase.

1 Introduction

Lithium intercalated $Li_xV_2O_5$ is attractive as cathode material for 3 V lithium rechargeable batteries. It is known that a host, V_2O_5 can intercalate reversibly a maximum of 2 Li^+ ions per V_2O_5. The variety in the valence of vanadium and the insertion of large amounts of Li^+ ion causes a complicated phase transition.[1,2]

The host oxide V_2O_5 has an orthorhombic structure and forms the VO_5 layers parallel to (001), consisting of edge- and corner-sharing VO_5 square pyramids. The bronzes α-$A_xV_2O_5$ (A=Li, Na, Ag, K, Cu; $0 \leq x \leq 0.10$) are isostructural. The chemical Li insertion in V_2O_5 gives rise to several structural modifications: α for $0 < x < 0.13$, ε for $0.32 < x < 0.88$, δ for $0.88 < x < 1.0$, and γ-phase for $x > 1.0$.[3] $Li_xV_2O_5$ prepared at high temperature by solid-state reaction shows the β/β'-phase in the range of $0.2 < x < 0.6$.[4,5] The composition limits in the solution are not firm and the neighboring phases are usually coexistent. $Li_xV_2O_5$ exhibits the almost same framework with only a slight puckering of V_2O_5 up to $x = 1.0$ and a good reversibility. However, beyond $x = 1.0$ the slightly modified V_2O_5 structure is not maintained and it is replaced by the γ-phase with remarkably distorted structure. Such phase transition is closely related to the Li^+ ionic motion. Intensive NMR studies[6–10] have been performed for several decades. They revealed the crystalline material maintains several structural phases depending on Li content as mentioned before. Although the solid-state reaction method is often used for sample synthesis, heating/melting process at high temperature easily causes the oxygen loss. It results in a non-stoichiometry in oxide and yields the multi-phase. In this study, the static-NMR and the MAS-NMR measurements were performed in order to elucidate the Li^+ ionic motion in the multi-phase and structural changes in $Li_xV_2O_5$. The temperature dependence of the line width and MAS-NMR spectra depending on Li content are discussed below.

2 Experiments

The samples of $Li_xV_2O_5$ (x = 0.4, 0.6, 0.8, 1.9, 1.2, and 1.4) have been prepared by the solid-state reaction method. Li_2CO_3 and V_2O_5 were mixed in appropriate molar ratio and ground into fine powders using mortar and pestle, and melted in a porcelain crucible and the melt was cooled slowly.[11]

The broadband ^{7}Li-NMR spectrum(abbreviated hereafter static NMR spectrum) was measured by use of a laboratory-made conventional pulse NMR spectrometer over the temperature range from 77 to around 600 K at 10.08 MHz. Each powdered polycrystalline sample was packed in 9 mm Pyrex tube. To confirm an existence of the multi-phase in sam-

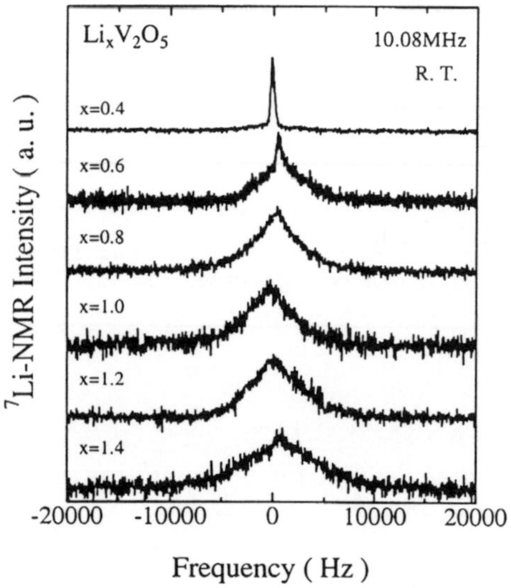

Figure 1. Static ^{7}Li-NMR spectrum of $Li_xV_2O_5$ at room temperature.

ples, ^{7}Li MAS(Magic Angle Spinning)-NMR measurement was performed by use of Bruker Avance 300 spectrometer with 4 mm probe. The spinning speed was 7 kHz. The MAS spectra at room temperature were measured at 116.672 MHz.

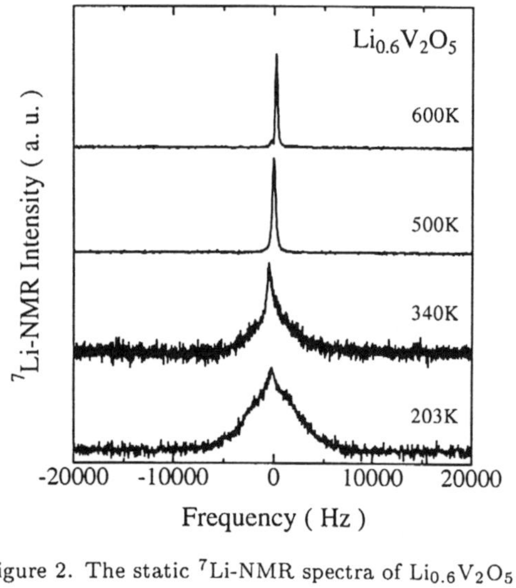

Figure 2. The static ^{7}Li-NMR spectra of $Li_{0.6}V_2O_5$.

3 Results

3.1 Static NMR Spectrum

A central transition line of ^{7}Li nucleus ($I = 3/2$) alone is observed in $Li_xV_2O_5$ resulting from the quadrupole broadening. Fig. 1 shows the static ^{7}Li-NMR spectra of $Li_xV_2O_5$ at room temperature. A broad central line observed for Li rich samples remarkably becomes narrow with the decrease of the Li content x, and a sharp line width is about 480Hz for $x = 0.4$.

The static NMR spectra show a considerable change in their line shapes with temperature. Fig. 2, 3, and 4 illustrate the static spectra for $x = 0.6$, 1.0, and 1.4 at several temperatures, respectively. The broad lines at low temperature change into sufficiently narrowed one at the highest temperature. The remarkable narrowing in the static spectra indicates the motional narrowing originating from the Li^+ ionic motion.

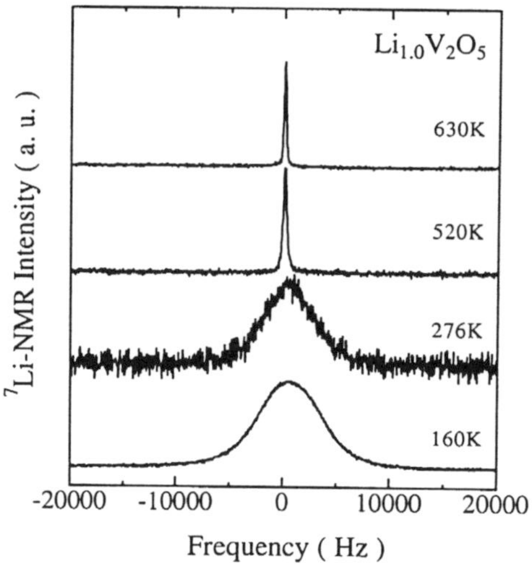

Figure 3. The static ^{7}Li-NMR spectra of $Li_{1.0}V_2O_5$.

3.2 MAS-NMR spectrum

No fine structures were observed in the static spectra. However, the solid-state high resolution NMR spectra reveal an existence of the multi-phases, because the environment surrounding Li nucleus is reflected in the NMR spectra. In facts, some information has been reported in previous XRD[12] and NMR[7,9] studies. In Fig. 5, the ^{7}Li MAS-NMR spectra of our samples show the changes in the line shape at room

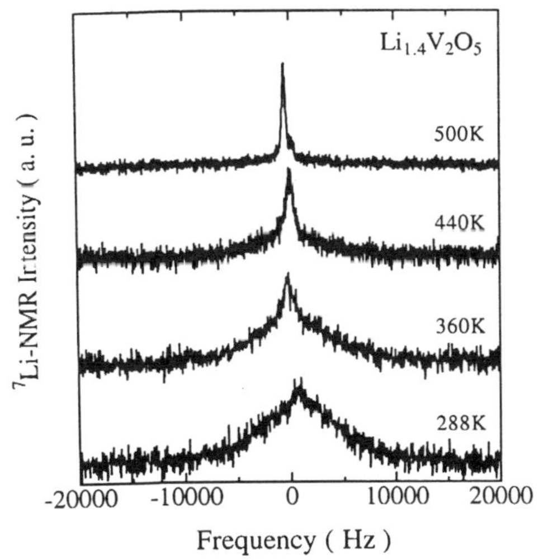

Figure 4. The static ^{7}Li-NMR spectra of $Li_{1.4}V_2O_5$.

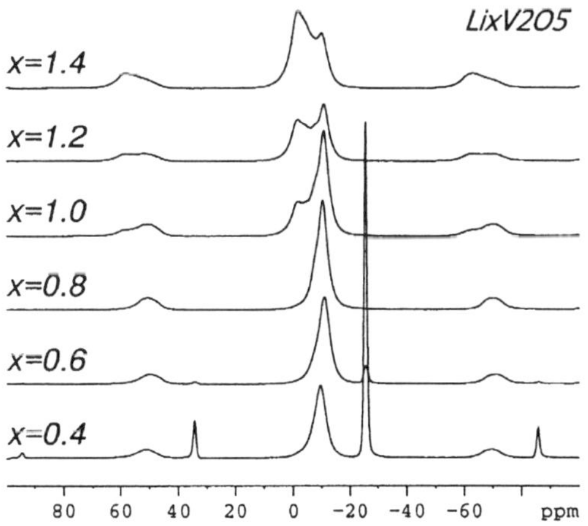

Figure 5. ^{7}Li MAS-NMR spectra for x = 0.4, 0.6, 0.8, 1.0, 1.2, and 1.4 of $Li_xV_2O_5$ at room temperature.

temperature. Two central lines with the spinning sidebands were observed around −9 and −25 ppm in the spectrum for $x = 0.4$ and 0.6. In the sample of $x = 0.6$, the peak around −25 ppm remarkably decreases and the peak around −9 ppm is dominant. The samples above $x = 0.8$ show asymmetric line shapes. A new peak appears on the left shoulder (around −1 ppm) of the central peak of $x = 1.0$ and grows up with the increase of x, becoming dominant in $x = 1.4$. The complicated spectra in the Li rich samples result in the line broadening because of superposition of coexisting phases.

A steep peak around −25 ppm for $x = 0.4$ would be ascribed to the β-phase as seen in Fig. 5, because the β-phase is reported to be a major one in the samples of $x = 0.4$ and 0.6 prepared at high temperature. It is sufficiently narrowed at room temperature and consistent with the narrowed width observed in the static spectra of $x = 0.4$. The β-phase is absolutely dominant in $x = 0.4$ and the other phases would contribute to a tail of the spectrum. The peak around −9 ppm observed in samples of $x = 0.4$ and 0.6 would be assigned to the ε-phase. The sample of $x = 0.6$ consists mainly of the ε-phase, because it is much larger than the peak from the β-phase around −25 ppm. A single spectrum for $x = 0.8$ would be mainly attributed to the δ-phase, because results of XRD[13] show an existence of the δ- and a small amount of the ε-phase. However, the difference between the ε- and δ-phase is elusive in these spectra. The new peak growing up above $x = 1.0$ would be assigned to the γ-phase because it appears above $x = 1.0$.[1] The result is consistent with a fact that the γ-phase is coexistent with the δ- and a small amount of ε-phase in $x = 1.0$.[12,13] The shoulder line for the samples above $x = 1.0$ corresponds to the line with larger chemical shift reported by Stallworth *et al.*[10] However, these shift are not consistent with our results. Actually, the varieties in the MAS-NMR data[7,9,10] for each phase are reported. It could be attributed to the fact that the phase content strongly depends upon fabrication procedure: chemical intercalation, electrochemical one, or high temperature reaction.

3.3 *The line width of $Li_xV_2O_5$*

Simple line patterns were observed in the MAS spectra below $x = 0.8$. We could reproduce the line shape with a single gaussian or lorentzian. Figs. 6(a), (b), and (c) show the decomposition of the static spectra for x = 0.4, 0.6, and 0.8, respectively. Upper and lower figures show the high and low temperature spectra, respectively. The spectrum for each sample shows a single lorentzian at the highest temperature, while each shows a broad gaussian spectrum at lower temperature. Especially the broad spectra can be represented by the superposition of a broad (I) and narrow (II) gaussian as seen in the low temperature spectum for $x = 0.6$. The line shapes for $x = 0.4$ and 0.8 are represented by a single gaussian or lorentzian over the entire temperature region. The line width in the samples above $x = 1.0$ is solely evaluated from

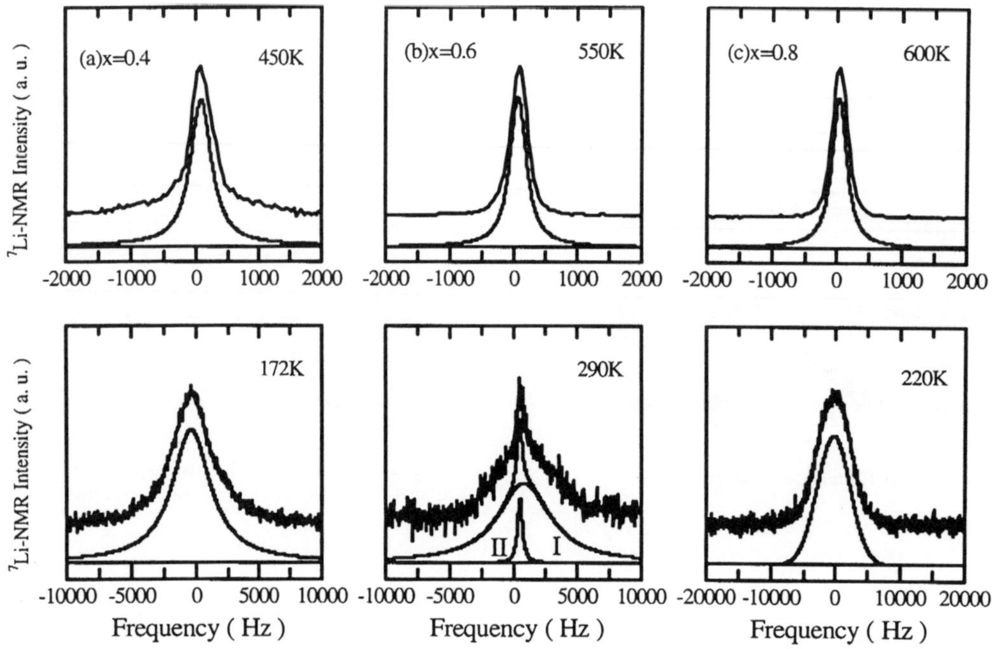

Figure 6. Typical spectra for $x = 0.4$(a), 0.6(b), and 0.8(c) in $Li_xV_2O_5$. Upper and lower are the observed spectra and calculated one, respectively. Solid lines are gaussian(or lorentzian) and superposed spectrum. The line width for $x = 0.6$ is decomposed into the broad and narrow line denoted by I and II.

the full width at half maximum of the spectrum, because the asymmetric line shape is complicated by the quadrupole splitting for $x = 1.2$ and 1.4. Fig. 7 shows the temperature dependence of the line width for the samples of $x = 0.4$ to 1.4. The line width of $x = 0.4$ is equal to the narrow line-II of $x = 0.6$. It means that the line-II would be ascribed to the β-phase. It would be consistent with the result of the MAS spectra. On the other hand, the line-I of $x = 0.6$ is in good agreement with the line width for $x = 0.8$ and 1.0. The line-I of $x = 0.6$ would correspond to the ε-phase. Almost same line width was observed for the samples with Li content above $x = 1.2$.

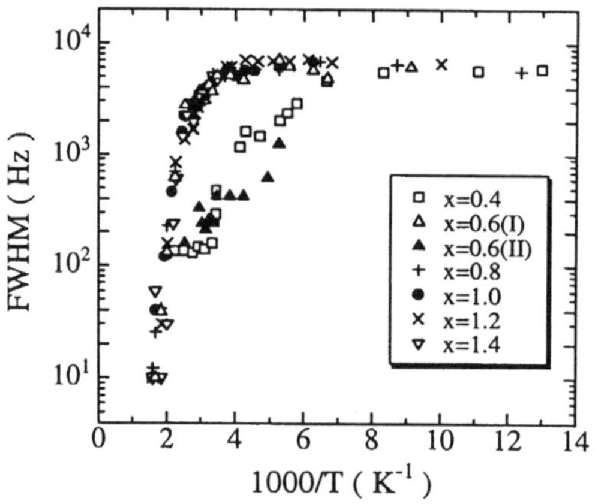

Figure 7. Arrhenius diagram representing the temperature dependence of the line width for $x = 0.4$, 0.6, 0.8, 1.0, 1.2 and 1.4 in $Li_xV_2O_5$

4 Discussion

The abrupt decrease in the temperature dependent line width for $Li_xV_2O_5$ would be attributed to the Li^+ ionic motion. According to a previous work[14], we assume that the motionally narrowed line width originated from diffusing Li nuclei obeys the following expression:

$$\Delta W(T) = A \exp\left(\frac{E}{k_B T}\right) + \Delta W_0$$

where E is the activation energy for the diffusion of Li^+ ions, A is the prefactor and ΔW_0 is the temperature independent line width. The first term expresses the line width due to the dipole-dipole coupling. The second term contains the other contributions including the external magnetic inhomogeneity and the quadrupole interaction. The linear variation of $\ln(\Delta W)$ vs. T^{-1} in the high temperature region allows the determination of the activation energy.

The activation energies for several phases in previous works[6,7,9] were evaluated from the data obtained below 400 K. As seen in Fig.7, however, the steep slopes above 400 K produce the larger activation energies than the previous values. The activation energies for $x = 0.4$ and the line-II of $x = 0.6$ are estimated to be 0.06 to 0.08 eV. The values are consistent with that of 1.55 kcal/mol (0.067 eV) for the β-phase in a previous work.[6] The activation energies for $x = 0.8$ and the line-I of $x = 0.6$ are evaluated to be 0.50 eV above 400 K and consistent with that of $x = 1.0$. The larger activation energy for $x = 0.8$ of the δ-phase at the high temperature region would be connected with the phase transition; the $\delta \leftrightarrow \varepsilon$-phase around 110 to 130 °C.[15] Because of this, the ε-phase is stable over wide range of $x \approx 0.3$ to 1.0 above 400K. Therefore, it means that the larger activation energies of 0.4 to 0.5 eV are ascribed to the ε-phase at high temperature.

The activation energies for the samples of $x =$ 1.2 and 1.4 above 400 K show almost same value of about 0.53 eV, which would be regarded as the averaged value over the typical phase: the δ, ε and γ-phases. On the other hand, the activation energies below 400 K are evaluated to be 0.08 to 0.10 eV. Their values are consistent with the reported value of 0.08 eV for the γ-phase.[7]

Li^+ ions migrate along the b-axis in the tunnels consisting of VO_5 pyramids in V_2O_5. There are many unoccupied sites in the samples with lower Li content. Li^+ ions can easily migrate through these sites. It leads to the relatively low activation energies in the low temperature region. The large number of Li insertion in V_2O_5 cause the increase of puckering of $[VO_5]_\infty$ layers. It leads to the phase transition from the β- to ε-, δ- and γ-phase. Nevertheless, the change in the activation energies is small in $Li_xV_2O_5$, because a diffusion path would be kept after undergoing the phase transition. Indeed, even in the γ-phase, the LiO_6 octahedron parallel to the b-axis forming by remarkable puckering of the VO_5 layers would provide a suitable environment for Li^+ ionic diffusion.[16]

In this system, the intrinsic activation energy should be defined above 400 K rather than at low temperature. The very low activation energy at low temperature seems to be unrealistic for the Li diffusion. The temperature range from room temperature to 400 K seem to be crossover region for several phases. The activation energies above 400 K are close to values from conductivity measurements.[17] The diffusion mechanism reflecting the large activation energy has not been understood yet at present.

5 Summary

The Li intercalated V_2O_5 prepared by solid-state reaction indicates the multi-phase region over wide Li content. In the β-phase, Li^+ ions can diffuse sufficiently at low temperature with the low activation energy. It could be distinct from the fact that the activation energies are 0.4 to 0.5 eV in the ε-phase at high temperature.

References

1. J. Galy, *J. Solid State Chem.* **100**, 229 (1992).
2. C. Delmas, H. Cognac-Auradou, J. M. Cocciantelli, M. Ménétrier, J. P. Doumerc, *Solid State Ionics* **69**, 257 (1994).
3. B. Pecquenard, D. Gourier, N. Baffier, *Solid State Ionics* **78**, 287 (1995).
4. A. D. Wadsley, *Acta Cryst.* **8**, 695 (1955).
5. J. Galy, J. Darriet, P. Hagenmuller, *Rev. Chim. Mine.* **8**, 509 (1971).
6. J. Gendell, R. M. Cotts, and M. J. Sienko, *J. Chem. Phys.* **37**, 220 (1962).
7. J. M. Cocciantelli, K. S. Suh, J. Senegas, J. P. Doumerc, J. L. Soubeyroux, M. Pouchard and P. Hagenmuller, *J. Phys. Chem. Solids* **53**, 51 (1992).
8. J. Hirschinger, T. Monglet, C. Marichal, P. Granger, J-M Savariault, E. Déramond, J. Galy, *J. Phys. Chem.* **97**, 10301 (1993).
9. J. M. Cocciantelli, K. S. Suh, J. Senegas, J. P. Doumerc, M. Pouchard, *J. Phys. Chem. Solids* **53**, 57 (1992).
10. P. E. Stallworth, F. S. Johnson, S. G. Greembaum, S. Passerini, J. Flowers, W. Smyrl, J. J. Fontanella, *Solid State Ionics* **146**, 3 (2002).
11. M. Vijayakumar, S. Selvasekarapandian, *Solid State Ionics* **148**, 329 (2002).
12. M. Vijayakumar, S. Selvasekarapandian, R. Kesavamoorthy, K. Nakamura, T. Kanashiro,

Mater. Lett. **57**, 3618 (2003).
13. M. Vijayakumar, Doctoral thesis, Bharathiar Univ., India , (2003).
14. T. G. Stoebe, R. D. Gulliver, T. O. Ogurtani, R. A. Huggins, *Acta. Mettallurgica* **13**, 701 (1965).
15. J. Galy, C. Satto, P. Sciau, and P. Millct, *J. Solid State Chem.* **146**, 129 (1999).
16. C. Satto, P. Sciau, E. Dooryhee, J. Galy, P. Mittet, *J. Solid State Chem.* **146**, 103 (1999).
17. K. Kuwabara, M. Itoh, K. Sugiyama, *Solid State Ionics* **20**, 135 (1986).

DIRECT DIFFUSION STUDIES OF SOLIDS USING RADIOACTIVE NUCLEAR BEAMS*

SUN-CHAN JEONG, ICHIRO KATAYAMA, HIROKANE KAWAKAMI, HIRONOBU ISHIYAMA, YUTAKA WATANABE, HIROARI MIYATAKE, EIKI TOJYO, MITSUHIRO OYAIZU, KATSUSHI ENOMOTO
Institute of Particle and Nuclear Studies, High Energy Accelerator Research Organization (KEK)
Oho 1, Tsukuba, Ibaraki 305-0801, Japan

MASAO SATAKA, SATORU OKAYASU, HIROYUKI SUGAI, SHIN-ICHI ICHIKAWA, KATSUHISA NISHIO
Japan Atomic Energy Research Institute
Shirane 2-4, Tokai, Ibaraki 319-1195, Japan

MASAHITO YAHAGI, TAKANORI HASHIMOTO
Faculty of Engineering, Aomori University
Kouhata 2-3-1, Aomori, Aomori 090-0943, Japan

KAZUNORI TAKADA, MAMORU WATANABE
National Institute of Materials Science
Namiki 1-1, Tsukuba, Ibaraki 305-0044, Japan

AKIHIRO IWASE
Faculty of Engineering, Aomori University
Kouhata 2-3-1, Aomori, Aomori 090-0943, Japan

YASUHARU SUGIYAMA
Nihon Advanced Technology
Shirane 2-4, Ibaraki 319-1195, Japan

KEK-JAERI Radioactive Nuclear Beam (RNB) facility, an ISOL based RNB facility in Japan, will be able to provide various short-lived radioactive isotopes for experiments with an energy ranging from 0.1 to 1.1 MeV/nucleon. As an interdisciplinary application of the RNB for materials science, two experiments have been proposed; diffusion studies in super ionic conductors, and β-Nuclear Magnetic Resonance (β-NMR) and Perturbed Angular Correlation (PAC) studies by implantation of polarized RNBs. The present status of the experimental and instrumental development for the experiments is discussed.

1. Introduction

Radioactive atoms have been used in materials science for many decades. Besides their classical application as tracers for diffusion studies, nuclear techniques (Mössbauer Spectroscopy, PAC, β-NMR, and Emission Channeling), primarily developed in nuclear physics for detecting particles or γ-radiations emitted during the decay of the radioactive atoms, are now being routinely used to gain microscopic information on the structural and dynamical properties of the bulk of materials via hyperfine interactions or emitted particles [1]. More recently these techniques have also been applied to study of complex bio-molecules, surfaces, and interfaces. With the advent of most versatile 'radioactive nuclear beam (RNB) factory' represented by the on-line isotope separator (ISOL) facility, the possibilities for such investigations have been greatly expanded during the last decade [2]. At Tandem Accelerator facility of JAERI-Tokai, we have been working on constructing a RNB facility [3], in which radioactive nuclei, produced with Tandem Accelerator beams, for example, 30 MeV proton beam, can be accelerated to 1.1 MeV/nucleon for

* This work is partially supported by the REIMEI Resource of Japan Atomic Energy Research Institute and by High Energy Accelerator Research Organization

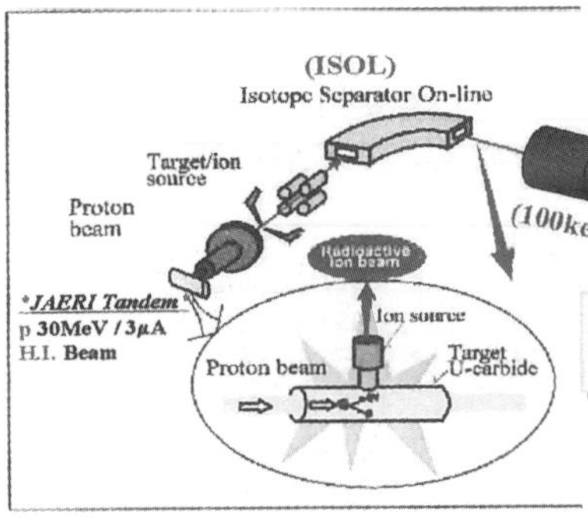

Fig.1. The principle of ISOL (isotope separator on-line) based RNB facility: radioactive nuclei are produced by nuclear reactions induced within a thick target by an external beam, e.g. UC_2 target by the irradiation of 30MeV proton from JAERI Tandem Accelerator. The target kept at a high temperature permits the fast diffusion of the reaction products into the ion source where they are ionized by plasma, or surface ionization. The singly charged ions are then extracted, mass separated in a magnetic field, and further, after being boosted to higher charge states, accelerated to the energies necessary for experiments.

ions with q/A of more than 1/10, where q and A are the charge state and the mass number of ions, respectively.

The energy is variable in the range from 0.1 to 1.1 MeV/nucleon, which allows us to implant the RNBs into specimens at a proper depth, avoiding the problem caused by the surface, e.g. diffusion barrier like oxide layers. Here, the production, the separation and the implantation of radioactive probes are integrated into one device, as shown in Fig.1. Typical ion species, under planning, include ^{8}Li($T_{1/2}$=0.8s), ^{18}F(110m), ^{20}F(11s), ^{111}In(2.8d), ^{143}Ce(1.7h) with intensities of 10^4 ~10^7 ions/s depending on nuclear species. We have been proposing two experiments using these RNBs. One is diffusion study in solids and the other is β-NMR or PAC study by implantation of polarized RNBs.

In the following, we will first describe our on-going experiment for measuring diffusion coefficients by a new method with a radiotracer beam of ^{8}Li emitting α-particles followed by β-decay. And then the experimental devices under development for diffusion studies using the conventional radiotracer method are presented. Finally, an experimental technique to polarize RNB will be introduced, which is now under development.

2. Diffusion experiments by short-lived RNBs

Though diffusion studies in solids with radioactive tracers have a history of more than half a century, there are some elements where no radioactive isotope has been used because radioactive tracers of lifetimes adequate for the application are not available. Among them Li and F elements are included. Our proposal employs ^{8}Li, ^{18}F and ^{20}F to be available in KEK-JAERI RNB facility. Since their lifetimes are too short to apply the conventional radiotracer method for the self-diffusion studies in solids, we have developed experimental methods to perform the experiment on line. There are two different schemes for the diffusion experiments depending on decay modes of the radioactive tracers, separately describing in the following.

2.1. ^{8}Li case: α-emitter

After being beta-decaying, ^{8}Li emits α-particles having a broad energy distribution with an average of 1.6 MeV and 600 keV in a full-width at half maximum (FWHM). In this case, the amount of incidental energy loss of the α-particles on their passage to the surface of the solid of interest depends on the position of the decaying emitter; the measured energies of the α-particles emerging from the solid are closely related to the decaying positions of the tracer. The time evolution of the energy spectra is therefore supposed to be a measure of the diffusivity of the tracer in the solid. The energy spectra are broadening with increasing diffusion time; the tracer diffusion coefficients could be simply obtained by the time-dependent widths of the measured energy spectra if the inherent energy of the emitted charged particles is well defined. In the present case, however, the inherent energy distribution of the α-particles is continuous and broad. Although the correspondence between the emitted position and measured energy of the charged particles is not so simple as above, the tracer diffusion coefficient could be obtained with the help of the simulation [4].

We have performed an experiment to measure the diffusion coefficients of Li in the sample of LiAl compound, by using the α-emitting radiotracer of ^{8}Li. The Li diffusion coefficients in LiAl have been measured in a wide range of temperature [5], and thus LiAl was adopted as an example of sample in the

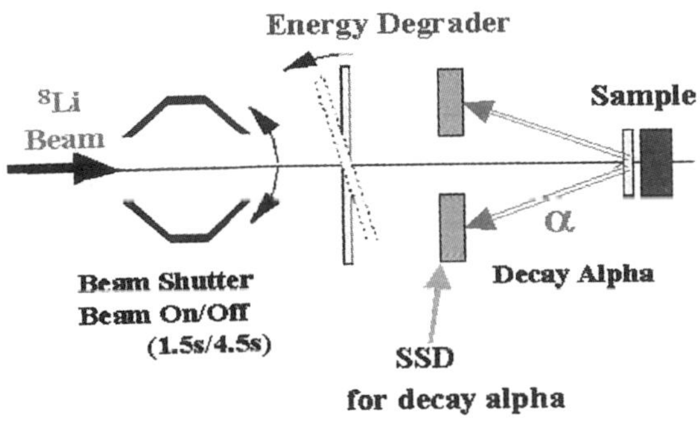

Fig.2. Experimental set-up of ^{8}Li diffusion study

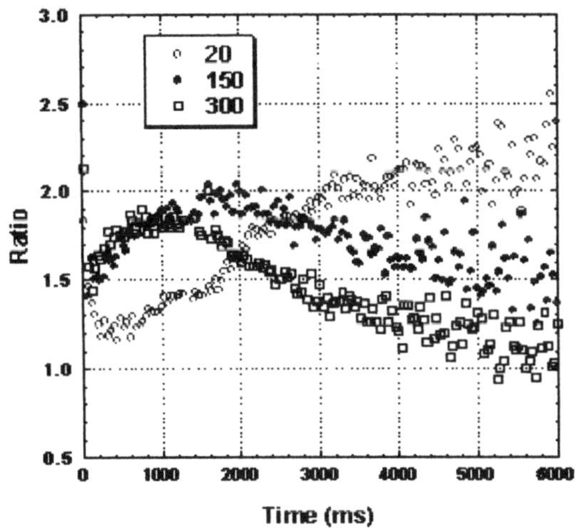

Fig.3. Normalized time spectra of α yields measured at the temperature of 20°C(○), 150°C(●) and 300°C(□), respectively. For the normalization, the time dependent curve of the activity of ^{8}Li during the time sequence for the measurement was used.

experiment for the experimental confirmation and calibration of the method.

The experimental set-up for the diffusion measurement is shown in Fig.2, installed nearly at the focal position of JAERI-RMS. The radioisotope of ^{8}Li was produced, by bombarding ^{7}Li of 24MeV on ^{9}Be target of 42μm in thickness. Separated by JAERI-RMS, the ^{8}Li beam of 14.6MeV with about 0.6MeV in FWHM was obtained with an intensity of 2500 particles/s by 30enA(electric nano-Ampere) of ^{7}Li. The energy of ^{8}Li for implantation was further reduced to 4MeV by the energy degrader installed upstream, corresponding to the average implantation-depth of about 12μm from the front surface of the sample. One time sequence for the measurement consisted of the time-duration of 1.5s for implantation (beam-on) and of 4.5s for subsequent diffusion (beam-off). The beam-on and off operation was carried out by the beam shutter. The sample was set at 20°C (room temperature) before starting the measurement. The energies of α– particles coming out of the sample were measured as a function of time by an annular solid-state detector (SSD). The sequence was repeated to obtain good statistics, where the time-zero is always at the beginning of the irradiation. We performed the measurement in the same way at the temperature of 150°C and 300°C, respectively.

The yield of α-particles with energies larger than 400keV showed different time-dependence according to the temperature of the sample. The observation implies that the time-dependence of α-yields is a good measure of the diffusivity of ^{8}Li in the sample, because the diffusivity depends on the temperature of the sample. By excluding the trivial time-dependence of the yields associated with the radioactive lifetime of ^{8}Li during the measurements, we obtained the time spectra only depending on the temporal evolution of the profiles of diffusing ^{8}Li at a temperature of the sample. In Fig.3, the time spectra are represented by the ratios, i.e., yields of α-particles divided by those of ^{8}Li existing in the sample at the time of interest. If ^{8}Li did not diffuse at all, the spectra should be constant over time.

By comparing the experimental time spectrum that is measured at a certain temperature, with results from the simulations, the diffusion coefficient could be finally obtained. The comparison is in progress.

The preliminary result has clearly shown the diffusion effect of ^{8}Li in LiAl. We will apply the method to measure the diffusion coefficients of $LiCoO_2$ compound, which is used as the positive electrode materials of most secondary Li ion batteries. The Li diffusion coefficients in $LiCoO_2$ have been measured by various electro-chemical ways so far. The values, however, are scattered over 6 orders of magnitudes [6]. Clearly, the measurement of the diffusion coefficients in a different way is highly desirable to settle down such disagreements, especially of importance in the recent general efforts to design the battery by simulations based on the first principle.

2.2 $^{18, 20}$F case: β-emitter

The other scheme is applicable to diffusion of β- and γ-emitter radioactive tracers. In this case, however, the energy loss of the radiation in solid is not sensitive to the position of the emitter in the present scale, say, ~10μm. Therefore, we follow the conventional method in conjunction with a serial sectioning technique [7]. Because of the short lifetime and deep implantation of the tracers as compared to the conventional method, we have developed an instrument to perform the experiment on line, in which RNB irradiation, diffusion, sectioning by ion beam sputtering, radioactivity measurement of sputtered specimen and measurement of sputtered depth, are performed in sequence. A schematic arrangement of subsequent processes and corresponding devices is shown in fig.4. We are planning to apply the instrument first to measure ^{18}F ion diffusion in super ionic conductor.

In parallel, we have performed a simulation in order to examine the feasibility of the experiment, by choosing a typical F ionic conductor, CaF_2, as an example of sample, whose diffusion coefficients range from 10^{-11} to 10^{-9} cm^2/s around the temperature of interest [8].

In the simulation, the sample is brought to high temperature and irradiated for 17s by the ^{18}F beam of 18 MeV with an energy spread of 2% in FWHM. The implantation depth, corresponding to the range of 18MeV ^{18}F in CaF_2, is about 8μm with a FWHM of 0.5μm estimated by TRIM code [9]. In the present case, the irradiation time corresponds to the diffusion annealing time in the conventional method, where the radioactive isotope of the diffusing element primarily deposited onto the sample surface diffuses inside the sample during the time. The sample is then sectioned in series for 600s with a sputtering rate of 0.5 μm/30s. Each section collected on the tape for 30s is transported to the detecting area, where the activity in each section is measured for 30s. This is one time-cycle for the measurement at a temperature (20 steps for 600s). During the sectioning and measuring time, the sample is sufficiently cooled down to the temperature where the diffusion is negligible.

In Fig. 5 are shown the simulated concentration depth profiles of ^{18}F at the end of the irradiation under a diffusion coefficient of $1x10^{-8}$ cm^2/s. The profile is supposed to be fixed during the following sectioning and measuring time. The primary profile in Fig.5 is the concentration of ^{18}F and slightly asymmetric due to the reflection at the surface as a usual boundary condition for diffusion; those ^{18}F diffused to the surface of the sample are simply reflected at the surface and again diffuse into the sample. Under the assumption of reflection, the irradiation time is critically limited by the largest diffusion coefficients to be measured in the experiment; the irradiation time is unreasonably long, and then the profiles at the end of the time are significantly distorted by diffusion between the implanted depth and the surface.

The measured profile in Fig.5 is the specific activity of ^{18}F at the corresponding depth where the decay-loss is taken into account according to the sectioning rate as mentioned above, i.e. 0.5μm/30sec. In the case of ^{18}F, the life-time is long enough as compared to the sectioning and measuring time and thus the decay-loss is at most 6% even when the serial sectioning is performed to the depth of 20μm, while the activity is just 0.3% of the primary yields, inversely proportional to the half lifetime. If the efficiency of the processes associated with the tape-collection of the sectioned sample and the measurement of the activity is assumed to be about 10%, the expected yields are further reduced by 0.1. Since the intensity of ^{18}F beam assumed in the simulation was $2x10^4$ particles/s, the yields expected in our future facility become more than about 20/0.5μm by referring to Fig.5. For smaller diffusion coefficients, a

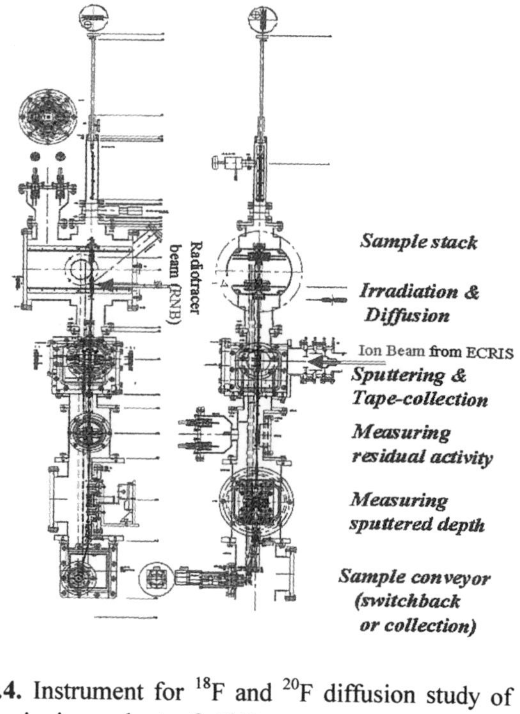

Fig.4. Instrument for ^{18}F and ^{20}F diffusion study of super ionic conductor β-PbF_2.

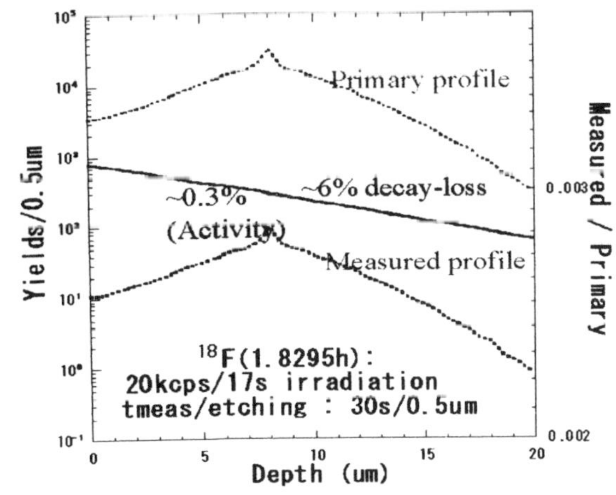

Fig.5. Simulated concentration-depth profile of ^{18}F at the end of the irradiation time under a diffusion coefficient of $1\times10^{-8}cm^2/s$ (Primary). The measured profiles, simply given by the specific radioactivity in the section of 0.5μm in thickness, are also presented after correcting the decay-loss at the corresponding time when the sputtering rate of 0.5μm/30s is assumed. The implantation rate during the irradiation time was assumed to be 2×10^4 particles/s.

longer irradiation time is possible. The activity of the implanted tracer accordingly increases and thus better statistics is expected.

The lower limit of the diffusion coefficient accessible by the present method is determined by the half-life time of the tracer. If we assume, as the criterion, the diffusion length of 1μm, two times larger than the width of the implanted depth profile, during the half-life time, the diffusion coefficient of 1×10^{-12} cm^2/s is expected to be measurable. Although more compromise of the measuring time sequence for better statistics is certainly necessary, the simulation reveals that the conventional radiotracer method can be applied for diffusion studies by using RNB with a half-life time of around 1h. It should be noted here that, for simplicity, we have assumed the energy of the beam for implantation to be 1MeV/nucleon. If we reduce the energy with the range less than 1μm, the applicable lifetime of tracers by our present device could be reduced to order of minutes in two regards: narrower spatial distribution of the primary depth profiles and shorter sputtering time to scan the depth of interest than considered so far.

3. Development of polarized RNB

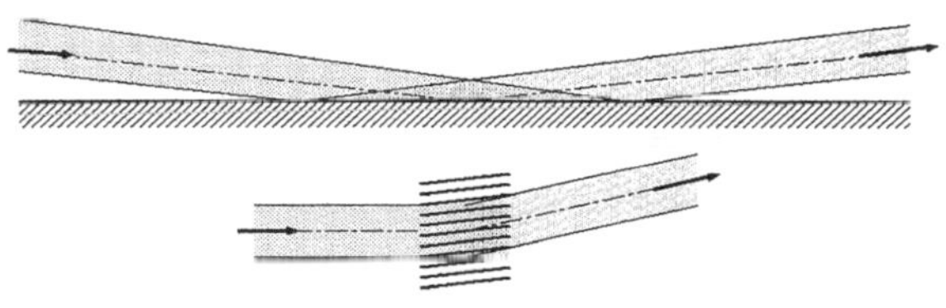

Fig.6. Schematic view of foil geometries with large tilted angle. Two geometries, typical (Upper panel) and folded (Lower panel), are respectively shown. The beam is incident from the left, and passes through the geometry after being specularly reflected on the foil surface.

Beams of polarized radioactive nuclei could open up the possibility of interesting applications in various fields. In addition to possible uses in nuclear β-decay studies to measure the nuclear properties of ground state, such beams would be well suited also for β-NMR and PAC experiments for materials science with short-lived radioisotopes. The combination of our RNB facility with the tilted-foil method would allow versatile production of short-lived radioisotopes, their subsequent polarization and then, after being implanted into the sample of interest, measurement of the hyperfine interaction of the nuclear moments with the fields of their surroundings using the β-NMR and PAC techniques.

The tilted-foil method uses the anisotropic atomic collision of incident ions with the conduction electrons at the exit surface of the foil. When the ion beams passing through a thin foil, tilted against the incident beam direction at an oblique angle, the electronic states of the outgoing ions are polarized [10]. Polarization is initially introduced in the orbital motion of the electrons by the surface interaction at the instant of exit of ions from the foil. During flight in free space, some of this electron spin (J) polarization is transferred to the nuclear spin (I) through hyperfine interaction. The direction of the polarization is well defined; **n** x **v** where **n** is the unit vector normal to the surface of the foil at the exit and **v** is the ion's velocity. By a successive passage of several such foils, interspersed with regions of free flight to allow a significant nuclear precession around the total angular momentum F=I+J in flight, the polarization effect is enhanced. In this way, rather sizable nuclear polarization, ~10%, for a wide variety of elements can be achieved [10].

The degree of the polarization increases with increasing tilted angle (θ), proportional to $sin\theta$. Though

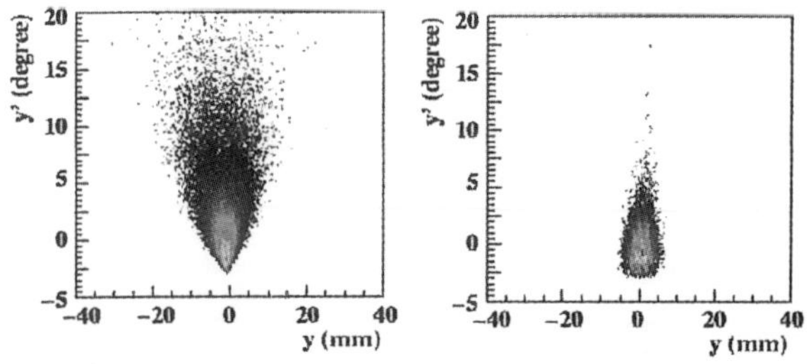

Fig.7. Expected beam emittance of scattered ions by the typical geometry (Left), and the folded geometry (Right).

the present energy of the RNBs is well suited to apply the tilted-foil method, larger tilted angle is desirable for higher polarization. However, in reality, the largest tilted angle is geometrically limited to about 60° because of the difficulty for preparing a tilted thin foil extended longitudinally to a large distance in order to cover the full trajectories of incident beams with a large tilted angle. On the other hand, geometry with the largest tilted angle could be easily realized if one used ion beam interaction at grazing incidence (IBSIGI) for polarization. In this case, the scattered ions from a solid surface are highly polarized. As shown in Fig.6 (upper panel), the geometry for grazing incidence is an extreme case of tilted foil geometry, while employing polarized ions reflected from the incident surface of the foil. In the typical IBSIGI geometry, as shown in the upper panel of Fig.6, the foil surface spatially extends to the longitudinal direction and thus the scattering points looking from the image point, i.e. position of the sample where the polarized ions are going to be implanted, are not well localized. As a more compact geometry, we have started to develop a folded IBSIGI geometry. In the folded geometry shown in the lower panel of Fig.6, foils with a largely reduced size are piled up along the vertical direction with respective to the beam direction. In this way the longitudinal extension of the target foil system could be largely reduced. Since the scattering points are relatively localized we can expect improved quality of scattered ions. In Fig.7 compared are the emittances of the polarized ion beams scattered from the foil surface in both geometries shown in Fig.6, well demonstrating the improved emittance in the folded foil geometry. Here, we have assumed $^{15}O^{1+}$ beams of 50keV with an emittance of 18π mm· mrad (±2mm x± 7mrad). This geometry is going to be tested under the collaboration between KEK, Kyoto Univ. and Tohoku Univ., by using an ISOL beam at CYRIC (cyclotron radioisotope center, Tohoku Univ.).*

In the IBSIGI application for the nuclear spin polarization, the folded geometry does not necessarily consist of very thin foils as required by the tilted-foil method. As a further development, we are going to make a folded thin foil geometry for the method.

4. Outlook

In KEK-JAERI RNB facility, various RNBs accelerated up to 1.1MeV/nucleon will be available from the beginning of 2004. Our several efforts to use the RNB for interdisciplinary applications have been introduced. Proposals to use the present RNBs for super ionic conductor study are welcome.

References

1. T. Wichert and M. Diecher, *Nucl. Phys.* **A693**, 327 (2001).
2. http://isolde.web.cern.ch/ISOLDE/, http://jkj.tokai.jaeri.go.jp/NuclPart/ExoticNucl/E-index.html/.
3. H. Miyatake et al., *Nucl. Insrt. And Meths.* **B204**, 747 (2003).
4. S.C. Jeong et al., *Jpn. J. Appl. Phys.* **42**, 4756 (2003).
5. J.C. Tarczon, W.P. Halperin, S.C. Chen and J.O. Brittain, *Mater. Sci. Eng.* **A101**, 99 (1988).
6. H. Sato, D. Takahashi, T. Nishina and I. Uchida, *J. Power Sources* **68**, 540 (1997).
7. F. Wener, A. Grude, R. Rummel, M. Eggersmann, T. Zumkley, N.A. Stolwijk and H. Merher, *Meas. Sci. Technol.* **7**, 632 (1996).
8. S. Chandra, *Super-Ionic Solids* (North-Holland, Amsterdam, 1981) p.120.
9. J.F. Ziegler, J.P. Biersack and U. Littmark, *The stopping and range of ions in solids* (Pergamon Press, New York, 1985) 1998-Version, Chap.8 .
10. H. Winter and R. Zomny, *Hyp. Int.* **22**, 237 (1985).

* M. Tanigaki (KUR, Kyoto Univ.) and T. Shinozuka (CYRIC, Tohoku Univ.).

X-RAY ABSORPTION SPECTROSCOPY OF LI ION BATTERY AND ELECTRONIC MATERIALS

J. B. METSON

Department of Chemistry, and MacDiarmid Institute of Materials Science and Nanotechnology
The University of Auckland, Private Bag 92019, Auckland, New Zealand. E-mail: j.metson@auckland.ac.nz

B. AMMUNDSEN

Pacific Lithium Ltd. 2 Mana Place, Manukau City, Auckland, New Zealand ,
E-mail: brett@pacificlithium.co.nz

Y. HU

Canadian Synchrotron Radiation Facility, Synchrotron Radiation Centre. University of Wisconsin Madison, Stoughton, Wisconsin, U.S.A. Email: yhu@uwo.ca

Layered lithium manganese oxides are of interest as potential cathode materials for Li ion secondary batteries. The manganese analogues of the widely used and well studied $LiCoO_2$ and are substantially cheaper and more environmentally friendly, but most formulations suffer from poor discharge characteristics and short cycle life due to capacity fade. This is attributed to alteration of the cathode to phases such as the spinel structured $LiMn_2O_4$. Monoclinic $LiMnO_2$ is a layered structure, isomorphous with $LiCoO_2$, however this monoclinic polymorph is only stable with structure regulating dopants such as Cr or Al in some of the manganese sites. In this study Al and Cr doped $LiMnO_2$ and a number of related compounds, have been prepared and examined by XPS, and XANES measurements of the transition metal L edges. In conjunction with XRD and Reitveld analysis, the high resolution L edge spectra reveal detail of the electronic structure of these materials, and provide substantially better definition of both manganese oxidation states, and the role of the dopant, in these compounds, than does conventional XPS. The absence of Mn^{3+} in the highly Cr doped materials means these are less prone to problematic Jahn-Teller effects.

1 Introduction

The selection of a cathode material for Li ion secondary batteries is a major determinant of battery capacity, voltage characteristics and longevity (cycle life). An increasingly mobile society places ever greater demands on such portable energy storage devices and thus requires ever better performance from such materials and the ability to package the system into thin and more flexible configurations. In the selection of cathode materials, $LiCoO_2$ has dominated, since the commercial introduction of Li secondary batteries in the early 1990's. However for reasons of cost, environmental impact and capacity limits, alternatives have long been sought. The most widely studied system has been $LiNiO_2$ /$LiCoO_2$, but a variety of transition metal oxide intercalation hosts have been explored. The list includes Mn, Al, Fe, Cr, and Cu, alone or typically in binary combinations. Belharouk et.al. [1] have recently reported studies on the $Li(Ni_{1/3}Co_{1/3}Mn_{1/3})O_2$ ternary system. Lithium iron and vanadyl phosphates have also been looked at as possible alternatives [2]. In some cases these materials have proceeded to commercial products.

Bruce et. al. outline the requirements for such a material [3] including the essential electronic properties, such as high capacity, adequate electronic conductivity, a low Fermi level coupled with high open circuit potential, and potential stability across a wide range of charge state (a flat discharge profile). Even more demanding are the structural properties of high and sustainable rates of Li intercalation/deintercalation, stability in contact with electrolyte and forming binders, and ease of fabrication.

The attractiveness of the Mn based materials in meeting these latter requirements are clear. The stable, spinel structured $LiMn_2O_4$ has been widely investigated but suffers from limited cycle life, poor characteristics in the discharge curve, and relatively low capacity [4]. However the $LiMnO_2$ stoichiometry offers a theoretical capacity of 285 mAh/g (based on the Mn^{3+}/Mn^{4+} couple), twice that of the spinel [5], but is far more complex to prepare. Monoclinic $LiMnO_2$ with the layered α-$NaFeO_2$ rocksalt type structure, isomorphic with $LiCoO_2$, is stabilised with dopants such as Al or Cr sharing some of the Mn sites [6-8]. However there are problems with cycle life in these materials, through eventual evolution to the spinel type discharge curve.

Studies have now demonstrated the performance advantages of the high Cr system Li_3CrMnO_5 first reported by Davidson *et al.* [9,10] which incorporates additional Li in the transition metal layers. However the capacity of this material (200 mAh/g) is beyond the stoichiometric capacity of the Mn^{3+}/Mn^{4+} couple and the dominance of the Cr^{3+}/Cr^{6+} couple in the electroactivity of this material has now been demonstrated [11].

The microstructural characterisation and identification of oxidation state changes in the cycling of these materials is critical to understanding performance, but can be extremely challenging. The broad nature of the transition metal 2p peaks, coupled with small binding energy shifts in some cases, has limited the effectiveness of conventional XPS in providing oxidation state information on the transition metal ions, especially the Mn^{3+}/Mn^{4+} shift [12], while on the other hand, $Cr^{3+}$and Cr^{6+} are readily resolved. The ability to resolve both atom positions and co-ordination in the Extended X-ray absorption Fine Structure (EXAFS) spectrum, and chemical state fingerprinting in the X-ray Absorption Near Edge (XANES) spectrum, makes this method ideal for resolving some of these questions.

A number of papers have reported the XANES spectra of the $LiCoO_2$ and $LiNiO_2$ and mixed Ni:Co systems in particular [13,14], however there has been little reported work in the $Li(Cr,Mn)O_2$ system. Horne [15] has reported Co and Ni data from $LiMn_{1.75}Co_{0.25}O_4$ and $LiMn_{1.75}Ni_{0.25}O_4$. We have made extensive measurements of the X-ray Absorption Near Edge Structure (XANES) of the transition metal 2p levels for the $LiCr_xMn_{1-x}O_2$ (x = 0.05) material, and a number of related standard materials and Ammundsen et. al. report more extensive studies of the higher Cr materials [11]. Li_2MnO_3 provides a standard monoclinic rocksalt type structure with Mn in the 4+ state. Similarly $LiCrO_2$ provides a layered hexagonal rocksalt structure with trivalent Cr in octahedral sites. Li_3MnCrO_5 provides a model of the mixed layered rocksalt structure with Li, Cr and Mn in octahedral co-ordination. This structure is of interest because it contains both Mn and Cr with the Cr presumed to be trivalent and the Mn tetravalent.

2 Experimental

LiM_xMn_{1-x} samples were prepared by reacting stoichiometric amounts of Li_2CO_3 with the Mn, Al and Cr oxides, at 1000 – 1050°C under flowing nitrogen as described previously [7]. The standards were again prepared by solid state methods using the stoichiometric ratios of the oxides and Li_2CO_3. XANES spectra were collected on the spherical grating monochromator (SGM) beamline of the Canadian Synchrotron Radiation Facility (CSRF) located at the Synchrotron Radiation Center (SRC), at the University of Wisconsin, Madison. All spectra were collected in the total electron yield (TEY) and fluorescence yield modes, although different depths are being analysed, typically the spectra are similar and the former provide better signal to noise. Because of the optimal range of operation of the monochromator, the Mn spectra were recorded on the second harmonic, while scanning the monochromator between 318 and 332 eV. Monochromator resolution, measured on the C K edge, is around 0.2 eV. The Cr spectra were recorded directly by scanning energies between 572 and 600 eV. All absorption spectra are presented normalised to I_0 which was monitored continuously during data collection using a 90% transmission grid.

SEM images were recorded on a Philips XL30 FEGSEM located in the Research Centre for Surface and Materials Science at the University of Auckland.

3 Results and Discussion

For $LiM_xMn_{1-x}O_2$ the monoclinic structure is only observed with specific dopants M occupying some of the Mn sites in the octahedral layer. Both Al^{3+} (0.53Å) and Cr^{3+} (0.62Å) are smaller than Mn^{3+} (0.64Å) and are considered to create zones of micro-strain in the lattice [7]. The typical Al-O bond length (1.92Å) is shorter than either the equatorial (1.94Å) or axial (2.3Å) Mn – O bonds in the Jahn Teller distorted octahedron. These strain/polarisation effects ultimately force the lattice into the monoclinic structure (Table 1). Reitveld analysis of the XRD patterns suggests this transition is partial at x = 0.05 in $LiAl_xMn_{1-x}O_2$ and complete above this level [7].

The limiting solubility of Al lies at around 10% and capacity is sacrificed by the substitution of the electro-inactive Al.

Table 1. Unit cell parameters (Å) for $LiAl_x Mn_{1-x}O_2$ and $LiCr_xMn_{1-x}O_2$ materials.

	a	b	c	β
$LiAl_{0.05}Mn_{0.95}O_2$	5.419(2)	2.802(1)	5.381(2)	115.89(2)°
$LiAl_{0.10}Mn_{0.90}O_2$	5.416(1)	2.804(1)	5.382(2)	115.91(2)°
$LiCr_{0.03}Mn_{0.97}O_2$	5.438(2)	2.808(1)	5.386(2)	115.98(2)°
$LiCr_{0.05}Mn_{0.95}O_2$	5.431(1)	2.807(1)	5.385(2)	115.98(2)°
$LiCr_{0.10}Mn_{0.90}O_2$	5.405(2)	2.815(2)	5.375(2)	115.61(2)°
$LiMnO_2$ (Ref.3)	5.439(3)	2.809(2)	5.395(4)	115.9(4)°

A performance comparison of the 5 and 10% materials, together with the spinel structured $LiMn_2O_4$, is shown in figure 1. The Cr substituted materials are of more interest as transition to the monoclinic structure is observed at x = 0.5 and there is complete solid solution between Cr and Mn. The Cr is electroactive, limiting any potential loss of capacity and indeed is the dominant electroactive species in the 50:50 substituted material discussed earlier [10]. The typical morphology of the $LiM_xMn_{1-x}O_2$ cathode particles retrieved after extended cycling in button cells is shown in relief and in cross-section in figure 2. Some adherent binder matrix and added carbon black is seen and the layered morphology is clearly visible in both images.

Figure 3 shows the Mn L-edge XANES spectra of the $LiCr_{0.05}Mn_{0.95}O_2$ material, along with the 50:50 Mn:Cr model compound Li_3CrMnO_5 and a Mn^{4+} standard in Li_2MnO_3. The spectra display the expected 2p doublet structure but the multiple scattering effects observed around the absorption edge produce a far richer and more informative spectrum than is observed in the typical XPS spectrum, where Mn^{3+} /Mn^{4+} shifts are reported between 0.3 and 0.7 eV and typical peak widths are greater than 3 eV [12,16,17].

The XANES spectra immediately resolve that the Mn in the 50:50 compound is predominantly tetravalent. A small shoulder at around 641 eV, preceding the sharp

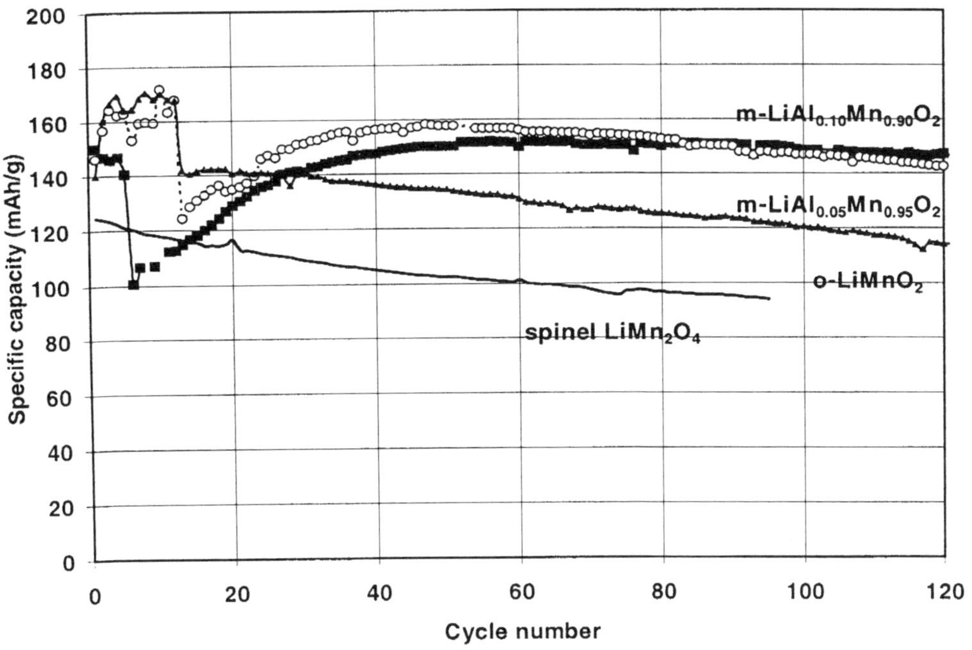

Figure 1. Performance comparison of the aluminium doped monoclinic $LiMnO_2$ with the undoped orthorhombic $LiMnO_2$ and the spinel structured $LiMn_2O_4$

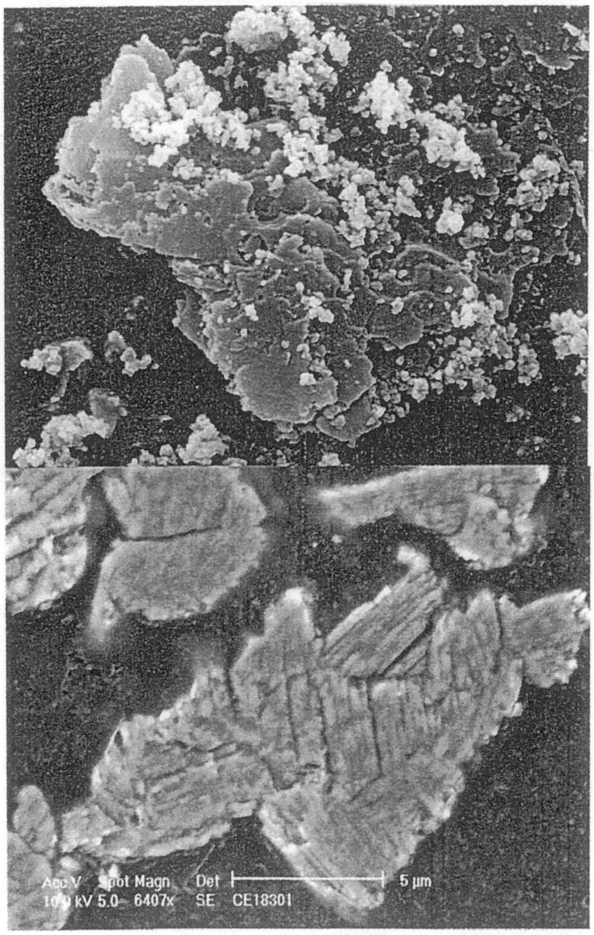

Figure 2. Top and cross-sectional views of $LiMnO_2$ cathode particles retrieved after cycling in a test cell. Note the visible layering of the structure.

threshold peak suggests there is some Mn^{3+} in the structure. However for the $LiCr_{0.05}Mn_{0.95}O_2$ structure, the spectrum is quite different. The threshold energy and the doublet move to lower energy by several eV and a rich fine structure is observed, particularly in the lower energy $2p_{3/2}$ peak. Thus the spectrum is consistent with Mn^{3+} as expected from the stoichiometry and X-ray structural analysis. XPS was not able to resolve Mn oxidation states in these structures.

Cr is more tractable in the XPS spectrum in that the ≈ 3eV binding energy shift between the 3+ and 6+ oxidation states in oxide materials is at least comparable to the observed linewidths in our non-monochromatised XPS instrument, and those in published spectra [18,19]. However again the XANES is less ambiguous, as seen in the spectra of the $LiCr_{0.05}Mn_{0.95}O_2$ phase and two reference compounds shown in figure 4, and in spectra published elsewhere [11]. Predictably, the spectra are consistent in confirming the dominance of Cr^{3+} in all the structures. For Li_3MnCrO_5 the confirmation of trivalent Cr is consistent with the Mn XANES spectra which indicate Mn^{4+} is dominant in this structure. In a separate study [11] the XANES spectra of the Cr L-edge of the Li_3MnCrO_5 material charged (Li depleted) to 4.4V, demonstrates the presence of Cr^{VI}, while the Mn remains essentially unmodified in the 4+ oxidation state.

This is particularly important in the context of cathode materials as the Jahn-Teller active Mn^{3+} is

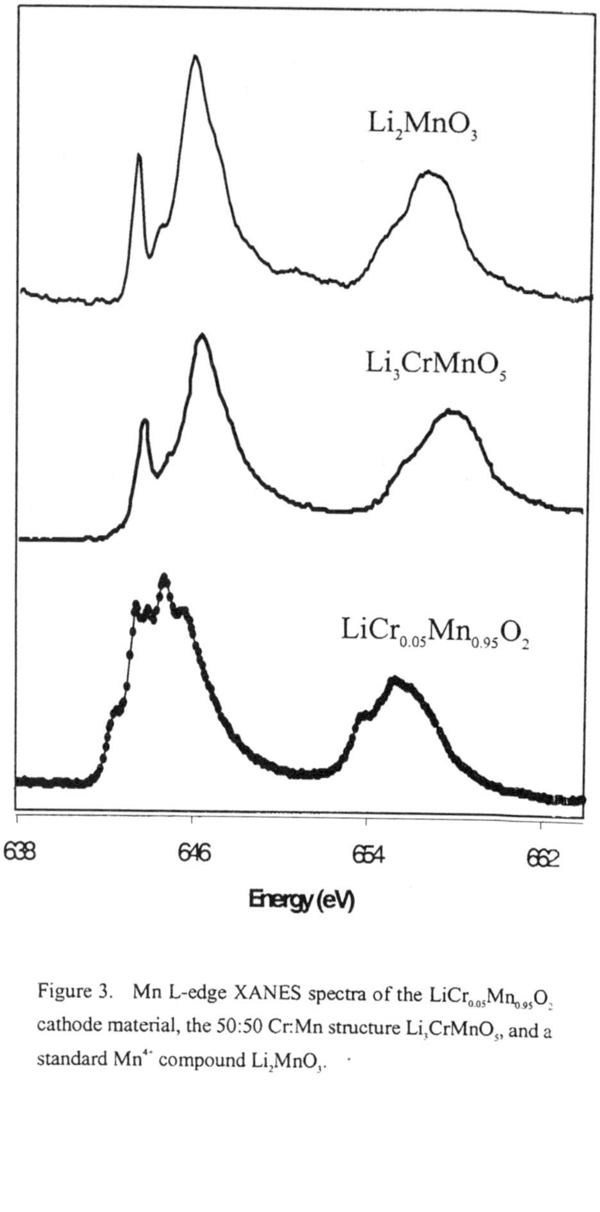

Figure 3. Mn L-edge XANES spectra of the $LiCr_{0.05}Mn_{0.95}O_2$ cathode material, the 50:50 Cr:Mn structure Li_3CrMnO_5, and a standard Mn^{4+} compound Li_2MnO_3.

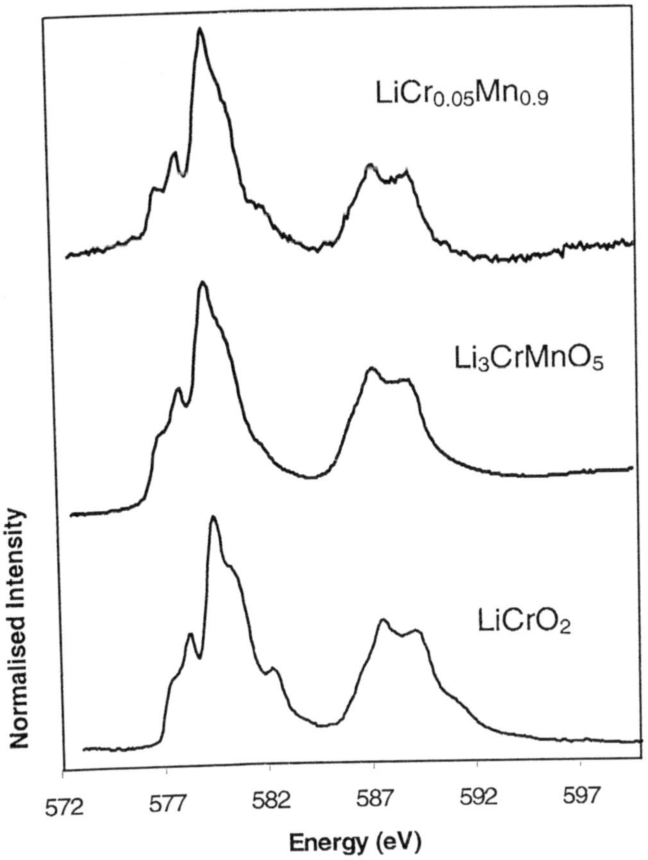

Figure 4. Cr L edge XANES spectra of the $LiCr_{0.05}Mn_{0.95}O_2$ cathode material, the 50:50 Cr:Mn structure Li_3CrMnO_5 , and a standard Cr^{3+} compound $LiCrO_2$.

problematic in encouraging the microstructural breakdown of the higher Mn materials where the Mn^{3+}/Mn^{4+} couple provides the charge balance during lithium cycling in and out of the structure [4,5].

4 Conclusions

Mn and Cr L-edge XANES spectroscopy has been used to examine the oxidation states of a number of $LiM_xMn_{1-x}O_2$ (M=Al, Cr) materials developed as potential battery materials for Li ion secondary batteries. The XANES spectra provide a detailed method of defining oxidation states, particularly for Mn, where the conventional XPS spectrum of the 2p levels struggles to provide adequate resolution. XANES spectra confirm the dominance of Mn^{3+} and Cr^{3+} in $LiCr_{0.05}Mn_{0.95}O_2$. This material provides superior performance in capacity and cycle life to the structurally analogous $LiAl_{0.05}Mn_{0.95}O_2$ which shows slow transformation to the spinel structure on cycling. In both cases the presence of the dopant elements on the manganese sites of the $LiMnO_2$ lattice stabilizes the layered monoclinic structure over the normal orthorhombic polymorph.

Identification of the electroactive species during cycling of the materials in test cells, has relied extensively on X-ray absorption Near Edge Spectroscopy (XANES). In XANES, the ability to examine both the surface sensitive electron yield and the fluorescence yield from depths of 10s of nanometres provides an especially powerful combination.

Acknowledgements

Part of this work is based on research conducted at the Synchrotron Radiation Center, University of Wisconsin - Madison, which is supported by the NSF under Award No. DMR-0084402. Access to the SGM line at the synchrotron ring was provided through the Canadian Synchrotron Radiation Facility.

References

1. I. Belharouk, Y-K Sun, J.Liu and K.Amine. *J.Power Sources*, **123**, p. 247-252 (2003).
2. B. M. Azmi, T. Ishihara, H. Nishiguchi and Y.Takita. *J.Power Sources*, **119-121**, p. 272-277 (2003).
3. P. G. Bruce, A. R. Armstrong and R. I. Gitzendanner. *J.Mat. Chem.* **9**, p.193-198 (1999).
4. B. Ammundsen and J. Paulsen, *Adv. Mater.* **13**, p.943-956 (2001).
5. G. G. Amatucci, C. N. Schmutz, A. Blyr, C. Sigala, A. S. Godz, D. Larcher, J. M. Tarascon, *J. Power Sources* **69**, p. 11 (1997).
6. Y.Shao-Horn, S.A.Hackney, A.R.Armstrong, P.G.Bruce, R. Gitzendanner, C.S.Johnson and M.M.Thackeray. *J.Electrochem Soc.* **146**, p.2404, (1999).
7. B.Ammundsen, J.Desilvestro, T.Groutso, D.Hassell, J.B.Metson, E.Regan, R.Steiner and P.J.Pickering, *Mater. Res. Soc. Symp. Proc.* 575, p. 479 (2000); B.Ammundsen, J.Desilvestro, T.Groutso, D.Hassell,

J.B.Metson, E.Regan, R.Steiner and P.J.Pickering, *J. Electrochem. Soc.* **147** (11) 4078 – 4082. (2000).

8. Y-M. Chiang, D.R.Sadoway, Y-I. Jang, B.Huang and H.Wang. *Electrochem. Solid State Lett.* **2**, p.107, (1999)
9. I.J.Davidson, R.S.McMillan, H.Slegr, B.Luyan, I.Kargina, J.J.Murray and I.P.Swainson. *J.Power Sources*. **81-82,** 406, (1999).
10. C.Storey, I.Kargina, Y.Grincourt, I.J.Davidson, Y. C. Yoo and D.Y.Seung. *J.Power Sources* **97-98**, p. 541-544 (2001).
11. B. Ammundsen, J. Paulsen, I. J. Davidson, R.-S. Liu, C.-H. Shen J.-M. Chen, L.-Y. Jang and J.-F. Lee, *J. Electrochem. Soc.* **149**, A431 (2002).
12. R.Benson and J.B.Metson. *European Conference on Applications of Surface and Interface Analysis.* H.J.Mathieu, B.Reihl and D.Briggs Eds. John Wiley and Sons. P. 373-376, (1996).
13. I Nakai, K.Takahashi, Y.Shirashi, T.Nakagome, F.Izumi, Y.Ishii, F. Nishikawa, T.Konishi. *J.Power Sources*, **68**, p.536-539 (1997).
14. L.A.Montoro, M.Abbate and J.M.Rosolen. *Solid State Letters*. **3** (9), p.410-412 (2000).
15. C.R.Horne *The Electrochemical Society Interface*, Spring 1998, p.61-62.
16. J.S.Foord, R.B.Jackman, G.C.Allen *Philos Mag A*. 49(5) p.657-663 (1984).
17. V.DiCastro, and G.J.Polzonetti. *J.Electron Spectrosc. Related. Phemon.* **48**, p.117-123 (1989).
18. D.Briggs and M.P.Seah (Eds) *Practical Surface Analysis,* 2nd Ed. Vol 1. John Wiley, Chichester (1990).
19. J.F Moulder, W.F.Stickle, P.E Sobol, and K.D.Bomben. *Handbook of X-ray Photoelectron Spectrsocopy.* Physical Electronics, Eden Prairie, MN. (1992).

HOLE BURNING SPECTROSCOPY AND SITE SELECTIVE SPECTROSCOPY FOR RARE EARTH IONS DOPED $LA_{0.5}LI_{0.5}TIO_3$

HIROSHI KOYAMA †

Institute of Multidisciplinary Research for Advanced Materials (IMRAM), Tohoku University, Katahira, Sendai, Japan 980-8577

SHINICHI FURUSAWA

Department of Electronic Engineering, Faculty of Engineering, Gunma University, Tenjin-cho, Kiryu, Gunmma, Japan 376-8515

TAKESHI HATTORI

Institute of Multidisciplinary Research for Advanced Materials (IMRAM), Tohoku University Katahira, Sendai, Japan 980-8577

We have observed the hole burning due to the optical pumping of the hyperfine levels of Eu^{3+} ion in LLT:Eu^{3+} and due to the local structural change in LLT:Pr^{3+}. It is considered that the hole burning in LLT:Pr^{3+} is due to the localized motion of Li ions near the equilibrium position because the hole decay has occurred below 21K. From luminescence spectra, it was found that there are four Eu^{3+} sites in LLT:Eu^{3+}. From site selective fluorescence spectra, it was found that all Eu^{3+} sites are in the C_{2v} symmetry or the lower than C_{2v} symmetry. The site symmetry for all Eu^{3+} sites is lower than symmetry which is expected from crystal structure of LLT. Both results suggest that Li ions are located at off-center position in LLT at low temperature.

1. Introduction

The site selective spectroscopy is useful to investigate the local structure in disordered materials which contain a small amount of Eu^{3+} ions as optical center because Eu^{3+} ion has relatively simple diagram of energy levels and they are sensitive to the local symmetry at the optically active ion in disordered materials [1,2]. In particular, the ground state $^7F_0$ and the principal emitting state $^5D_0$ show no Stark splitting. This fact simplifies the analysis of the experimental data. In the case which the $^7F_0 \leftrightarrow ^5D_0$ absorption band is excited by means of monochromatic light, high site selectivity is accomplished. Additionally, most convenient dye lasers used Rhodamine 6G can be use for the fluorescence and absorption experiment of Eu^{3+} ion because transition between the $^7F_0$ and $^5D_0$ level are as same as the wavelength region of laser dye.

Persistent spectral hole burning (PSHB) which is one of the site selective spectroscopy has been applied to investigate the structural and dynamical properties of disordered materials. PSHB by motion of ions has been observed in ionic conductors doped with rare earth ions such as Eu^{3+} ion or Pr^{3+} ion. In resent studies, the potential energies for the light induced local motion of ions have been determined from the analysis of the thermal decay-profile of the persistent hole in ionic conductors by several authors [3,4]. On the other hand, the homogeneous line width can be measured using to PSHB in the inhomogeneous broadening of disordered materials [5]. Because the homogeneous line width becomes wide in the disordered materials. It is reflected the disordered nature [6,7]. We reported that there are two states in which the degree of ordered is differ from each other in YSZ and these ordered states are responsible to the concentration dependence of ionic conduction in YSZ [8].

The perovskite-type lanthanum lithium titanates ($La_{0.5}Li_{0.5}TiO_3$(LLT)) is a high lithium conductor. The ionic conductivity σ is about 10^{-3} S/cm at room temperature [9]. In LLT, ionic conduction arises from transportation by Li ions through A-site vacancies.

† Present address: Nanomaterials Laboratory, National Institute for Materials Science, Namiki 1-1, Tsukuba, Ibaraki, Japan, 305-0044, E-mail: KOYAMA.Hiroshi@nims.go.jp

However, the Li ionic conduction in LLT has not been understood perfectly. In this study, we study PSHB and luminescence spectra of LLT-doped with Eu^{3+} and Pr^{3+} ions in order to know information of localized Li site. We report the observation of PSHB in LLT: Eu^{3+} and LLT: Pr^{3+} and discuss the mechanism of the hole burning in LLT. Additionally, we apply the site selective fluorescence spectroscopy to investigate the relationship the hole burning and the local symmetry in LLT.

2. Experimental

Single crystals are made by cooling from the melt at 1560°C in platinum crucible using to Molybdenum furnace. A cooling rate from 1560°C to 1520°C is 0.5°C/hour for LLT: Eu^{3+}, and 0.33°C/hour for LLT:Pr^{3+}. A cooling rate from 1520°C to room temperature is 100°C/hour. Sample surfaces were polished for optical measurements. The content of rare earth ions was kept for 1mol%.

X-ray diffraction powder (XRD) patterns were recorded with Cu Kα radiation in a RIGAKU RINT-V diffractmetor. Data were taken with a 1 degree divergence slit, receiving slit of 0.30mm. The 2-theta range analyzed was 10-90 degree, with a step scan of 0.01 and a counting time of 0.3s for a step.

Fig.1 shows the powder X-ray diffraction patterns which are recorded at room temperature for LLT:Eu^{3+} and LLT:Pr^{3+}. For both samples, XRD patterns could be indexed with the double perovskite cell ($a=a_p$, $c\sim 2a_p$) in the space group P4/mmm similar to previous studies [9,10].

The experimental set-ups and the methods of the hole burning experiment and the site selective fluorescence experiment are also the same as previous works [1,3,4]. At the hole burning experiment, a CW ring dye laser using Rhodamine 6G with frequency jitter

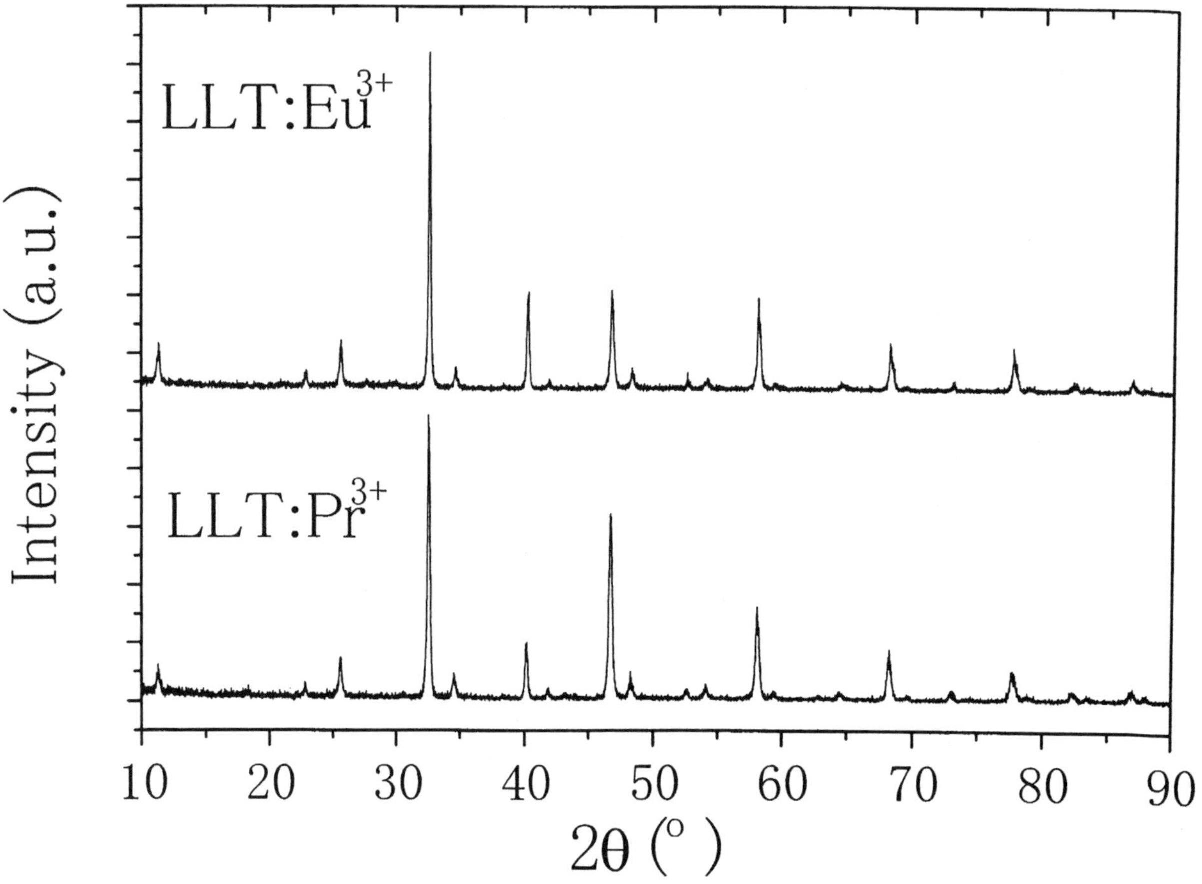

Fig.1 Powder X-ray diffraction (XRD) pattern for LLT:Eu^{3+} (upper) and LLT:Pr^{3+} (lower) at room temperature.

width of 1MHz was used as a light for the burning of hole and for the detecting of a hole by the scanning of the laser frequency. We made hole burning experiment by using the $^3H_4 \leftrightarrow ^1D_2$ transition of Pr^{3+} for LLT:Pr^{3+}, and by using the $^7F_0 \leftrightarrow ^5D_0$ transition of Eu^{3+} for LLT:Eu^{3+}.

3. Result and discussion

Fig.2 shows a luminescence spectrum which is due to the $^5D_0 \rightarrow ^7F_0$ transition of Eu^{3+} ions for LLT:Eu^{3+} under the excitation at 308nm. There are four Eu^{3+} sites (A, B, C and D) which the local crystal field differs from each other because both the $^5D_0$ state and the $^7F_0$ state are non-degenerate. Since the $^5D_0 \rightarrow ^7F_0$ transition of Eu^{3+} ions observed is a zero phonon transition, there are absorption bands at same wavelength. Under the excitation at 576.5nm (B band) or at 578.9nm (C band), we can observe the persistent hole. These holes were vanished 30-40 seconds after from stopping of irradiation for a burning light. Holes had antiholes and side holes as shown in Fig.3.

In the previous studies [5,11], three type mechanisms have been reported as production of long-lived holes. The first is an optical pumping mechanism which is observed for $^7F_0 \rightarrow ^5D_0$ transition of Eu^{3+} ions in crystals, glasses and some ionic conductors. In this case, antiholes and side holes are observed in any case and hole is stable only at the lower temperature than 15K. Additionally, hole has a short decay time as compared with holes which are due to other two mechanisms. The second is an optically induced

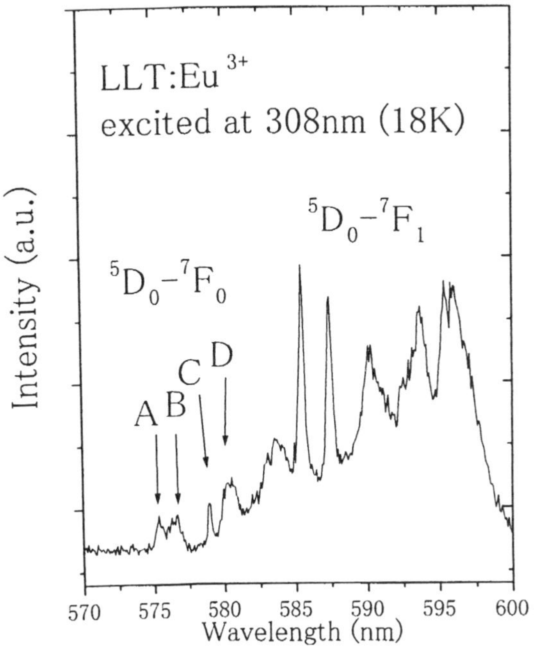

Fig.2 a luminescence spectrum under the excitation at 308nm for LLT:Eu^{3+} (18K).

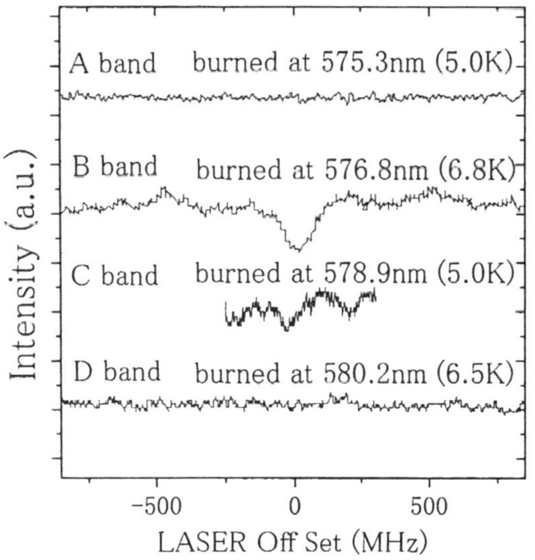

Fig.3 Hole spectra for LLT:Eu^{3+}. Laser power is 30mW/mm^2. Burning time is 300s.

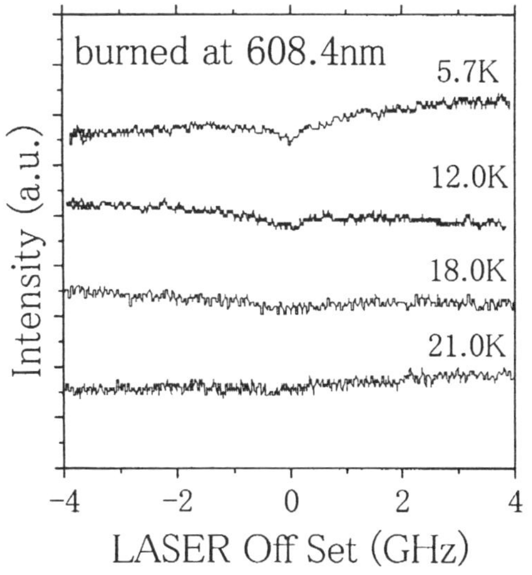

Fig.4 Hole spectra for LLT:Pr^{3+} at several temperature. Burning wavelength is 608.4nm.Laser power is 30mW/mm^2. Burning time is 600s.

rearrangement of local structure of materials. In this case, hole has no antiholes and side holes. A decay time of the hole is from half hour to infinite. The third is an ionization of rare earth ion. Hole by this mechanism can be observed at even room temperature in several materials. From the character of holes in LLT:Eu^{3+}, we concluded that holes are due to the optical pumping of the hyperfine levels of Eu^{3+} ion.

Fig.4 shows hole spectra for LLT:Pr^{3+} at several temperature. This hole can be burned between 608nm and 609nm in absorption band of the $^1D_2 \rightarrow {}^3H_4$ transition of Pr^{3+} ions. The hole can be observed 30min after from stopping of irradiation for a burning light and hardly be observed at above 21K. This hole has no antiholes. These results suggests that the mechanism of the hole production in LLT:Pr^{3+} is a local structural change by a light irradiation.

The local structural change in LLT:Pr^{3+} by light irradiation is not a hopping of Li ions between A-sites in perovskite structure. If so, the hole should be observed at over 21K because activation energy for ionic conduction of LLT is 0.40eV [9]. Therefore, this hole burning is due to the localized motion by Li ions near the equilibrium position, probably between off-center position. Because this hole could be observed below 21K, Li ions in the off-center position are responsible for the hole burning in LLT:Pr^{3+}.

Fig. 5 shows the temperature dependence of the hole width (Full Width at Half Maximum (FWHM)) in LLT:Pr^{3+} at low temperature. We represented the hole by a Lorentzian curve in order to obtain the hole width from the hole spectrum for LLT:Pr^{3+}. The hole width was determined from the calculated Lorentzian curve. The temperature dependence of the hole width (Γ_h) was fitted with a function of $\Gamma_h = \alpha T^{\beta}$ by least squares method, as shown in Fig. 5 by solid line. The temperature dependence of Γ_h shows as $\Gamma_h \sim T^{1.1}$ in LLT:Pr^{3+}. The homogeneous line width ($\Gamma_{homo} \sim \Gamma_h/2$) depends on the temperature as $\Gamma_{homo} \propto T^{1\sim2}$ in disordered materials such as glasses. In ionic conductors, the similar temperature dependence of the homogeneous line width was observed at low temperature for YSZ:Pr^{3+} [8,12]. The temperature dependence of Γ_h shows a disordered nature of LLT. It is considered that the disorder nature in LLT arises from Li ions which occupy the off-center position near lattice point.

Fig.6 shows the site selective spectra of the $^5D_0$-$^7F_1$ transition of Eu^{3+} ions for LLT:Eu^{3+}.

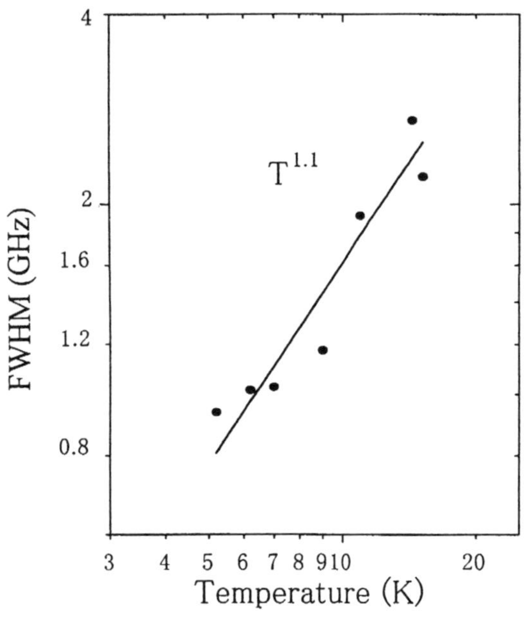

Fig.5 Temperature dependence of the hole width (FWHM) for LLT:Pr^{3+}. Solid line is a calculated line.

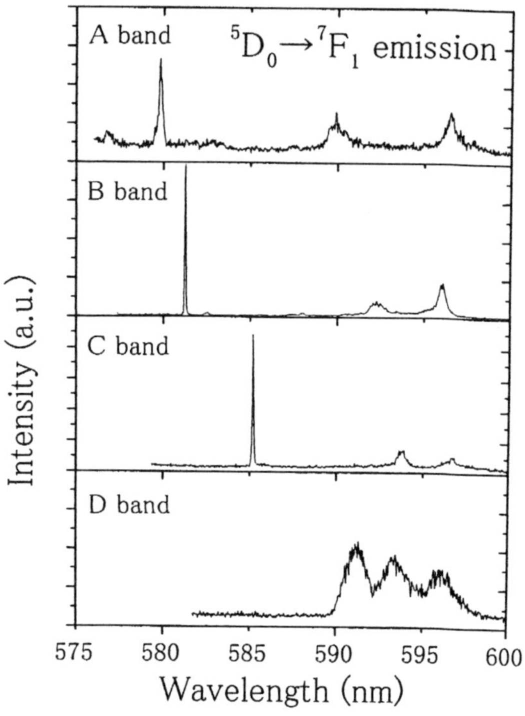

Fig.6 Site selective fluorescence spectra of the $^5D_0$-$^7F_1$ transition of Eu^{3+} ions for A, B, C and D band in LLT:Eu^{3+} at 10K.

Degeneracy of the $^7F_1$ state of Eu^{3+} ions is three. This state splits to some Stark levels under the low site symmetry. The number of the Stark levels of the $^7F_1$ state is one for O_h symmetry, two for D_{4h} or C_{4v} symmetry, three for C_{2v} or lower symmetry. Since all site selective spectra of the $^5D_0$-$^7F_1$ transition of Eu^{3+} ions have three peaks, it is suggested that the degeneracy of the $^7F_1$ state is completely removed at all sites in LLT. The result suggests that the local symmetry at all Eu^{3+} sites is C_{2v} or lower than C_{2v}.

Fig.7 shows the site selective spectra of the $^5D_0$-$^7F_2$ transition of Eu^{3+} ions. Degeneracy of the $^7F_2$ state of Eu^{3+} ions is five. The $^7F_2$ state splits to five Stark levels under C_{2v} or lower than C_{2v}. For C_{2v} symmetry, the number of allowed optical transition of the $^5D_0 \rightarrow {}^7F_2$ transition of Eu^{3+} ions is four among five Stark levels of the $^7F_2$ state. Under C_2 or C_1 symmetry, all transition of the $^5D_0 \rightarrow {}^7F_2$ transition of Eu^{3+} ions are allowed.

In A and B sites, five peaks are observed for the spectra of the $^5D_0$-$^7F_2$ transition of Eu^{3+} ions. Symmetry of these sites is C_2 or C_1. In C site, we can observe three peaks only because the luminescence from Eu^{3+} ions occupied D site overlapped in the wavelength region over 615nm. From the spectrum of the $^5D_0$-$^7F_1$ transition of Eu^{3+} ions, we concluded that the site symmetry of C site is C_{2v} or lower than C_{2v}. In D site, the spectrum has two broad inhomogeneous bands, namely we cannot distinguish obviously the number of peaks for the spectrum. However, since this broad inhomogeneous band is often observed in Eu^{3+} ion doped disordered materials such as glasses. It is indicated that the site symmetry of D site is most low symmetry like as glasses.

Site symmetry of all Eu^{3+} sites is lower than that is expected from crystal structure of LLT. Because the primitive unit cell of LLT is the cubic perovskite structure, the site symmetry is O_h in the case which the A-site ions are all same kind ions. The site symmetry of A-site is D_{4h} symmetry in the case which La ion and Li ion are perfectly ordered parallel to c-axis. However, the site symmetry for all Eu^{3+} sites is lower than these symmetry judging from the site selective spectra for LLT:Eu^{3+}.

It is considered that two origin of this low site symmetry in LLT. One is the mixed arrangement of Li ions and La ions at A-site for perovskite structure. In the case of the mixed arrangement of Li ions and La ions at A-site, the site symmetry can be reduced to the C_{2v}. The result for the site selective spectra of the $^5D_0$-$^7F_1$ transition of Eu^{3+} ions suggested that the low site symmetry is due to the mixed arrangement of A-site ions around Eu^{3+} ion.

The other is the occupation off-center position around lattice point by Li ions. In this case, the site symmetry can be reduced from C_{2v} to C_2 or C_1. Recently, it is reported that there is a quasi-equilibrium position between lattice points for Na ions in Na-β"alumina by the analysis of thermal decay profile of PSHB [3]. The result of site selective spectra for LLT:Eu^{3+} suggests that there are off-center positions which can be occupy by Li ions.

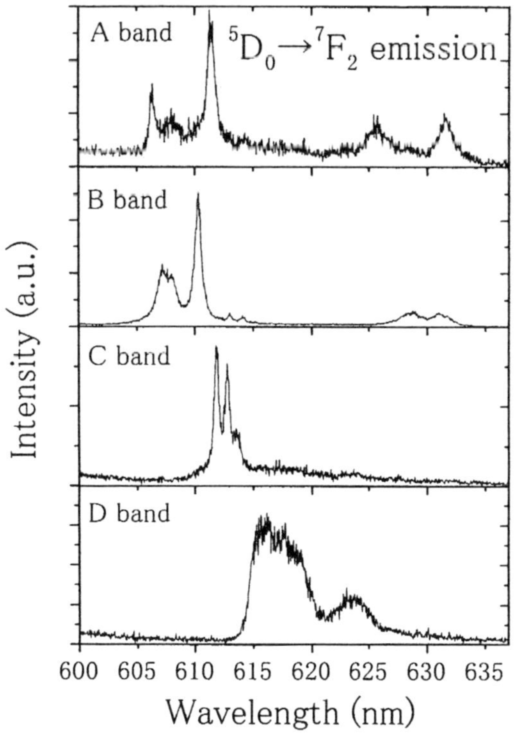

Fig.7 Site selective fluorescence spectra of the $^5D_0$-$^7F_2$ transition of Eu^{3+} ions for A, B, C and D band in LLT:Eu^{3+} at 10K.

4. Summary

We have observed the hole burning in LLT:Eu^{3+} and LLT:Pr^{3+}. The hole in LLT:Eu^{3+} is due to the optical pumping of the hyperfine levels of Eu^{3+} ion. The hole in LLT:Pr^{3+} is due to the localized motion of Li ions near the equilibrium position by a light irradiation because of the fact that the hole can be observed below 21K.

From a luminescence spectrum by exciting at 308nm for LLT:Eu^{3+}, it was found that there are four Eu^{3+} sites which the local crystal field differ from each other. We have applied the site selective spectroscopy for four Eu^{3+} sites in LLT:Eu^{3+}. All Eu^{3+} sites are in the C_{2v} symmetry or the lower than C_{2v} symmetry. Both results suggest that Li ions are located at off-center positions in LLT.

Acknowledgements

This work was partly supported by CREST of JST (Japan Science and Technology).

References

1. H.Yugami, A.Koike, T.Suemoto and M.Ishigame, Phys. Rev. **B44**, 9214 (1991).
2. M.A.Saltzberg and G.C.Farrington, *J. Solid State Chem.* **83**, 272 (1989).
3. T.Hattori, R.Yagi, K.Aso and M.Ishigame, *Solid State Ionics* **136/137**, 409(2000).
4. S.Matuo, H.Yugami and M.Ishigame, *Phys. Rev.* **B64**, 24302 (2001).
5. R.M.Macfarlane and R.M.Shelby, *Opt. Commun.* **45**, 46 (1983).
6. T.Okuno, K.Tanaka, K.Koyama, M.Namiki and T.Suemoto, *J. Lumin.* **58**, 184 (1994).
7. K.W.Jang and R.S.Weltzer, *Phys. Rev.* **B52,** 6431 (1995).
8. H.Koyama and T.Hattori, *Science and Technology of Advanced Materials* **4,** 131 (2003).
9. Y.Inaguma, C.Liquan, M.Itoh, T,Nakamura, T.Uchida, H.Ikuta and M.Wakihara, *Solid State Commun.* **86,** 689 (1993).
10. J.L.Fourquet, H.Duray and M.P.Crosnier-Lopez, *J. Solid State Chem.* **127,** 284 (1996).
11. R.M.Macfarlane and R.M.Shelby, *Opt. Lett.* **9,** 533 (1984).
12. K.Tanaka, T.Okuno, H.Yugami, M.Isigame and T.Suemoto, *Opt. Commun.* **86,** 45 (1991).

RELATION BETWEEN STRUCTURE AND LITHIUM ION CONDUCTIVITY IN $La_{4/3-y}Li_{3y}Ti_2O_6$

YANHUI ZOU

Department of Physics, Faculty of Science, Ehime University, 2-5 Bunkyocho, Matsuyama, Ehime, Japan.
E-mail: zouyanhui@phys.sci.ehime-u.ac.jp

NAOKI INOUE

Department of Physics, Faculty of Science, Ehime University, 2-5 Bunkyocho, Matsuyama, Ehime, Japan.
E-mail: naokinoue@phys.sci.ehime-u.ac.jp

The solid solutions $La_{4/3-y}Li_{3y}\square_{2/3-2y}Ti_2O_6$ (y=0.09~0.33) have been investigated by complex impedance spectroscopy, X-ray diffraction (XRD) and nuclear magnetic resonance (NMR) methods. The ionic conductivity showed maximum value around y=0.21, and kept high values in high y concentration. The XRD patterns showed single phase for all concentration. Rietveld's profile analysis was applied to refine XRD results. The crystal structure is orthorhombic for y=0.09~0.15 and tetragonal for y=0.17~0.33. The ^7Li NMR spectra showed main central peak with a Lorenzian shape for y≦0.21. The central peak was divided into two parts for y≧0.23. Narrow intense peak was a mobile component due to mobile ions, and a small broad central peak was due to less mobile lithium ions which contributed to immobile component. We report in this study the Li^+ locations and occupancy, activation energy, concentration and positions of vacancies, change of ionic conduction mechanism from 2-D to 3-D related to bottleneck size, and explain the reason why the high values of conductivity showed in high y concentration.

1. Introduction

The solid state lithium ion conductors have received considerable attention due to potential applications in solid state devices: rechargeable batteries, electrochemical sensor, electrochemical display, etc. The highest lithium ionic conductivity has been reported for $La_{4/3-y}Li_{3y}\square_{2/3-2y}Ti_2O_6$ (LLTO) with perovskite-like structure, values as high as 10^{-3} S/cm at room temperature [1~3]. In this compound, La^{3+}, Li^+ and vacancies are located in A-sites and B-sites are fully occupied by TiO_6 octahedral. The high ionic conductivity is considered to originate from the presence of vacancies in A-site, numerous equivalent sites for Li^+ to occupy and capacity of Li^+ to move freely through the bottleneck formed by four adjacent TiO_6 octahedra in A-site. Many studies have been performed on this material. The effective carrier concentration and site percolation were found to be predominant factors in conductivity [4]. On the other hand, high value of conductivity in sample with high lithium concentration was explained by assuming the existence of unusual position for Li^+ [5]. The structural analysis led to a tetragonal structure based on $P4/mmm$ space group. Recently the transition from $Pmmm$ to $P4/mmm$ dependent on the lithium concentration was reported [2~7]. The ^7Li nuclear magnetic resonance (NMR) revealed a particular behavior of the relaxation time T_1 [8, 9]. However, the occurrence of high value of conductivity at high lithium concentration, the 2-D or 3-D conduction mechanism and vacancy positions are still debatable and need further study. In this report, the conductivity, conduction mechanism, crystallographic factors such as occupation, vacancy concentration and its position, bottleneck size and local structure which mainly influence the ionic conduction are discussed.

2. Experimental Procedure

The samples $La_{4/3-y}Li_{3y}\square_{2/3-2y}Ti_2O_6$ (y=0.09~0.33) were prepared by conventional solid-state reaction method. The starting materials La_2O_3, Li_2CO_3 and TiO_2 were mixed and heated for a few hours at 300, 500, 800 and 1150℃. After that the powder was ground, pressed into pellets of 10mm diameter and 5-6mm thickness under a pressure of 250 kg/cm^2 and then sintered at 1350℃. During sintering temperature was increased by a rate of 10℃/min from 250℃ to 1350℃ and then sample was cooled down at the same sweep rate to 550℃. Sintered pellets coming from the same preparation run have been used for the powder X-ray diffraction, conductivity and NMR measurements.

The powder X-ray diffraction (XRD) patterns were recorded at room temperature with $CuK\alpha$ radiation and $(\theta/2\theta)$ Bragg geometry with angular range 10°~120°.

Electrical conductivity measurements were carried out by complex impedance spectroscopy using automatically controlled HP4285A precision LCR meter. All measurements were performed on cylindrical pellets with evaporated gold electrodes. The frequency and temperature range were 100Hz~20MHz and 30 ℃ ~112℃, respectively.

^{7}Li NMR spectra at 7T were taken in a solid Fourier transform spectrometer with CMX300 at room temperature. The standard sample was Li_2CO_3 solid solution.

3. Results and Discussion

3.1. *Ionic conductivity*

The bulk conductivity data extracted from the impedance complex plane are shown with some literatures values in Fig.1. The ionic conductivity increases with increasing y concentration and temperature. The maximum ionic conductivity value was obtained around y=0.21 and the same value was at higher concentration. The problems are why the conductivity shows maximum and saturation effect. Especially, high value of conductivity in y=0.33 can not be explained only by vacancy concentration □2/3-2y. We will discuss this aspect in detail in following sections.

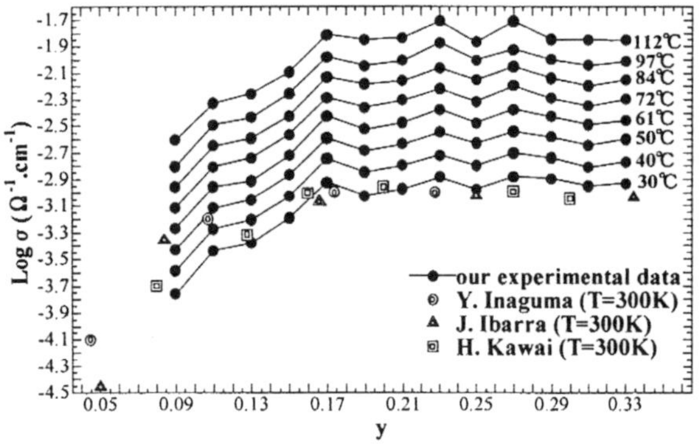

Fig.1. Conductivity against y. The solid circles are our experimental data, and other points are some literature values [2,3,7,10].

3.2. *X-ray diffraction and Rietveld analysis*

The XRD patterns show a single phase for all compositions. Rietveld's profile analysis method was applied to refine the XRD results and it was carried out in orthorhombic structure with space group *Pmmm* (y=0.09~0.15) and tetragonal structure with space group $P4/mmm$ (y=0.17~0.33) as proposed by Ibarra etc. [7]. Table 1 shows atomic positions and coordinates for orthorhombic structure *Pmmm* and tetragonal structure $P4/mmm$. Fig.2 presents Rietveld's refinement profile observed and calculated patterns for y=0.21 and they are in good agreement.

Table 1. Position parameters in the structure.

	Orthorhombic Pmmm		Tetragonal P4/mmm	
Atom	Position	Coordinate	Position	Coordinate
La1, Li1	1a	0,0,0	1a	0,0,0
La2, Li2	1c	0,0,1/2	1b	0,0,1/2
Ti	2t	1/2,1/2,z	2h	1/2,1/2,z
O1	1f	1/2,1/2, 0	1c	1/2,1/2,0
O2	1h	1/2,1/2,1/2	1d	1/2,1/2,1/2
O3	2r	0,1/2,z	4i	0,1/2,z
O4	2s	1/2,0,z	...	...

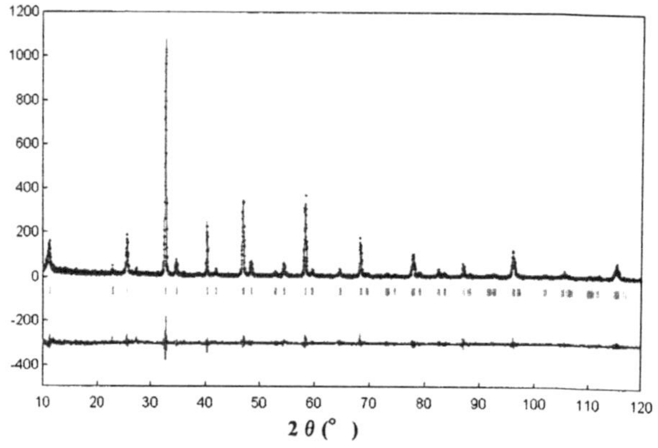

Fig.2. Rietveld's refinement profile of LLTO.

The refinement was done with following models: (1) model 1: XRD intensity is given only by La^{3+} content and no consideration of vacancy position. Li^+ contribution to XRD intensity is neglected due to its low scattering factor. (2) model 2: All vacancies are concentrated in 1b/1c-site. Li^+ content is determined by the formulas, Li1=1-La1 and Li2=1-□-La2. (3) model 3: 75% vacancies are distributed in 1b/1c-site, and 25% in 1a-site. Li^+ content is determined by the formulas, Li1=1-25%□-La1 and Li2=1-75%□-La2. The La^{3+} occupation in A-site by model 1 quite agreed with model 2 in y=0.09~0.21 as shown in Fig.3. Therefore, in this study we chose assumption model with full vacancies occupation in 1b/1c-site in low concentration (y=0.09~0.21). Model 3 in y=0.23~0.33 also showed good agreement with model 1. But as □2/3-2y is very small in y=0.25~0.33, there is a lot of ambiguity. It means vacancy concentration doesn't give any effect for the Li^+ content calculation. However, we assumed at present state that three quarters vacancies distributed in 1b-site and one quarter in 1a-site in high concentration (y=0.23~0.33). With this model the refinement was performed and the structure parameters were obtained. Fig.4 shows the lattice parameters against Li^+ concentration according to our data and others [2,7,9,11~13]. It can be seen that the lattice parameters of a, b and c axes show crystal structure change at y=0.15 and decrease slightly with increasing y concentration. Fig.5 shows A-site ions occupation dependent on y concentration. La^{3+} is mainly assigned to 1a-site and clearly decreases with increasing y content, contrarily the occupancy shows no evident change in 1b/1c-site at low concentration, and slightly increases at high concentration. Li^+ in 1a-site slightly increases and linearly increases in 1b/1c-site with increasing y concentration. Li^+ contents in y=0.25~0.33 couldn't be determined as previously described.

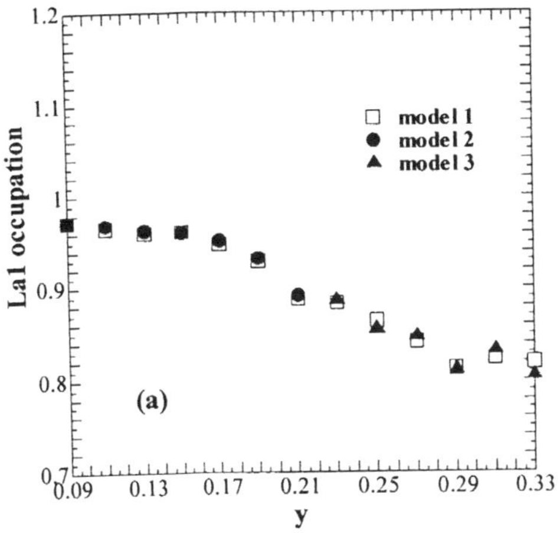

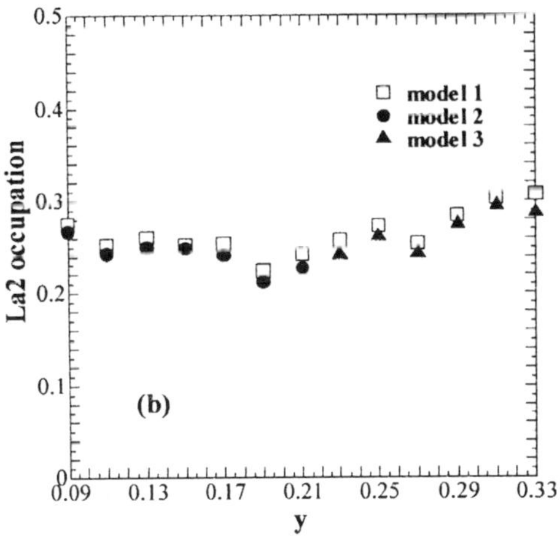

Fig.3. La^{3+} occupation at A-sites according to three models, (a) La1, (b) La2.

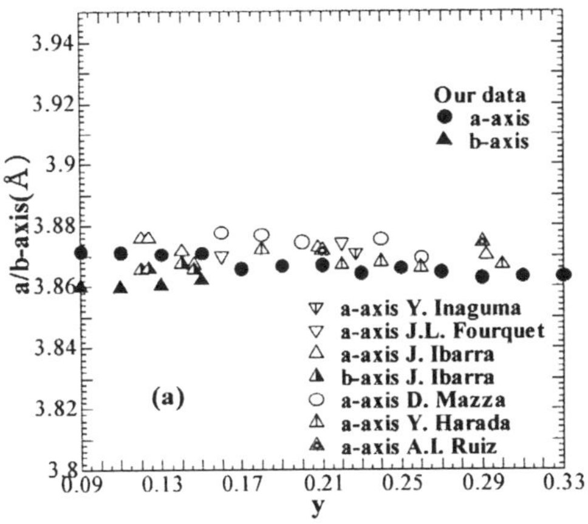

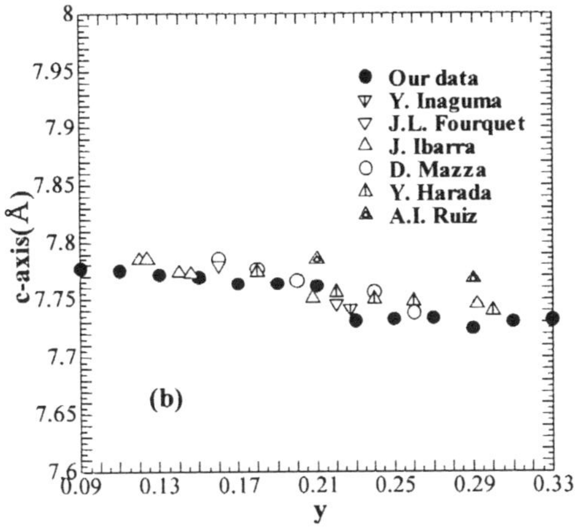

Fig.4. Lattice parameters against y, (a) a and b-axis, (b) c-axis.

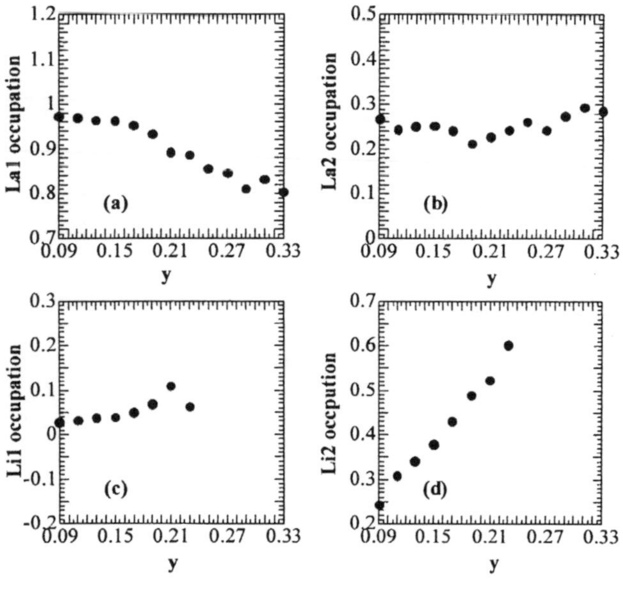

Fig.5. A-site ions occupation, (a) La1, (b) La2,(c) Li1, (d) Li2.

In this structure, Li^+ migration is considered through the bottleneck formed by four TiO_6 octahedra. Therefore, ionic conductivity should be strongly depended on the B-site ions positions. Fig.6 shows O^{2-} and Ti^{4+} positions, and distances between O^{2-} and 1a, O^{2-} and 1b/1c in the structure. In orthorhombic structure (y=0.09~0.15) O^{2-}2s is shifted to 1a-site and O^{2-}2r stands on the intermediate position from 1a-site and 1c-site. In tetragonal structure (y=0.17~0.33) O^{2-}4i is slightly shifted to 1a-site. Ti^{4+} is close to 1b/1c-site in all compositions. The bottleneck size is defined by the smallest cross-sectional area of the interstitial pathway that is constructed by four O^{2-} as shown in Fig.7. Here, 1b-1b, 1a-1a and 1a-1b in tetragonal structure define the bottleneck size via the pathway from 1b to 1b, 1a to 1a and 1a to 1b, respectively. In orthorhombic structure 1b is 1c. Fig.8 shows the bottleneck size against y concentration. As can be seen, the sizes of 1b-1b and 1c-1c bottleneck are larger than 1a-1a bottleneck size in low concentration. 1b-1b bottleneck decreases with increasing y concentration, becomes comparable with 1a-1a and 1a-1b bottleneck sizes in high concentration. Therefore, Li^+ exhibits high occupancy at 1b and 1c-sites and may easily migrate in 1b-1b and 1c-1c at low concentration, which is 2-D ionic conduction mechanism. At high concentration, the Li^+ and vacancy which have occupations in both 1a and 1b-sites can choose three comparable pathways 1b-1b, 1a-1b and 1a-1a, therefore, the ionic migration becomes 3-D ionic conduction mechanism.

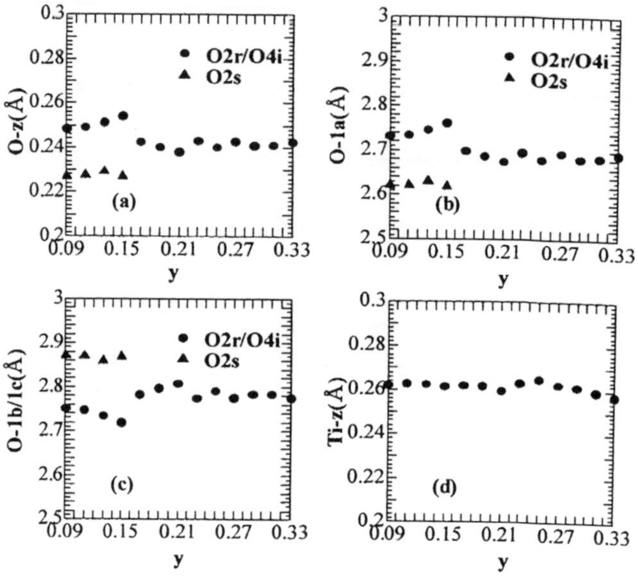

Fig.6. O^{2-} and Ti^{4+} positions in the structure, (a) O-z, (b) O-1a, (c) O-1b, (d) Ti-z.

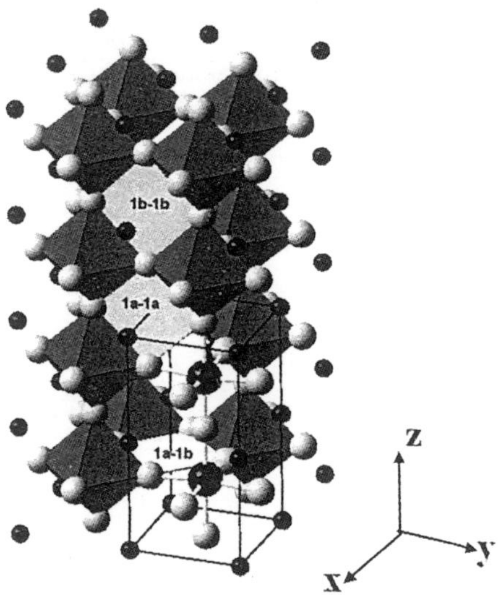

Fig.7. Crystal structure of LLTO in tetragonal unit cell.

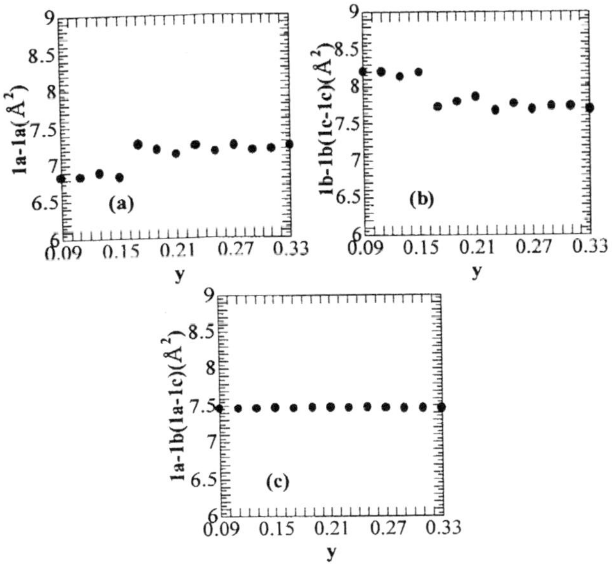

Fig.8. Bottleneck size in the structure, (a) 1a-1a, (b) 1b-1b(or 1c-1c), (c) 1a-1b(or1a-1c).

3.3. *NMR results*

^{7}Li NMR spectra were recorded with static method dependent on y concentration. They show main central peak with a Lorenzian shape for y=0.09~0.21 and two compositions for y=0.23~0.33 in Fig.9. Using Lorenzian fitting the main central peak in y=0.23~0.33 is easily divided into two parts: a narrow large intense peak (mobile component) and a small broad central peak (immobile component) as shown in Fig.10. We indicate this is the first observation of main peak dependent on y concentration. Hirakoso [14] and co-workers have recently observed the main similar peak for α, β form dependent on sintering temperature in $La_{0.56}Li_{0.33}TiO_3$. The mobile lithium ions which give the sharp main peaks are located in highly symmetrical sites with weak static quadrupole interaction. Fig.11 and Fig.12 show the spectral area and the spectral width against y concentration, respectively. The area of mobile component increases with increasing y concentration and almost keeps constant in high concentration (y=0.23~0.33). The area of immobile component is zero in y=0.09~0.21, and appeared in y=0.23~0.33. Here the area corresponds to the lithium content. The spectral width of mobile component is very narrow compared with broad component. That means the mobile component plays role of carrier in this structure. The lithium immobile component can not migrate into the lattice sites and occupy in unusual sites, such as 2e, 2f and 2g of $P4/mmm$. Therefore, the vacancies increase in 1a and 1b-sites in y=0.23~0.33. Fig.13 shows the vacancy concentration estimated by initial formula □(2/3-2y) in open circles and slow components in solid circles in y=0.23~0.33. If we assume the conductivity is proportional to mobile components that are Li^+ mobile content and related to activation energy, we can estimate the conductivity by Li^+ mobile content and the least square value of activation energy. Fig.14 shows the calculated conductivity against y concentration at 30℃. The calculated solid circles and line are in good agreement with our experimental data and some literature values.

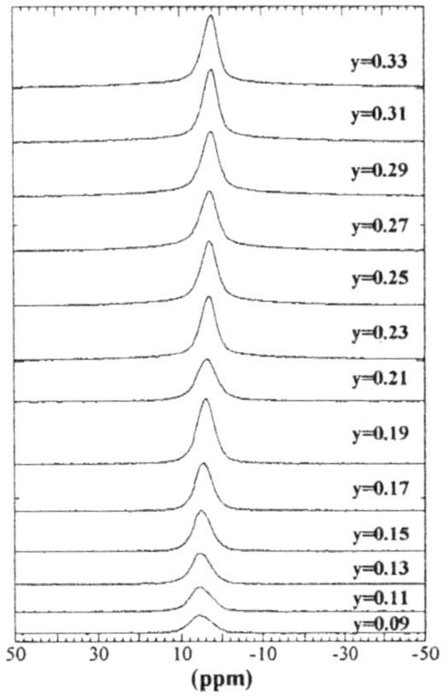

Fig.9. Static spectra of ^{7}Li NMR.

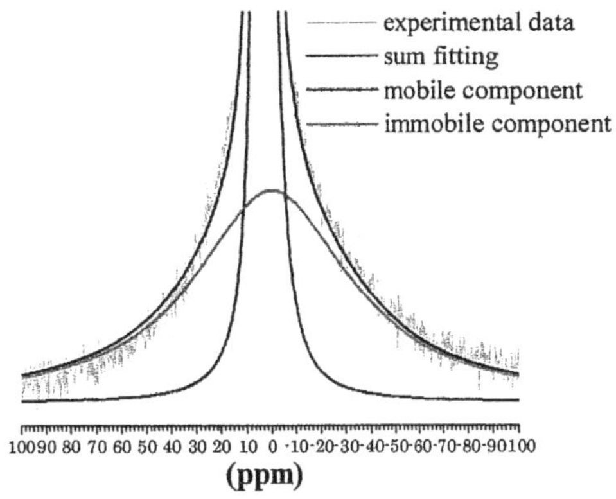

Fig.10. Lorenzian fitting of ^{7}Li NMR spectrum.

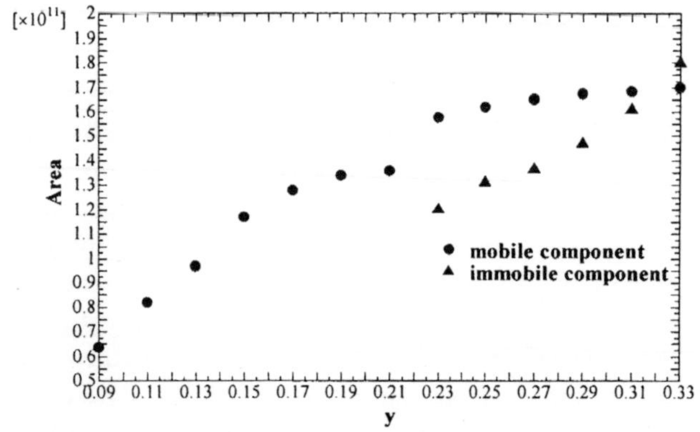

Fig.11. Area under peak for two components in static NMR spectra.

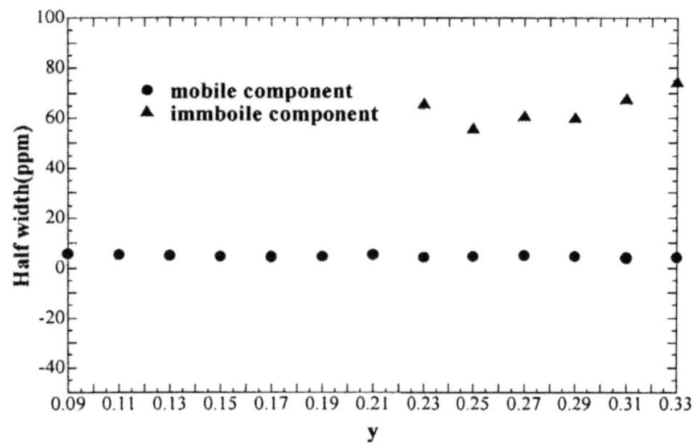

Fig.12. Half width of two components of NMR spectra.

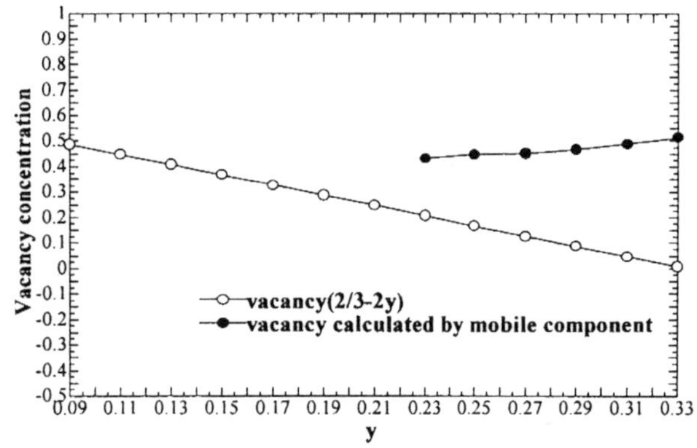

Fig.13. Vacancy concentration in the structure.

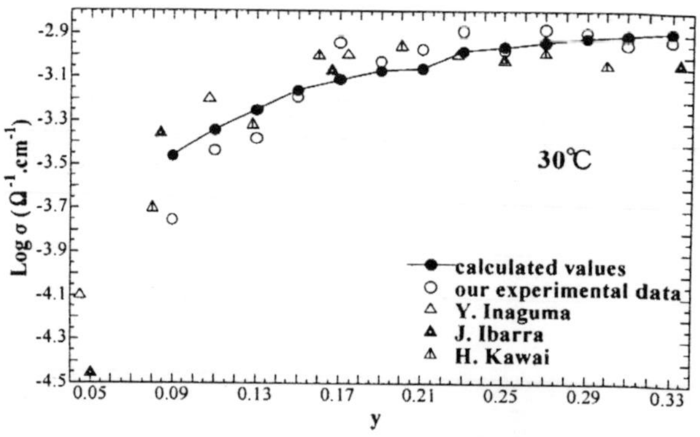

Fig.14. Calculated conductivity and experimental values against y.

4. Conclusion

The impedance spectroscopy showed increase of ionic conductivity with increasing Li^+ concentration and temperature. The maximum value of conductivity was obtained around y=0.21 and saturated in high y concentration. Our research was focused on providing evidence (i) why the conductivity showed the maximum and high values at high concentration, (ii) what is the conduction mechanism in the structure. The XRD patterns showed a single phase for all concentrations. Rietveld's profile analysis method was applied to refine XRD results. The refinement was done with three models. The crystal structure is orthorhombic with *Pmmm* space group for y=0.09~0.15, and tetragonal with $P4/mmm$ space group for y=0.17~0.33. Lattice parameters of a-axis and c-axis in orthorhombic structure were larger than those in tetragonal structure, and gradually decreased with increasing y concentration. In orthorhombic structure b-axis is smaller than a-axis. 1a-site was nearly fully occupied by La^{3+} and 1b-site was preferred by Li^+. With increasing y concentration La^{3+} occupation clearly decreased in 1a-site and showed no evident change in 1b/1c-site. Vacancies were found only in 1b/1c-site for y=0.09~0.21 and became randomly distributed at high concentration. The Li^+ contents were calculated based on the consideration of La^{3+} and vacancy contents. Li^+ slightly increased in 1a-site and linearly increased with increasing y concentration in 1b/1c-site. The bottleneck is constructed by four adjacent TiO_6 which depend on the B-site ions positions. In orthorhombic structure (y=0.09~0.15) O^{2-}2s is shifted to 1a-site and O^{2-}2r stands on the intermediate position from 1a-site and 1b-site. In the tetragonal structure (y=0.17~0.33) $O^{2-}4i$ is slightly shifted to 1a-site. The ionic conduction should be related with the larger 1b-1b and 1c-1c bottleneck sizes and full vacancy concentration in 1b/1c-site for y≦0.21, and three comparable bottleneck sizes 1b-1b, 1a-1b and 1a-1a with random distribution of vacancy concentration for y≧0.23. Therefore, it was concluded that the Li^+ ionic conduction was 2-D in y≦0.21 and changed to 3-D ionic conduction mechanism in y≧0.23, and which is the first description of conductivity mechanism in LLTO studies.

The ^7Li NMR spectra showed main central peak with a Lorenzian shape for y≦0.21. Using Lorenzian fitting, the central peak was divided into two parts for y≧0.23. Narrow large intense peak is a mobile component due to mobile ions in the structure, and a small broad central peak is attributed to less mobile lithium ions which contributed to immobile component. The mobile component increased with increasing y concentration, saturated at high concentration at a constant value. The immobile component was zero for y≦0.21 and appeared for y=0.23~0.33. Li^+ giving rise to the mobile component play carrier role in the structure. Contrarily, Li^+ contributing to the immobile component can not migrate into the lattice sites, and probably occupy unusual sites. As the result, vacancy concentration increased and vacancies were randomly distributed in 1a and 1b sites. Using the mobile lithium ion content and activation energies, we estimated the y concentration dependence behavior of conductivity. The calculated conductivity was in good agreement with the experimental results. That is the first explanation of the ionic conductivity behavior in LLTO.

Acknowledgments

We thank Integrated Center for Sciences and Center for Cooperative Research and Development Ehime University for supporting NMR and XRD experiments.

References

1. A.G. Belous, G.N. Novitskaya, S.V. Polyanetskaya, and Yu.I. Gornikov, Izv. Akad. Nauk SSSR. Neorg. Mater. **23**, 470 (1987).
2. Y. Inaguma, L, Chen, M. Itoh, T. Nakamura, T. Uchida, H. Ikuta and M. Wakihara, Solid State Commun. **86**, 689 (1993).
3. Y. Inaguma, L. Chen, M. Itoh and T. Nakamura, Solid State Ionics **70/71**, 196 (1994).
4. Y. Inaguma and M. Itoh, Solid State Ionics **86-88**, 257 (1996).
5. J.A. Alonso, J. Sanz, J. Santamaria, C. Leon, A. Varez and M.T. Fernadaz-Diaz, Angew. Chem. Int. Ed. **39**, 619 (2000).
6. A.D. Robertson, S. Garcia-Martin, A. Coats and A.R. West, J. Mater. Chem. **127**, 1405 (1995).
7. J. Ibarra, A. Varez, C. Leon, J. Santamaria, L.M. Torres-Martineg and J. Sang, Solid State Ionics **134**, 219 (2000).
8. J. Emery, O. Bohnke, J.L. Fourquet, J.Y. Buzare, P. Florian and D. Massiot, J. Phys.: Condens. Matter. **14**, 523 (2002).
9. J.L. Fourquet, H. Duroy and M.P. Crosnier-Lopez, J. Solid State Chem. **127**, 283 (1996).
10. H. Kawai and J. Kuwano, J. Electrochem. Soc. **78**, 141 (1994).
11. A.I. Ruiz, M.L. Lopez, M.L. Veiga and C. Pico, Solid State Ionics **112**, 291 (1998).
12. D. Mazza, S. Ronchetti, O. Bohnke, H. Duroy and J.L. Fourquet, Solid State Ioincs **149**, 81 (2002).
13. Y. Harada, T. Ishigaki, H. Kawai and J. Kuwano, Solid State Ionics **108**, 407 (1998).
14. Y. Hirakoso, Y. Harada, J. Kuwano, Y. Saito, Y. Ishikawa and T. Eguchi, Key Engineering Materials **169-170**, 209 (1999).

Theoretical Investigation of Lithiation of Intermetallic Anode Materials: InSb, Cu_2Sb and η'-Cu_6Sn_5

S. Sharma* and C. Ambrosch-Draxl
Institute for Theoretical Physics, Karl-Franzens-Universität Graz, Universitätsplatz 5, A-8010 Graz, Austria.
(Dated: May 16, 2004)

In this work the mechanism of Li insertion/intercalation in the anode materials InSb, η'-Cu_6Sn_5 and Cu_2Sb is investigated by the means of first-principles total-energy calculations. The average insertion/intercalation voltage and volume expansion for transition from InSb to Li_2InSb, Cu_2Sb to Li_2CuSb and Cu_6Sn_5 to Li_2CuSn are calculated and are found to be in good agreement with the experimental values.

Keywords: Batteries, Li-ion, Anode, FPLAPW.

1. Introduction

Light weight and compact, lithium-ion batteries [1] are ideal energy storage devices for the use in appliances like laptops, mobile phones and electric vehicles. But there are concerns about the safety of the lithium-ion batteries in their present form, because, the most common material used as anodes in these batteries is graphite and its lithiated potential is very close to that of the lithium. Intermetallic compounds present an attractive alternative to graphite as anode materials in Li ion insertion batteries due in particular to the high capacity, an acceptable rate capability and operating potentials well above the potential of metallic lithium. In search of new anode materials one of the desirable qualities is that the structural and volumetric distortion of the host anode material on lithiation are small. This was found to be achieved if there exists a strong structural relation [2–4] between the host and its lithiated products. Some of the compounds which show such structural relations are Cu_6Sn_5[5–7], InSb [2, 4, 8–10], Cu_2Sb [3], MnSb, Mn_2Sb [11], SnSb [12], GaSb [10] *etc.* These compounds can be divided into three main structure types, NiAs, ZnS and Cu_2Sb. In these materials, one of the components (Cu, In, Mn, Sn, Ga) is less active and is extruded on lithiation, while the other component provides the structural skeleton for further lithiation.

Since it is desirable for an ideal anode material to reversibly insert/intercalate Li ions while maintaining the structural stability, the knowledge of the exact amount of Li accommodated within the structure before the extrusion of the less active component is extremely important. To this extent we perform total energy calculations for studying the mechanism of Li insertion/intercalation in one representative compound from each of the structure types. These are InSb which belongs to the zinc blende structure, η'-Cu_6Sn_5 with NiAs type structure and Cu_2Sb which is the prototype for the third class of materials.

*Electronic address: `sangeeta.sharma@uni-graz.at`

2. Methodolody

Total energy calculations are performed using the full potential linearized augmented planewave FPLAPW method implemented in the WIEN2k code [13]. The scalar relativistic Kohn-Sham equations are solved in a self-consistent scheme. For the exchange-correlation potential we use the local density approximation (LDA). All the calculations are converged in terms of basis functions as well as in the size of the k-point mesh representing the Brillouin zone.

For all the energy differences reported in the present work optimized crystal structures are used, i.e. the atomic positions and the unit cell volumes have been varied until the global energy minimum has been found.

3. Results

3.1. Crystal structures

In order to study the lithiation mechanism in detail we first need to look at the crystal structures of the host anode materials for possible voids or lack of them. InSb exists in the zinc blende structure with In atoms at (0,0,0) and the Sb atoms occupying four tetrahedral sites: (0.25,0.25,0.25), (0.75,0.25,0.75), (0.75,0.75,0.25) and (0.25,0.75,0.75). This kind of structural arrangement leaves four tetrahedral sites vacant, namely (0.75,0.25,0.25), (0.25,0.25,0.75), (0.25,0.75,0.25) and (0.75,0.75,0.75). Along with this there is another void in the structure at (0.5,0.5,0.5).

Cu_2Sb crystallizes in a tetragonal structure with space group P4/nmm and 2 formulae units per unit cell. The Cu atoms occupy the sites (0.75,0.25,0), (0.25,0.75,0.0), (0.25,0.25,0.27) and (0.75,0.75,0.73) while the Sb atoms are situated at (0.25,0.25,0.70) and (0.75,0.75,0.3). There exists a small void in the structure at (0.5,0.5,0.5).

Unlike InSb and Cu_2Sb the unit cell for η'-Cu_6Sn_5 is not as simple. The chemical unit cell (CUC) has the space group C2/c with four different kinds of Cu atoms at sites 4e, 4a and 8f and three different kinds of Sn atoms at sites 4e and 8f. There are a total of 4 formula-units (44 atoms) per CUC [14]. Thus in order to clearly see the voids for possible insertion/intercalation we refer the reader to look at the total crystalline valence charge density in Fig. 5 of Ref. 7. With this information we go on to study how well the voids in the various crystal types accommodate Li atoms.

3.2. Lithiation Mechanism

On lithiation of the host anode materials there are two distinct probable processes: 1) The incoming Li ions go to one of the voids mentioned above. 2) A substitutional reaction takes place, i.e., the less active component (In or Cu) is expelled and is replaced by Li ions.

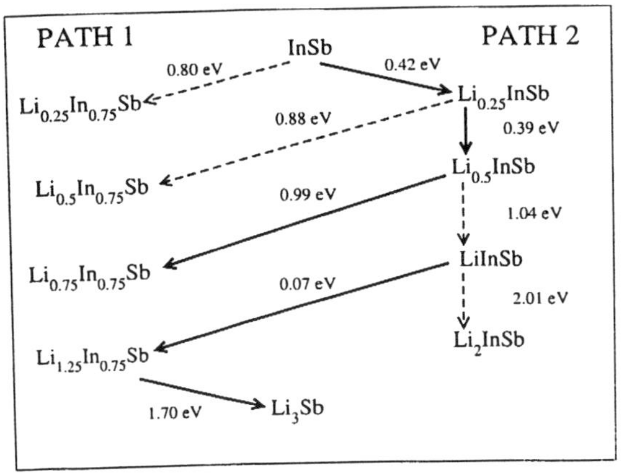

FIG. 1: Possible lithiation reaction paths for InSb. The energies are given in eV per formula unit. The energetically favourable path is marked by the bold arrows.

In this regard, recently the electrochemistry of Li insertion in InSb was studied experimentally and contradictory results were published: Vaughey *et al.* [2] reported that 2 Li ions can be incorporated within the InSb structure with very small volume expansion (5.6%) and without In extrusion. They proposed the reaction mechanism to be

$$(x + y)\mathrm{Li} + \mathrm{InSb} \rightarrow \mathrm{Li}_{x+y}\mathrm{In}_{1-y}\mathrm{Sb} \qquad (1)$$

for $0 \leq x \leq 2$ and $0 \leq y \leq 1$. This finding was further supported by Johnson *et al.* [8] On the contrary, Hewitt *et al.* [9]. claimed that x reached a maximum value of 0.27 before In extrusion. In yet another work by Tostmann *et al.* [15], this value of x was found to be close to 1 before In extrusion. In light of the existing controversy there is a need for clarification of exact lithiation path. Using the first-principles total-energy calculations we point out the energetically favourable reaction path in Fig 1. The energy required to go from one structure to another is written on the arrows where the energetically favourable reaction path is indicated by bold arrows. It is found that initially it is energetically favourable for Li atoms to incorporate in the voids of the InSb structure (path 2). After the formation of $\mathrm{Li}_{0.5}\mathrm{InSb}$ substitutional reaction is energetically favourable, but the energy difference between the two paths is very small. If this energy barrier of 0.05 eV could be over come under certain experimental conditions, then formation of LiInSb would lead to a large volume expansion of 14.4 %. On further lithiation energetics clearly favour substitutional reaction. Our suggested reaction mechanism best agrees with the previous suggestions by Hewitt *et al.* [9]. The unit cell volume expansion for transition from InSb to $\mathrm{Li}_2\mathrm{InSb}$ is calculated to be 15.4 %, while the experimental [2] data is reported to be 5.6%. Our calculated volume expansion for transition from InSb to $\mathrm{Li}_{0.5}\mathrm{InSb}$ is 5.3 %. The calculated average intercalation voltage (AIV) for a transition from InSb to LiInSb is 1.83 V, while for transition from InSb to $\mathrm{Li}_{0.5}\mathrm{InSb}$ is 0.82 V. The experimental value [8] of the plateau in the voltage profile in the range 0.62V - 0.75V, depending upon the sample being a single crystal or ball milled. As pointed above, the initial incoming Li atoms have the freedom to occupy either position (0.75,0.25,0.25) or (0.5,0.5,0.5). We find that the probability for Li insertion at the position (0.75,0.25,0.25) is higher than at (0.5,0.5,0.5). The difference in energy between the two structures, $\mathrm{Li}_{0.25}\mathrm{InSb}$ with Li at (0.75,0.25,0.25) and $\mathrm{Li}_{0.25}\mathrm{InSb}$ with Li at (0.5,0.5,0.5), is 0.05 eV per formula unit.

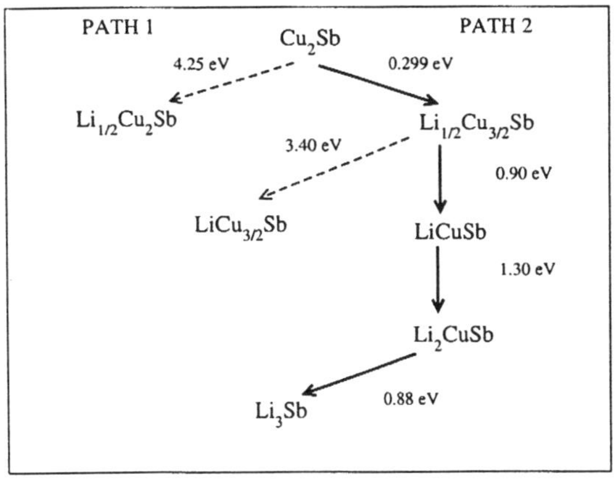

FIG. 2: Possible lithiation reaction paths for $\mathrm{Cu}_2\mathrm{Sb}$. The energies are in eV per formula unit. The energetically favourable path is marked by bold arrows.

The lithiation reaction path for $\mathrm{Cu}_2\mathrm{Sb}$ is shown in Fig. 2. Unlike InSb, in this case the Cu atoms are expelled from the structure very early on in the reaction. In fact, it seems impossible to insert/intercalate Li ions into the structure without Cu extrusion. These results are in accordance with the reaction path previously suggested by Fransson *et al.* [3]. The first incoming Li atoms replace the Cu atom at site (0.25,0.25,0.70). This leads to a structural reorganization such that an ideal face-centered-cubic CuSb arrangement is formed. Further lithiation leads to the extrusion of one more Cu atom to form LiCuSb which is a zinc blende structure with the Cu atom at (0,0,0), the Sb atom at (0.25,0.25,0.25) and the Li atom at (0.5,0.5,0.5). Further lithiation leads to the formation of Li_2CuSb structure with an extra Li atom at (0.75,0.25,0.25) compared to the LiCuSb structure. The theoretically optimized structure of Li_2CuSb shows a volume expansion of 26.4% per formula unit compared to the initial compound Cu_2Sb. This is in good agreement with the experimental value of 25.2%. The calculated value of the AIV for going from Cu_2Sb to Li_2CuSb is 0.98 V, which is slightly higher than the experimental value of 0.82 V [3].

In situ X-ray diffraction (XRD) studies have shown that, during the electrochemical reaction of potential anode material η'-Cu_6Sn_5, with lithium Li_2CuSn is formed [16–18]. The ideal reaction for this phase transformation is described by the following reaction:

$$Cu_6Sn_5 + 10Li \longrightarrow 5Li_2CuSn + Cu. \qquad (2)$$

Further lithiation of the Li_2CuSn structure results in further extrusion of Cu from the structure, to finally form $Li_{4.4}Sn$. If the material is fully lithiated to form $Li_{4.4}Sn$, the cycling stability is dramatically affected and the ability to insert lithium reversibly gradually declines. The detailed mechanisms of the reaction above (Eq. 2), for example the positions of the initial Li in the η'-Cu_6Sn_5 structure and the starting point of Cu extrusion from the structure are not known. Although there have been suggestions regarding the mechanism of the phase change from η'-Cu_6Sn_5 to Li_2CuSn [16, 19], there exists no experimental or theoretical calculation to back these suggestions.

By total energy calculations it was found that the energy requirement for the insertion of a Li atom at site (0.907,0.625,0.0408) (labeled A in Fig. 5 of Ref. 7) is minimum as compared to any other void in the structure. As can be seen from Fig. 3 the energy required for the substitutional reaction is 2.63 eV per formula unit. Thus energetics clearly favours the insertion/intercalation reaction. On complete relaxation of the structure with eight Li atoms per CUC (i.e. formation of $Li_2Cu_6Sn_5$) at the eight crystallographically symmetric A sites, a very small volume expansion of 4.6% is calculated. This volume expansion could not be seen experimentally [7].

There are two scenarios for further lithiation, one where lithium is inserted without any Cu extrusion and the other where the lithium insertion is accompanied by extrusion of the Cu atoms. It was found that the first case (*i.e.* formation of $Li_3Cu_6Sn_5$) is accompanied by a volume expansion of 47.3 % and costs large amount of energy (11.83 eV per formula unit). While, the substitutional reaction costs very small energy of 0.27 eV per formula unit. We, therefore, suggest that after the insertion of two Li atoms per Cu_6Sn_5 formula unit, the Cu atoms start to extrude from the structure accompanied by structural rearrangements. This picture is supported by our XRD measurements which does not show any detectable structural change up to 2 Li atoms per Cu_6Sn_5 formula-unit but indicates a major structural rearrangement beyond that [7]. The calculated value of the AIV for the transformation from Cu_6Sn_5 to Li_2CuSn is 0.378V which is in good agreement with the experimental value of the plateau in the voltage profile [7].

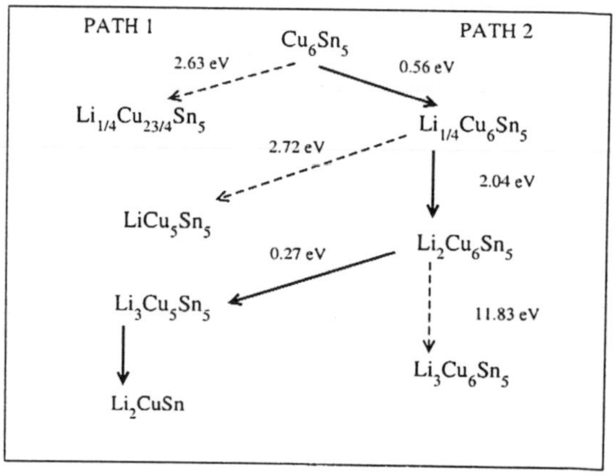

FIG. 3: Possible lithiation reaction paths for η'-Cu_6Sn_5. The energies are in eV per formula unit. The energetically favourable path is marked by bold arrows.

4. Conclusions

To conclude, we have investigated the three prototype structures, η'-Cu_6Sn_5, InSb and Cu_2Sb, for the use as anode materials in Li-ion batteries. From the first-principles total-energy calculations the energetically favourable lithiation path for these materials has been determined. The three compounds studied in the present work show different behaviour. For InSb insertion/intercalation reaction is found to be more favourable uptill the formation of $Li_{0.5}InSb$. Beyond this substitutional reaction is favoured by the energetics, while for Cu_2Sb the substitutional reaction is favoured by the energetics right from the begining. Total-energy and force calculations show that after the insertion of two Li atoms per Cu_6Sn_5 formula-unit the Cu atoms are extruded form the structure leading to large structural rearrangement.

The calculated values of volume expansion and AIV can be directly compared with experiments. The volume expansion on going from InSb to Li_2InSb is calculated to be 15.4% which does not agree with experimental value of 5.6%, on the other hand our calculated volume expansion for transition InSb to $Li_{0.5}InSb$ is much closer to this value. The AIV for the transition InSb $\rightarrow$ $Li_{0.5}InSb$

is 0.82 V and for InSb → LiInSb is 1.8 V. The volume expansion for the transition $Cu_2Sb \rightarrow Li_2CuSb$ is calculated to be 26.5 % which is in very good agreement with the experimental data (26.5%). The AIV for this transition is about 19.5 % higher than the experimental value. The AIV of the η'-Cu_6Sn_5 to Li_2CuSn transformation has been calculated to be 0.38 V in excellent agreement with the experimental result of $\sim$ 0.4V.

Acknowledgements

We thank the Austrian Science Fund (project P16227) for financial support. SS would also like to thank K. Edström and M. M. Thackeray for useful discussions and suggestions during the course of this work.

References

[1] J.-M. Tarascon and M. Armand, Nature **414**, 359 (2001).
[2] J. T. Vaughey, J. O'Hara, and M. M. Thackeray, Electrochem. Solid-State Lett. **3**, 13 (2000).
[3] L. M. L. Fransson, J. T. Vaughey, R. Benedek, K. Edström, J. O. Thomas, and M. M. Thackeray, Electrochem. Commun. **3**, 317 (2001).
[4] A. J. Kropf, H. Tostmann, C. S. Johnson, J. T. Vaughey, and M. M. Thackeray, Electrochem. Commun. **3**, 244 (2001).
[5] E. Nordström, S. Sharma, E. Sjöstedt, L. Fransson, L. Häggström, and K. Edström, Hyperfine Interact. **136**, 555 (2001).
[6] R. Benedek and M. M. Thackeray, J. Power Sources **110**, 406 (2002).
[7] S. Sharma, L. Fransson, E. Sjöstedt, L. Nordströsm, B. Johansson, and K. Edström., J. Electrochem. Soc. A **150**, 330 (2003).
[8] C. S. Johnson, J. T. Vaughey, M. M. Thackeray, T. Sarakonsri, S. A. Hackney, L. Fransson, K. Edtröm, and J. O. Thomas, Electrochem. Commun. **2**, 595 (2000).
[9] K. C. Hewitt, L. Y. Beaulieu, and J. R. Dahn, J. Electrochem. Soc. A **148**, 402 (2001).
[10] J. T. Vaughey, C. S. Johnson, A. J. Krof, R. Benedek, M. M. Thackeray, H. Tostmann, T. Sarakonsri, S. Hackney, L. Fransson, K. Edström, et al., J. Power Sources **97**, 194 (2001).
[11] L. Fransson, J. Vaughey, K. Edström, and M. Thackeray, J. Electrochem. Soc. A **150**, 86 (2003).
[12] H. Li, L. H. Shi, W. Lu, X. J. Huang, and L. Q. Shen, J. Electrochem. Soc. **148**, 915 (1999).
[13] P. Blaha, K. Schwarz, G. K. H. Madsen, D. Kvasnicka and J. Luitz, WIEN2k, Vienna University of Technology An Augmented Plane Wave + Local Orbitals Program for Calculating Crystal Properties revised edition 2001.
[14] A. K. Larsson, L. Stenberg, and S. Lidin, Acta. Cryst. B **50**, 636 (1994).
[15] H. Tostamann, A. J. Kropf, C. S. Johnson, J. T. Vaughey, and M. M. Thackeray, Phys. Rev. B **66**, 014106 (2002).
[16] K. D. Kepler, J. T. Vaughey, and M. M. Thackeray, Electrochem. Solid State Lett. **2**, 307 (1999).
[17] D. Larcher, L. Y. Beaulieu, D. D. MacNeil, and J. R. Dahn, J. Electrochem. Soc. **147**, 1658 (2000).
[18] L. Fransson, E. Nordström, K. Edström, L. H. ström, J. T. vaughey, and M. M. Thackeray, J. Electrochem. Soc. A **149**, 736 (2002).
[19] M. M. Thackeray, J. T. Vaughey, A. J. Kahaian, K. D. Kepler, and R. Benedek, Electrochem. Commun. **1**, 111 (1999).

DISPERSION OF PERMITTIVITY IN IONIC AND MIXED CONDUCTORS

JÓZEF R. DYGAS*

Faculty of Physics, Warsaw University of Technology, Koszykowa 75, 00-662 Warszawa, Poland

Increase of the ac conductivity with increasing frequency, observed in ionic and mixed electronic/ionic conductors, is accompanied by dispersion of permittivity. In a range of materials this dispersion is well described by the Cole-Cole relaxation function, which gives the power law frequency dependence of conductivity in the limit of high frequencies. The relaxation strength $\Delta\varepsilon$, the dc conductivity σ_0 and the relaxation frequency ω_C obey an empirical relation: $\sigma_0 = P\varepsilon_0\Delta\varepsilon\omega_C$, where P is a coefficient of the order of 1. Experimental results for ionic conductors: single crystals of BICUVOX, glasses from the LiO-SnO_2-TiO_2-P_2O_5 system and polymer electrolytes – lithium salt dissolved in PEO, as well as mixed electronic/ionic conductor – lithium manganese spinel, agree with the proposed model. Value of the coefficient P is characteristic for given material and provides information about the length scale and time scale of local, non-random motion of hopping charge carriers.

1. Introduction

The universal ac response of systems with ionic, electronic or polaronic hopping charge carriers has been characterized by Jonscher [1] as limiting power law frequency dependence of the conductivity observed when the ac conductivity increases above its dc value. Some authors confined presentation of the ac response to logarithmic plots of conductivity versus frequency, both in case of experimental data [2] and theoretical models [3] or computer simulations [4]. Analysis was often based on relation between the dc conductivity, σ_0, and the frequency dependent real part, $\sigma'(\omega)$, of the complex ac conductivity, which is roughly given by:

$$\sigma'(\omega) = \sigma_0\left[1 + (\omega/\omega_O)^n\right] , \qquad (1)$$

where the power law exponent n ranged between 0 and 1. The onset frequency ω_O of the conductivity dispersion was interpreted as the rate of charge carrier hopping [5,6].

Analysis of the complex conductivity $\sigma^*(\omega)$ data by inspection only of the real part $\sigma'(\omega)$, disregards peculiarities of the imaginary part, which are revealed when the complex conductivity is expressed by the complex dielectric function, $\varepsilon^*(\omega)=\varepsilon'(\omega)-\mathrm{j}\varepsilon''(\omega)$:

$$\sigma^*(\omega) = \sigma_0 + \mathrm{j}\omega\varepsilon_0\varepsilon^*(\omega) , \qquad (2)$$

where ε_0 is permittivity of vacuum, j denotes imaginary unit.

Simple expression for the complex dielectric function, which gives the power law frequency dependence of conductivity according to Eq. (1), is:

$$\varepsilon^*(\omega) = \frac{\sigma_0}{\varepsilon_0\omega_O\cos(n\pi/2)}\left(\mathrm{j}\frac{\omega}{\omega_O}\right)^{n-1} \qquad (3)$$

Dielectric function of Eq. (3) cannot reproduce a finite change $\Delta\varepsilon$ of the dielectric constant $\varepsilon'(\omega)$, which often accompanies the onset of increase of the real part of conductivity with increasing frequency. Dispersion of the dielectric constant $\varepsilon'(\omega)$ and peak of the dielectric loss $\varepsilon''(\omega)$ at frequencies near the onset of conductivity dispersion have been observed in glasses [7]. Empirical relation between the dc conductivity, σ_0, the finite change of the dielectric constant (relaxation strength), $\Delta\varepsilon$, and the frequency of relaxation, ω_C, was formulated by Barton [8], Nakajima [9] and Namikawa [10]:

$$\sigma_0 = P\omega_C\varepsilon_0\Delta\varepsilon , \qquad (4)$$

where P is a dimensionless coefficient of the order of 1.

The BNN relation has recently been applied as basis for temperature-frequency scaling of the ac conductivity of ionic glasses [11], disordered solids [12] and generally ion-conducting solids [13]. An

*E-mail: jrdygas@if.pw.edu.pl, fax: 004822 6282171

increase of dielectric constant with decreasing frequency, leading to saturation at a constant value, $\varepsilon_S = \varepsilon_\infty + \Delta\varepsilon$, was observed in crystalline ionic conductors at frequencies higher than the range, where the interfacial impedance of ion-blocking electrodes dominated the impedance [14,15]. Similar observations were made for electronic conductors, in which charge carriers are small polarons [16].

In this work, parameters of the ac response are obtained by nonlinear least squares fitting of a model, which incorporates dispersion of permittivity, to impedance spectra measured at various temperatures. Results are compared for four different materials: single crystalline oxide ion conductor $Bi_2Cu_{0.1}V_{0.9}O_{5.35}$, amorphous lithium ion conductors: inorganic glass of the system Li_2O-SnO_2-TiO_2-P_2O_5 and polymer electrolyte based on poly(ethylene oxide) as well as mixed electronic/ionic conductor $LiMn_2O_4$ with hopping polarons as major charge carriers.

2. Measurement and Analysis of Impedance Spectra

Impedance measurements were made in the frequency range from 10^{-2} to 10^7 Hz. Since the onset frequency of conductivity dispersion is proportional to the value of dc conductivity, measurements at relatively low temperature, when conductivity is low, are best suited for study of permittivity dispersion. Analyzers capable to measure impedance of high absolute value were used: Solarton 1260 augmented by Keithley 428 current amplifier [17] or Novocontrol Alpha-N, both covering the range up to 10^{13} Ω. Opposing faces of solid samples were polished and covered with sputtered gold or platinum films serving as electrodes. Polymer electrolyte foils were placed between gold plated stainless steel disks equipped with controlled spring loading. Good contact between polymer and metal was obtained after softening of polymer during the first heating run.

Complex impedance spectra were analyzed by non-linear least squares fitting of the model impedance function to data. For an experimental spectrum consisting of m complex values (impedance or admittance) $Y_i^* = Y_i' + jY_i''$ measured at frequency f_i, i=1,2, ... m and the model response function $F_Y^*(f,X)$ of frequency and r parameters: $X = \{x_1, x_2 ... x_r\}$, deviation of model from data is: $\Delta Y_i^* = Y_i^* - F_Y^*(f_i, X)$. The minimized objective function was a weighted sum of squared real and imaginary parts of deviations:

$$Q = \sum_{i=1}^{m} \left[w_{1,i} (\Delta Y_i')^2 + w_{2,i} (\Delta Y_i'')^2 \right] \quad (5)$$

with modulus weights $w_{1,i} = w_{2,i} = |Y_i|^{-2}$ or weights based on variance of random errors sampled by repetitive measurements of impedance at each frequency [18, 19]. Quality of fit was judged by the minimum value Q_{min} of the objective function and by width of confidence limits for estimated parameters.

3. Model of the ac Response

It is proposed that the complex conductivity, Eq. (2), can be modeled by a superposition of the dc conductivity σ_0 and the frequency dependent contribution given by the complex dielectric function in form of the Cole-Cole relaxation function [20]:

$$\varepsilon^*(\omega) = \varepsilon_\infty + \frac{\Delta\varepsilon}{1 + (j\omega/\omega_C)^{1-n}}, \quad (6)$$

where ε_∞ is the high frequency value of dielectric constant. In the limit of high frequency, $\omega \gg \omega_C$, the real part of conductivity given by Eq. (2) and Eq. (6) follows the power law:

$$\sigma'(\omega) \cong \sigma_0 + \varepsilon_0 \Delta\varepsilon \cos(\pi n/2) \omega_C^{1-n} \omega^n \quad (7)$$

The frequency of permittivity relaxation, ω_C, is not identical with the onset frequency of conductivity dispersion, ω_O, and frequency dependence of conductivity near the onset of dispersion is different than in Eq. (1). Comparison of Eq. (7) and Eq. (1) results in relation between ω_O and ω_C:

$$\omega_O = \left[\frac{\sigma_0}{\varepsilon_0 \Delta\varepsilon \omega_C \cos(\pi n/2)} \right]^{1/n} \omega_C \quad (8)$$

Using coefficient $P = \sigma_0/(\varepsilon_0 \Delta\varepsilon \omega_C)$ defined by Eq. (4):

$$\omega_O = \left[\frac{P}{\cos(\pi n/2)} \right]^{1/n} \omega_C \quad (9)$$

Thus, only when $P \cong 1$, as often found in glasses, the onset frequency and the frequency of relaxation are approximately equal: $\omega_O \cong \omega_C$. Generally, $P > 1$ and $\omega_O > \omega_C$.

In order to reproduce flat dielectric loss observed at high frequencies in amorphous ionic conductors, a further term was added to Eq. (6)

$$\varepsilon_{\mathrm{L}}^{*}(\omega) = A_{\mathrm{L}}\left(\mathrm{j}\omega/1\,\mathrm{Hz}\right)^{m-1}, \qquad (10)$$

where typical value of the exponent is $m \cong 0.8$.

In polymer electrolytes and in lithium-manganese spinel an additional dipolar relaxation was seen at high frequencies which was modeled by a second Cole-Cole term added to dielectric function of Eq. (6), with the relaxation strength $\Delta\varepsilon_{\mathrm{D}}$ not related to the dc conductivity.

For ionic conductors an equivalent circuit representing interfacial impedance of ion-blocking metal electrodes was placed in series with the resistance representing the dc conductivity. Lithium manganese spinel did not exhibit electrode polarization.

4. Results

4.1. *Single crystal of oxide ion conductor BICUVOX*

Oxide ion conductors, derived from $Bi_4V_2O_{11}$ by partial substitution of vanadium by other metal, exhibit high ionic conductivity at low temperature, when the high temperature γ phase is stabilized [21]. The crystal structure of $Bi_4V_2O_{11}$ is built of alternating $VO_{3.5}^{2-}$ oxygen-deficient sheets and $Bi_2O_2^{2+}$ layers perpendicular to the c-axis [22]. Single crystals of BICUVOX $Bi_2Cu_{0.1}V_{0.9}O_{5.35}$ are highly anisotropic. Conductivity in the direction parallel to the layers of the structure is larger by a factor of about 100 than in the direction of the c-axis [23]. Conductivity of single crystals at temperature below about 770 K was found to depend on thermal history. After rapid cooling the conductivity was high, while annealing at temperatures between 600 and 730 caused slow decrease of conductivity leading to values lower even by a factor of the order of 1000 after several days of heat treatment. Similar decrease of conductivity upon annealing was also observed for polycrystalline samples of the same composition and explained in terms of ordering of oxygen vacancies [24], which eventually led to transition to the low conductivity α phase. The ac response of single crystal rapidly cooled down from 770 K, measured with the electric field parallel to layers of the crystal structure, is presented here.

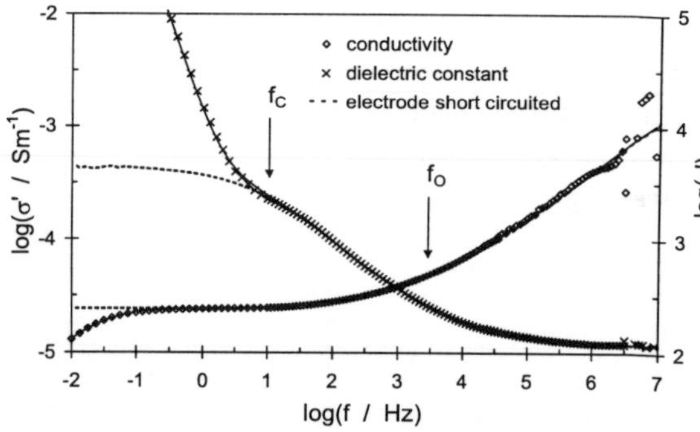

Figure 1. Dispersion of conductivity and dielectric constant in BICUVOX at 345 K (electric field parallel to structure layers). Continuous lines represent the fitted response of the model: σ_0=2.42×10^{-5} S/m, ε_∞=116, $\Delta\varepsilon$=4340, n=0.46, P=9.3. Dashed lines represent simulated response when the electrode part of equivalent circuit of Fig. 2 is short-circuited.

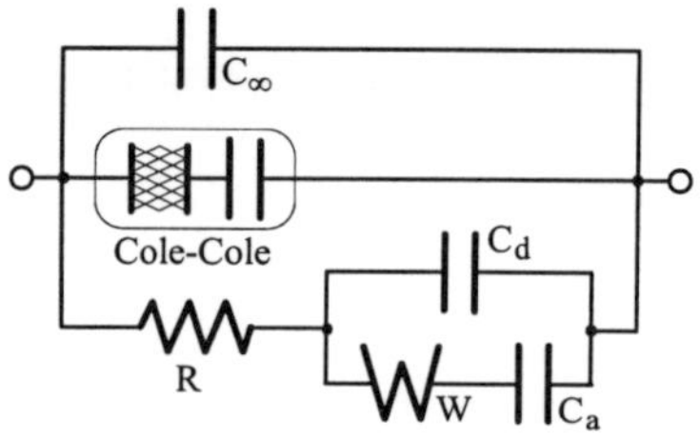

Figure 2. Equivalent circuit used to model the ac response of BICUVOX single crystal. The dielectric function of Eq. (6) is represented by capacitor C_∞ and parallel branch composed of capacitor and constant phase element (cross-hatched symbol). Capacitors: C_d, C_a, and Warburg element, W, represent impedance of electrodes.

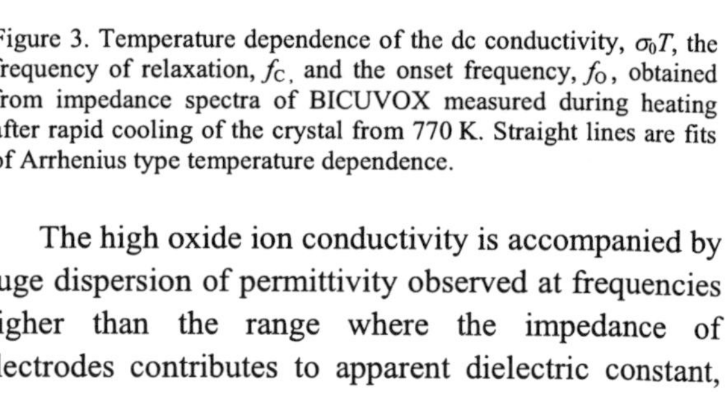

Figure 3. Temperature dependence of the dc conductivity, $\sigma_0 T$, the frequency of relaxation, f_C, and the onset frequency, f_O, obtained from impedance spectra of BICUVOX measured during heating after rapid cooling of the crystal from 770 K. Straight lines are fits of Arrhenius type temperature dependence.

The high oxide ion conductivity is accompanied by huge dispersion of permittivity observed at frequencies higher than the range where the impedance of electrodes contributes to apparent dielectric constant, see spectrum measured at 345 K in Fig. 1. Model used to fit the impedance spectrum contains section, which represents contact between sputtered platinum electrode and single crystal – Fig. 2. High values of capacitance: $C_d \cong 14\ \mu F/cm^2$, $C_a \cong 300\ \mu F/cm^2$, clearly indicate interfacial accumulation of ions. Saturation of permittivity at low frequencies is better seen, when contribution of the electrodes is removed, see the simulated curve in Fig. 1. Strength of relaxation is very high, $\Delta\varepsilon/\varepsilon_\infty \cong 37$. The relaxation frequency, $f_C=\omega_C/2\pi$, falls within the range of conductivity plateau. Dispersion of conductivity becomes significant only at much higher frequencies, $\omega_O/\omega_C \cong 240$, and may be considered as a tail of permittivity dispersion. The high value of the ratio of characteristic frequencies is associated with high value of the BNN coefficient $P \cong 10$, in accordance with Eq. (9).

The dc conductivity as a function of temperature, see Fig. 3, follows the Arrhenius dependence:

$$\sigma_0(T) = \frac{B}{T}\exp\left(-\frac{E_{\sigma T}}{k_B T}\right) \tag{11}$$

with activation energy, $E_{\sigma T}=0.72$ eV.

The onset frequency follows the Arrhenius dependence:

$$\omega_O(T) = \Omega_O \exp\left(-\frac{E_O}{k_B T}\right) \tag{12}$$

with activation energy, $E_O=0.75$ eV, approximately equal to that of the dc conductivity. The pre-exponential factor (attempt frequency) is $\Omega_O \cong 1.5\times10^{15}\ s^{-1}$.

Temperature dependence of the relaxation frequency, ω_C, is also of Arrhenius type, but with slightly lower activation energy, $E_C=0.65$ eV. Deviations of values of the relaxation frequency from the fitted straight line in Fig. 3 are larger than in the cases of the dc conductivity and the onset frequency, which reflects uncertainty of values estimated by fitting of the model to impedance spectra. Values of the coefficient P are between 8 and 19, increasing as temperature increases from 320 K to 485 K.

When the onset frequency is used as an estimate of the effective jump rate [5,6], the dc conductivity can be expressed as:

$$\sigma_0 = \frac{\gamma N a^2 q^2}{\alpha k_B T}\omega_O\ , \tag{13}$$

where N is concentration of charge carriers. Correlation factor can be taken $\gamma=1$, charge of oxide ion is $q=2e=3.2\times10^{-19}$ C and $\alpha=4$ for two dimensional transport. Assuming jump distance to be equal to separation between two adjacent oxygen positions in the equatorial vanadium plane, $a=0.2845$ nm [25], concentration of mobile oxygen vacancies was estimated. Averaging of estimates obtained for temperatures between 300 K and 500 K resulted in $N \cong 4.5\times10^{27}\ m^{-3}$. This value is in good agreement with concentration of vacancies $N=5.45\times10^{27}\ m^{-3}$ calculated based on crystal structure determination for similar compound $Bi_2Co_{0.1}V_{0.9}O_{5.35}$ [26]. Thus the assumption that ω_O is a good estimate of jump rate was confirmed.

4.2. *Li^+ ion conducting oxide glasses*

Glasses based on lithium titanium phosphate exhibit high Li^+ ion conductivity and good chemical stability in air [27]. In the investigated glasses of general formula $50Li_2O{:}xSnO_2{:}(10\text{-}x)TiO_2{:}40P_2O_5$, two forms of permittivity dispersion were identified [28]:

1. at the onset of conductivity dispersion, the finite change of the dielectric constant was modeled by the relaxation function, Eq. (6), with exponent $n \cong 0.35$;
2. at higher frequencies conductivity followed power law frequency dependence with exponent $m \cong 0.82$, as described by flat dielectric loss of Eq. (10).

Example of conductivity and dielectric constant spectrum, together with the fitted response of the model, is given in Fig. 4. In order to demonstrate existence of the two dispersions, response curves of partial models are also plotted in Fig. 4: without contribution of the flat dielectric loss or without contribution of the relaxation function. The initial increase of the real part of conductivity and decrease of the dielectric constant with increasing frequency are well reproduced by model without contribution of flat loss. At frequencies higher than the onset frequency, a discrepancy between measured conductivity and the response of model without flat loss becomes visible. At high frequencies, response of the model including flat dielectric loss, but without relaxation contribution, is in satisfactory agreement with the data.

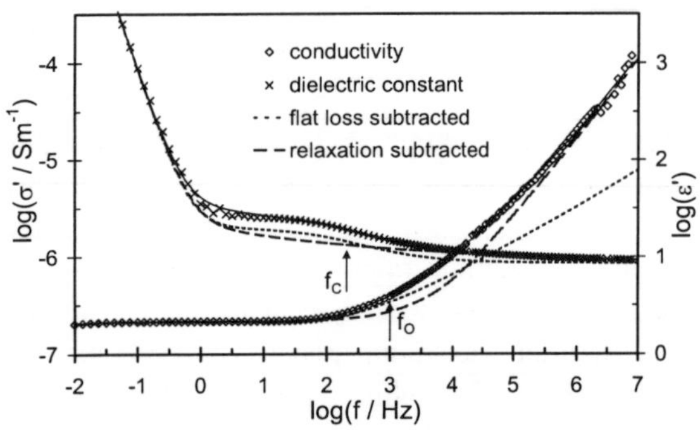

Figure 4. Dispersion of conductivity and dielectric constant in $50Li_2O{:}4SnO_2{:}6TiO_2{:}40P_2O_5$ glass at 267.5 K. Continuous lines represent fitted response of the model: σ_0=2.14×10^{-7} S/m, ε_∞=8.3, $\Delta\varepsilon$=11.5, n=0.39, P=1.6, A_L=16.8, m=0.825. Dashed lines represent response of the model when the flat dielectric loss is set to zero, $A_L=0$, or the relaxation strength is set to zero, $\Delta\varepsilon$=0.

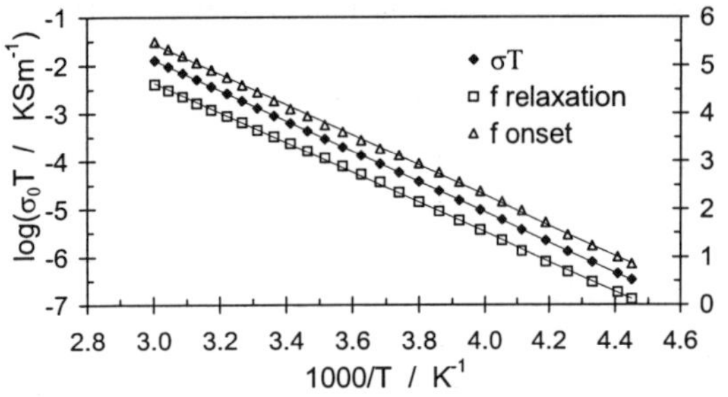

Figure 5. Temperature dependence of the dc conductivity, $\sigma_0 T$, the frequency of relaxation, f_C, and the onset frequency, f_O, obtained from impedance spectra of $50Li_2O{:}4SnO_2{:}6TiO_2{:}40P_2O_5$ glass. Straight lines are fits of Arrhenius type temperature dependence.

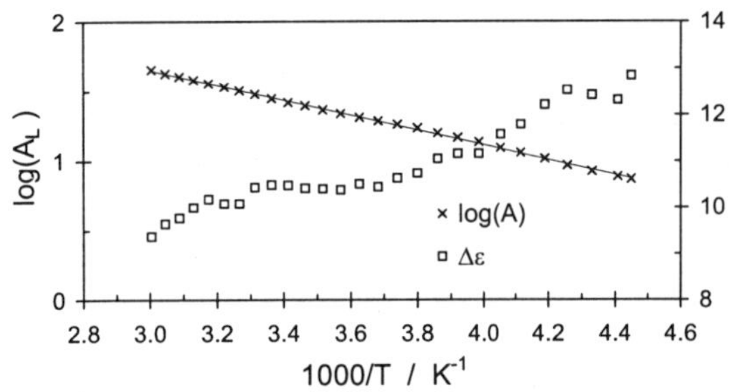

Figure 6. Temperature dependence of the relaxation strength $\Delta\varepsilon$ and the magnitude of flat dielectric loss A_L obtained by fitting model to impedance spectra of $50Li_2O{:}4SnO_2{:}6TiO_2{:}40P_2O_5$ glass.

The dc conductivity follows Arrhenius temperature dependence with $E_{\sigma T}$=0.630 eV, see Fig. 5. Activation energy of characteristic frequencies is nearly the same as that of the dc conductivity: the onset frequency E_O=0.627 eV and the relaxation frequency E_C=0.619 eV. Activation energy of the flat dielectric loss is E_L=0.106 eV – see Fig. 6. The obtained values of activation energy obey relation:

$$E_L = (1-m)E_{\sigma T} \qquad (14)$$

Thus the flat dielectric loss undergoes the temperature-frequency scaling together with the dc conductivity and the charge carrier relaxation.

The estimated values of the strength of relaxation, $\Delta\varepsilon$, decrease from 13 to 9 with increasing temperature, see Fig. 6, and are just a little larger than the high frequency value of the dielectric constant, ε_∞=8.3.

4.3. *Polymer electrolytes PEO-LiTFSI*

Polymer electrolytes, obtained by dissolution of salts in poly(ethylene oxide), PEO, undergo partial crystallization with formation of crystalline PEO or crystalline PEO-salt complexes [29]. Ionic conductivity decreases with increasing content of the crystalline phase. The ac response is strongly dependent on the multi phase microstructure [30].

The lithium salt with large and flexible anion: lithium bis(trifluoromethanesulfone)imide (LiTFSI) $LiN(CF_3SO_2)_2$ inhibits crystallization of the system when dissolved in PEO at molar ratio of O to Li between 8 and 12. Although crystallization of PEO_{10}LiTFSI has been observed by drift sensitive impedance spectroscopy at temperatures between 263 and 274 K [30], it is possible to obtain amorphous polymer electrolyte by rapid cooling from melt. The results presented here were obtained during heating run after polymer was rapidly cooled from 357 to 208.5 K. A decrease of conductivity by a factor about 100, recorded at constant temperature of 263.5 K, signaled crystallization. Impedance spectra measured prior to crystallization are characteristic of amorphous system.

At temperatures below 220 K, the dc conductivity is lower than the detection limit, which corresponds to sample resistance of about 10^{14} Ω. The ac response, see Fig. 7, comprises relaxation associated with the C-O bond dipoles within PEO molecular chains and a flat dielectric loss. This two contributions are modeled by the Cole-Cole function with exponent n_D=0.58 and the constant phase element with exponent m=0.77, Eq. (10), respectively. Strength of dipole relaxation, $\Delta\varepsilon_D$, constitutes only a fraction of the high frequency dielectric constant: $\Delta\varepsilon_D/\varepsilon_\infty \cong 0.17$. Frequency of the dipole relaxation increases slowly with increasing temperature, see Fig. 10. Magnitude of the flat dielectric loss, A_L, also increases with temperature.

At temperatures above 220 K, both the ionic conductivity and the dispersion of permittivity associated with mobile ions can be identified in the ac response. The full model of the ac response comprises five parallel contributions: the high frequency permittivity, the relaxation of bond dipoles, the flat dielectric loss, the relaxation of mobile ions and the ionic conductivity with electrode polarization; see Fig. 9. The Cole-Cole function with exponent n=0.38, modeling the carrier relaxation, reproduces only the onset of dispersion of the real part of conductivity, while the steeper frequency dependence above the onset frequency is due to the flat dielectric loss term and the dipole relaxation, see Fig. 8.

The dc conductivity and the frequency of charge carrier relaxation follow very similar temperature dependence, see Fig. 10. Above the glass transition temperature this dependence can be well fitted by the Vogel-Tamman-Fulcher (VTF) function:

$$F(T) = S\exp\left(-\frac{B}{T - T_0}\right) \qquad (15)$$

Values of parameters B and T_0 estimated for the dc conductivity, σ_0, for the relaxation frequency, f_C, and fore the onset frequency, f_O, are approximately equal.

Coefficient P of the BNN relation assumed values between 1.1 and 1.9, increasing with rising temperature. The ratio of the onset frequency to the relaxation frequency, ω_O/ω_C increased from 2 to 9, following increase of the coefficient P.

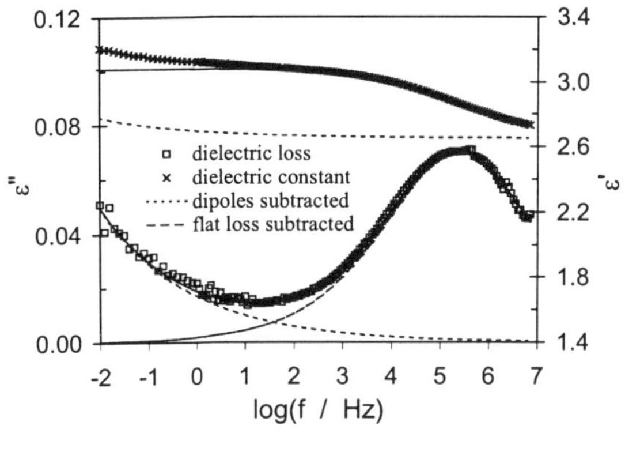

Figure 7. Dielectric function of PEO_{10}:LiTFSI at 211 K. Continuous lines represent model comprising top three branches of the circuit of Fig. 9: ε_∞=2.7, $\Delta\varepsilon_D$=0.44, n_D=0.58, A_L=0.075, m=0.77. Dashed lines represent response of the model when flat loss is zero, A_L=0, or dipolar relaxation is subtracted, $\Delta\varepsilon_D$=0.

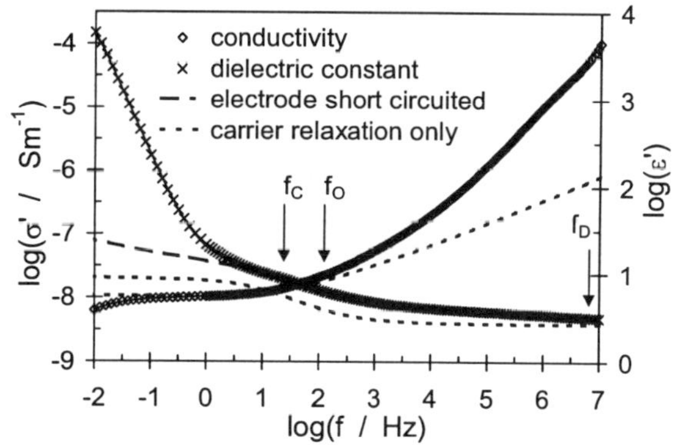

Figure 8. Dispersion of conductivity and dielectric constant in PEO_{10}:LiTFSI at 244 K. Continuous lines represent fitted response of the model: ε_∞=2.7, $\Delta\varepsilon_D$=0.71, n_D=0.58, A_L=8.1, m=0.77, σ_0=1.06×10^{-8} Sm^{-1}, $\Delta\varepsilon$=6.2, n=0.38, P=1.6. Dashed lines – response of the model with electrode part short circuited or when only ε_∞, σ_0 and $\Delta\varepsilon$ contributions are nonzero.

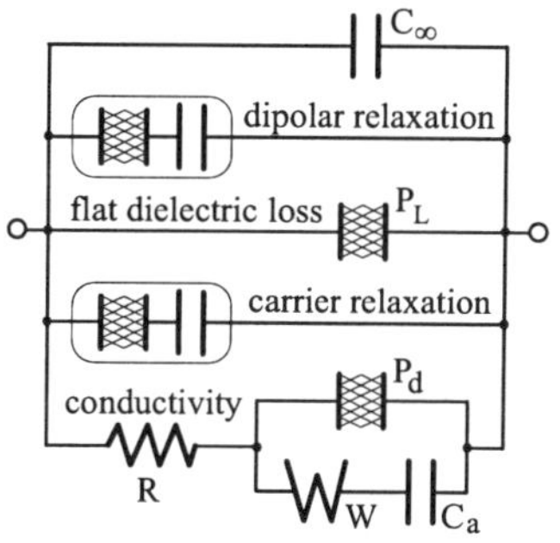

Figure 9. Equivalent circuit representation of a model of the ac response of amorphous polymer electrolyte.

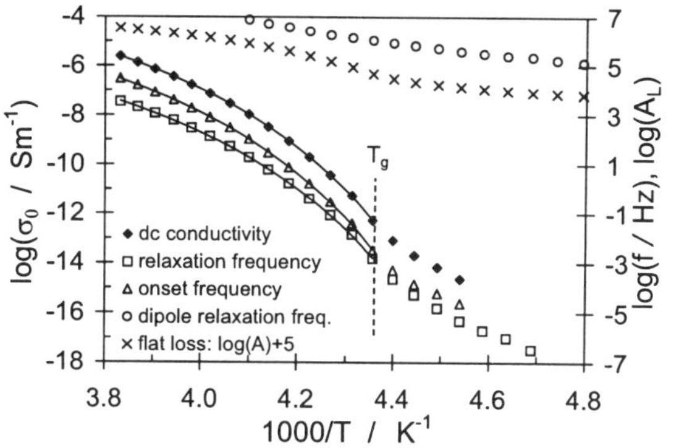

Figure 10. Temperature dependence of the dc conductivity, σ_0, frequency of relaxation, f_C, onset frequency, f_O, relaxation frequency of dipoles, f_D, and magnitude of flat dielectric loss, A_L, in PEO_{10}:LiTFSI measured during heating run after rapid cooling from melt. Continuous lines represent fitted VTF function.

4.4. *Lithium manganese spinel - polaronic conductor*

Lithium-manganese oxides are promising materials for cathode in rechargeable lithium ion batteries. In stoichiometric spinel, $LiMn_2O_4$, manganese ions coexist in two valence states Mn^{3+} and Mn^{4+} in equal proportion, so the chemical formula can be written as $Li(Mn^{3+}Mn^{4+})O_4$. The electrons at the partially occupied e-orbital of manganese ions remain localized, trapped in local lattice vibrations. The electronic conductivity is due to thermally activated hopping of small polarons between mixed valence manganese ions in neighbor sites [31].

The stoichiometric spinel undergoes a phase transition from cubic to orthorhombic symmetry near room temperature. At low temperatures, a partial charge ordering exists, leading to a superstructure composed of columns of Mn^{3+} type ions running along the [001] direction, which are surrounded by Mn^{4+} ions [32]. In nanocrystalline lithium manganese spinel, prepared by sol-gel method and sintered at 1073 K [33], change of the crystal symmetry is accompanied by a decrease of conductivity by a factor about 10. Conductivity exhibits thermal hysteresis [34].

Constant value of the real part of admittance, representing the dc conductivity, was measured for $LiMn_2O_4$ over several decades of frequency: from 10 mHz up to the onset of dispersion in the kHz region. Absence of the electrode polarization indicates that ionic contribution to the electrical conductivity is negligible. The impedance spectra do not indicate charge carrier blocking at grain boundaries.

Dispersion of permittivity in the low temperature phase of the $LiMn_2O_4$ spinel comprises two relaxation regions, see Fig. 11, which can be modeled by two Cole-Cole type relaxation functions, see Fig. 12.

The low frequency relaxation is associated with increase of dielectric constant, $\Delta\varepsilon \cong 12$. The relaxation frequency, f_C, follows the same temperature dependence as the dc conductivity, see Fig. 13. Values of activation energy are: E_σ=0.37 eV and E_C=0.36 eV. The BNN relation, Eq. (4), is obeyed with coefficient P≅7. While the onset of conductivity dispersion is associated with the high frequency tail of the low frequency relaxation, the onset frequency, f_O, is about two orders of magnitude higher than the relaxation frequency, f_C, in agreement with the large value of coefficient *P*, see Eq. (9).

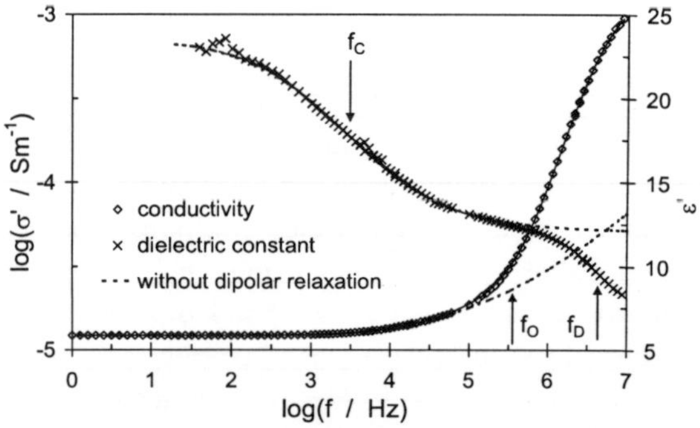

Figure 11. Dispersion of conductivity and dielectric constant in $LiMn_2O_4$ at 213 K. Continuous lines represent fitted response of the model shown in Fig. 12: σ_0=1.22×10^{-5} S/m, ε_∞=7.0, $\Delta\varepsilon$=11.8, n=0.45, P=6.9, $\Delta\varepsilon_D$=5.1, n_D=0.11. Dotted lines represent model with strength of dipolar relaxation set to zero.

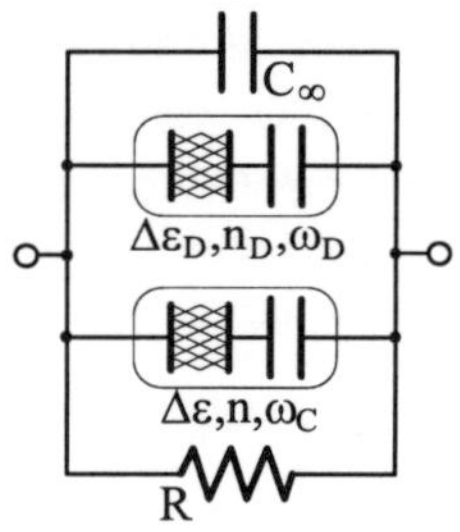

Figure 12. Equivalent circuit used to model the ac response of $LiMn_2O_4$ spinel. Two Cole-Cole type relaxation branches represent the charge carrier relaxation (low frequencies) and relaxation of dipolar structures by local polaron hopping (high frequencies).

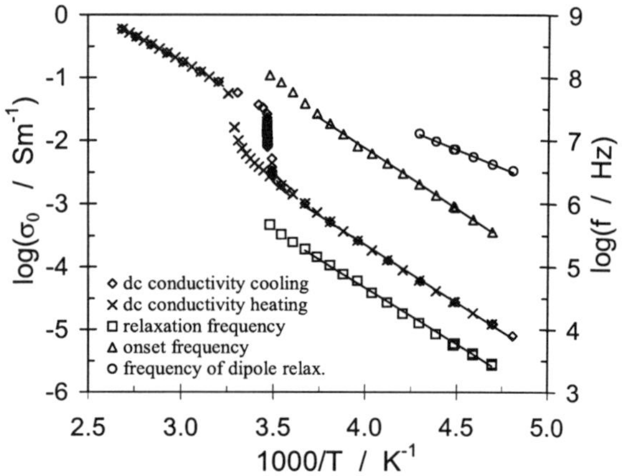

Figure 13. Temperature dependence of the dc conductivity, the frequency of relaxation, f_C, the onset frequency, f_O, and the relaxation frequency of dipolar structures, f_D, in $LiMn_2O_4$ spinel undergoing phase transition.

The rise of conductivity continues sharply at frequencies above the onset frequency. Slope of the log(σ') vs log(f) plot is larger than 1 in the low frequency tail of the high frequency relaxation. This relaxation is nearly of the Debye type, exponent of the Cole-Cole function, n_D=0.11, is close to zero. The rise of conductivity levels off at frequencies higher than the relaxation frequency f_D. These features distinguish clearly the high frequency dipolar relaxation from the case of flat dielectric loss – compare Figs. 4 and 11.

The low frequency relaxation seen in lithium manganese spinel is of the same nature as the charge carrier relaxation observed in ionic conductors. The thermally activated hopping of small polaron from Mn^{3+} to neighbor Mn^{4+} site constitutes charge transport mechanism quite similar to the thermally activated hopping of ion to neighbor vacant site.

The high frequency relaxation is not directly related to the dc conductivity. Its activation energy, E_D=0.23 eV, is distinctly lower than that of the dc conductivity. This relaxation has been observed only in the low temperature phase of the stoichiometric spinel. Tentatively, it can be assigned to relaxation of dipolar configurations of Mn^{3+} and Mn^{4+} ions formed as a result of charge ordering. Rearrangement of dipolar moment may be due to local hopping of small polarons, which involve lower potential barrier than potential barriers encountered by the charge carrying polarons on the conduction pathway. The relaxation strength is quite high, $\Delta\varepsilon_D$=5.1, in accordance with participation of significant number of Mn^{3+} - Mn^{4+} pairs in forming active dipoles.

5. Discussion

The presented experimental results show that the permittivity dispersion accompanying the conductivity by hopping ionic or polaronic charge carriers exhibits marked universality. Near the onset of conductivity increase with increasing frequency, the Cole-Cole type complex dielectric function accurately fits the measured spectra. This characteristic dispersion can be well identified in the experimental spectra when other contributions to the ac response are taken into account, when present. These are: the flat dielectric loss in the case of amorphous ionic conductors and the dipolar relaxation found in polymer electrolyte and in low temperature phase of lithium manganese spinel. The new form of universality is of great importance since it concerns the frequency range where deviation of the real part of conductivity from the dc value just begins. This corresponds to time scale characteristic for crossover from long range transport to charge carrier motion probing the local environment. Charge carriers undergo rearrangements with respect to oppositely charged counter-ions or immobile charged defects, which leads to configurations possessing electric dipole moment. Polarization by charge carriers leads to increase of the dielectric constant at frequencies below the onset of conductivity dispersion. The inverse of the relaxation frequency, ω_C, is thus the mean relaxation time, τ_{rel}, of dipolar configurations formed by charge carriers. Process of decay of such polarization may be termed charge carrier relaxation. Initial increase of the real part of conductivity over the dc value is governed by the high frequency tail of charge carrier relaxation.

Within the universality of charge carrier relaxation, characteristic parameters of the ac response of various hopping systems should be analyzed with the aim of establishing correlation with the structure of conductivity pathways and mechanism of transport. The nonlinear least squares analysis, applied in the present study, permits quantitative estimation of parameters of the model leading to identification of similarities and differences between the studied systems.

Fitted values of exponent of the Cole-Cole dielectric function, n, are close to 0.38 for amorphous lithium ion conductors: inorganic glass and polymer electrolyte. Higher values of exponent were obtained for single crystal of BICUVOX, oxide ion conductor with two dimensional conduction pathway, n=0.46, and for lithium manganese spinel, where columnar ordering Mn^{3+} and Mn^{4+} hosts of polarons takes place at low temperatures, n=0.45. Characteristic dependence of the power law exponent on the dimensionality of the conduction space was noted by Sidebottom [35]. The exponent, derived from the slope of log(σ') vs. log(f) plots in the range of frequency well above the onset of dispersion, was found to be smaller for systems of lower dimensionality [35]. In the present study, the steeper rise of the real part of conductivity at frequencies above the onset of dispersion is associated with the flat dielectric loss term in the model, not with the charge carrier relaxation term. Although the flat dielectric loss seems to be related to the dc conductivity, as evidenced by the joint temperature-frequency scaling observed in the case of glasses, in the present model it constitutes contribution independent of the charge carrier relaxation term.

The power law exponent of the flat dielectric loss term was close to m=0.8 in the studied lithium ion conducting glasses and polymer electrolyte. The fitted values of exponent are clearly distinct from those typical for the nearly constant loss (NCL), which, as reported for various amorphous ionic conductors [36-38] (and also crystalline [39]), are nearly equal 1. The nearly constant loss has not been identified in the experimental dielectric spectra of ionic and mixed conductors included in this study.

Recently, Roling et al. [40,41] showed that a well-defined transport length can be extracted from the dispersion of permittivity. They introduced a mean square displacement of the center of charge of mobile carriers due to non-random hopping:

$$\left\langle \tilde{R}^2(t) \right\rangle = \left\langle R^2(t) \right\rangle - 6D'_\sigma(0)t \ , \qquad (16)$$

where the contribution of long-range diffusion, given by the conductivity diffusion coefficient, D'_σ, was subtracted. Using the linear response theory, the infinite time limit of this displacement can be related to the strength of permittivity relaxation, $\Delta\varepsilon$, associated with charge carriers [41]:

$$\left\langle \tilde{R}^2(\infty) \right\rangle = \frac{6k_B T \varepsilon_0}{Nq^2} \Delta\varepsilon \ , \qquad (17)$$

where N and q have the same meaning as in Eq. (13).

Here a new proposal is put forward to combine Eq. (17) and the standard equation for the dc conductivity, where the hopping rate is identified with the onset frequency of conductivity dispersion, Eq. (13) (α=6 for three-dimensional transport). Using the two equations, concentration of charge carriers can be eliminated and as the result a ratio of the long time limit of mean square charge displacement to the jump distance is obtained:

$$\Gamma^2 = \frac{\left\langle \tilde{R}^2(\infty) \right\rangle}{a^2 \gamma} = \frac{\varepsilon_0 \Delta\varepsilon}{\sigma_0} \omega_O = \frac{1}{P} \frac{\omega_O}{\omega_C} \qquad (18)$$

The relation between the onset frequency, ω_O, and the relaxation frequency, ω_C, Eq. (9) based on the Cole-Cole representation of the charge carrier relaxation, gives:

$$\Gamma^2 = \frac{\left\langle \tilde{R}^2(\infty) \right\rangle}{a^2 \gamma} = \left[P^{n-1} \cos(\pi n/2) \right]^{-1/n} \qquad (19)$$

This proposal offers new interpretation of the coefficient P of the Barton-Nakajima-Namikawa relation, Eq. (4), as a measure of the length scale, Γ, of nonrandom hopping expressed in units of jump distance. In particular, when the exponent of the relaxation function, Eq. (6), is $n \cong 0.5$, as for example in the case of BICUVOX single crystal, then $\Gamma^2 \cong 2P$.

Different values of the coefficient P, obtained for the studied ionic and polaronic conductors, indicate different range of the non-random hopping. Value of the coefficient P, by means of Eq. (9), determines also the ratio of the onset frequency, ω_O, to the relaxation frequency, ω_C. When the exponent of the relaxation function is $n \cong 0.5$, then this ratio is $\omega_C/\omega_O \cong 2P^2$. Ratio of the two characteristic frequencies is equal to ratio of the relaxation time for rearrangement of dipolar configuration of charge carriers, τ_{rel}, to the average time between jumps of given ion, τ_{hop}:

$$\frac{\omega_O}{\omega_C} = \frac{\tau_{rel}}{\tau_{hop}} = \left[\frac{P}{\cos(\pi n/2)} \right]^{1/n} \qquad (20)$$

The parameter Γ may also be interpreted as an average length (expressed in units of elementary jump distance) of relative charge shift, caused by local ordering of hopping carriers, which leads to formation of electric dipole.

The results, corresponding to the spectra presented in Figs. 1, 4, 8 and 11, are summarized in Table 1.

Table 1. Characteristic parameters of the charge carrier dispersion of permittivity in the studied systems.

System	$\Delta\varepsilon/\varepsilon_\infty$	n	P	ω_O/ω_C	Γ
BICUVOX	37	0.46	9.3	240	5.1
Li^+ ion glass	1.4	0.38	1.63	5.9	1.9
$PEO_{10}LiTFSI$	2.3	0.38	1.57	5.4	1.8
$LiMn_2O_4$	1.7	0.45	6.9	132	4.4

6. Conclusions

Dispersion of permittivity, which accompanies conductivity by hopping ions or electrons (polarons) in crystalline and amorphous ionic and mixed ionic/electronic conductors, can be modeled by the Cole-Cole dielectric function. Parameters of the function: strength of relaxation, $\Delta\varepsilon$, relaxation frequency, ω_C, and the exponent, n, which characterizes shape of the dispersion, can be estimated by nonlinear least squares fitting of the model to the measured complex impedance spectrum. The coefficient linking the dc conductivity with the charge carrier relaxation $P=\sigma_0/(\varepsilon_0\Delta\varepsilon\omega_C)$ provides valuable information about the length scale and time scale of nonrandom hopping, expressed respectively in units of jump length and units of average time between jumps of a hopping carrier.

Values of coefficient P significantly larger than one are found for the single crystal of oxide ion conductor, BICUVOX, and for the nanocrystalline polaronic conductor, lithium manganese spinel. Large P indicates relaxation time of the local dipolar configurations extending over several hundred elementary jumps as well as the spatial range of nonrandom hopping on the order of several (4-6) elementary jump lengths. In the case of amorphous Li^+ ion conductors: lithium titanium phosphate glass and PEO based polymer electrolyte, coefficient P has values somewhat less than 2. Small P indicates limited extent of non-random hopping: displacement of about two jump lengths and the relaxation time corresponding to about six jumps of a carrier.

Proper analysis of permittivity dispersion relies on identification of all contributions to the ac response, which include flat dielectric loss in the case of amorphous conductors and relaxation of electric dipoles not related directly to charge carriers, if those are present in the studied material.

Acknowledgements

Experimental results reviewed here were obtained in joint effort with colleagues, whom the author is deeply indebted: W. Bogusz, Z. Florjańczyk, M. Kopec, F. Krok, P. Kurek, M. Marzantowicz, K. Pietruczuk, E. Zygadło-Monikowska at Warsaw University of Technolgy, M.W. Breiter, G. Fafilek at University of Technology Vienna, I. Abrahams, E. Hadzifejzovic at University of London, Z. Kaszkur, D. Lisovytskiy, J. Pielaszek at Insitute of Physical Chemistry Polish Academy of Sciences, M. Molenda, R. Dziembaj at Jagiellonian University Kraków.

This work was supported in part by Committee for Scientific Research under grant PBZ KBN 013/T08/12.

Invitation and support from organizers of The 1st ernational Discussion Meeting on Superionic nductor Physics is gratefully acknowledged.

References

1. A.K. Jonscher, *Nature* **267**, 673 (1977).
2. C. Cramer, *Ber. Bunsenges. Phys. Chem.* **100**, 1497 (1996).
3. K. Funke, *Prog. Solid State Chem.* **22**, 111 (1993).
4. B. Rolling, *Phys. Chem. Chem. Phys.* **3**, 5093 (2001).
5. D.P. Almond and A.R. West, *Solid State Ionics* **9&10**, 277 (1983).
6. D.L. Sidebottom, P.F. Green and R.K. Brow, *J. Non-Cryst. Solids* **222**, 354 (1997).
7. M. Tomozawa in *Treatise on Materials Science and Technology*, vol.12, eds. M. Tomozawa and H. Doremus (Academic Press, New York, 1977).
8. J. L. Barton, *Verres et Refr.* **20**, 328 (1966).
9. T. Nakajima in *1971 Ann. Rep. Conf. on Electrical Insulation and Dielectric Phenomena* (Nat. Acad. of Sci., Washington, 1972), p.168.
10. H. Namikawa, *J. Non-Cryst. Solids* **18**, 173 (1975).
11. D.L. Sidebottom, *Phys. Rev. Lett.* **82**, 3653 (1999).
12. J.C. Dyre and T.B. Schrøder, *Rev. Mod. Phys.* **72**, 873 (2000).
13. D.L. Sidebottom, B. Rolling and K. Funke, *Phys. Rev.* B **63**, 024301 (2000).
14. K.E.D. Wapenaar and J. Schoonman, *Solid State Ionics* **5**, 637 (1981).
15. J.B. Bates and J.C. Wang, *Solid State Ionics* **28-30**, 115 (1988).
16. L. Murawski and R.J. Barczyński, *J. Non-Cryst. Solids* **185**, 84 (1995).
17. J.R. Dygas and M.W. Breiter, *Electrochim. Acta* **41**, 993 (1996).
18. J.R. Dygas, W. Bogusz and F. Krok, *Proc. 4th Int. Symp. Electrochemical Impedance Spectroscopy*, Rio de Janeiro, Brazil, August 1998, p. 9.
19. J.R. Dygas and M.W. Breiter, *Electrochim. Acta* **44**, 4163 (1999).
20. K.S. Cole and R.H. Cole, *J. Chem. Phys.* **9**, 341 (1941).

21. F. Abraham, J.C. Boivin, G. Mairesse and G. Nowogrocki, *Solid State Ionics* **40-41**, 934 (1990).
22. F. Abraham, M.F. Debreuille-Gresse, G. Mairesse and G. Nowogrocki, *Solid State Ionics* **28-30**, 529 (1988).
23. P. Kurek, J.R. Dygas and M.W. Breiter, *J. Electroanal. Chem.* **378**, 77 (1994).
24. J.R. Dygas, P. Kurek and M.W. Breiter, *Electrochim. Acta* **40**, 1545 (1995).
25. I. Abrahams, F. Krok and J.A.G. Nelstrop, *Solid Sate Ionics* **90**, 57 (1996).
26. A. Kežionis, W. Bogusz, F. Krok, J.R. Dygas, A. Orliukas, I. Abrahams and W. Gębicki, *Solid State Ionics* **119**, 145 (1999).
27. I. Abrahams and E. Hadzifejzovic, *Solid State Ionics* **134**, 249 (2000).
28. J.R. Dygas, K. Pietruczuk, F. Krok, E. Hadzifejzovic, I. Abrahams, *Molecular Phys. Rep.* **35**, 150 (2002).
29. F.M. Gray, *Polymer Electrolytes* (RSC Materials Monographs, Cambridge, 1997).
30. J.R. Dygas, B. Misztal-Faraj, Z. Florjanczyk, F. Krok, M. Marzantowicz and E. Zygadlo-Monikowska, *Solid State Ionics* **157**, 249 (2003).
31. J.B. Goodenough, A. Manthiran and B. Wnetrzewski, *J. Power Sources* **43-44**, 269 (1993).
32. G. Rousse, C. Masquelier, J. Rodriguez-Carvajal and M. Hervieu, *Electrochem. Solid State Lett.* **2**, 6 (1999).
33. R. Dziembaj, M. Molenda, D. Majda and S. Walas, *Solid State Ionics* **157**, 81 (2003).
34. D. Lisovytskiy, Z. Kaszkur, N.V. Baumer, J. Pielaszek, M. Molenda, R. Dziembaj, J. Marzec, J. Molenda, J. Dygas, M. Kopeć and F. Krok, *Molecular Phys. Rep.* **35**, 36 (2002).
35. D.L. Sidebottom, *Phys. Rev. Lett.* **83**, 983 (1999).
36. W.K. Lee, J.F. Liu and A.S. Nowick, *Phys. Rev. Lett.* **67**, 1560 (1991).
37. A.S. Nowick, A.V. Vaysleyb and Wu Liu, *Solid State Ionics* **105**, 121 (1998).
38. D.L. Sidebottom and C.M. Murray-Krezan, *Phys. Rev. Lett.* **89**, 195901 (2002).
39. A. Rivera, J. Santamaría, C. León, J. Sanz, C.P.E. Varsamis, G.D. Chryssikos and K.L. Ngai, *J. Non-Cryst. Solids* **307-310**, 1024 (2002).
40. B. Roling, C. Martiny and K. Funke, *J. Non-Cryst. Solids* **249**, 201 (1999).
41. B. Roling, C. Martiny and S. Brückner, *Phys. Rev.* B **63**, 214203 (2001).

DIFFUSE X-RAY SCATTERING AND MOLECULAR DYNAMICS STUDIES OF K-HOLLANDITE AT HIGH TEMPERATURES

Yuichi Michiue[1], Mamoru Watanabe[1], Yoshito Onoda and Shinzo Yoshikado[2]

[1]Advanced Materials Laboratory, National Institute for Materials Science,
1-1 Namiki, Tsukuba, Ibaraki 305-0044, Japan
Fax: 029-860-4662, E-mail: MICHIUE.Yuichi@nims.go.jp
[2]Department of Electronics, Doshisha University, 1-3 Tataramiyakodani, Kyotanabe 610-0321, Japan

Abstract

One-dimensional ionic conductor $K_xMg_{x/2}Ti_{8-x/2}O_{16}$ (x=1.55) with the hollandite structure was studied by diffuse X-ray scatterings (295 - 673 K) and molecular dynamics simulations (200 - 800 K). Ionic correlations between K ions were significant up to at least 673K and observed peaks were indexed by the two basis vectors; one is corresponding to the period of unit separation of K ions (nc^*) and another is of the host framework (c^*) (n: K ion concentration, c^*: dimension of the reciprocal lattice). Temperature dependence of the K ion distribution along the tunnel from simulations was in good agreement with that from X-ray diffraction experiments previously reported. Static structure factors based on the simulation were in qualitative agreement with those from the diffuse X-ray scattering over the temperature range measured.

Keywords: hollandite, one-dimensional ionic conductor, static structure factor, diffuse X-ray scattering, molecular dynamics method

1. Introduction

Intensive studies have been made on the ionic correlation in solids, which is one of the most important factors affecting the ion-transport property in superionic conductors. It is known that in K-hollandite, a one-dimensional (1-d) superionic conductor, the cation-cation interaction is so strong as to be comparable to the periodic potential due to the framework structure. The system is characterized by K ions tending to arrange with the uniform separation to reduce repulsive interactions between themselves, concurrently perturbed by the framework potential. This model was originally obtained through the effort to explain the profile in the diffuse X-ray scattering [1]. Besides Bragg peaks, the diffuse scattering is localized in planes perpendicular to the c^* axis in a rotation photograph of $K_xMg_{x/2}Ti_{8-x/2}O_{16}$ (x=1.55) with 1-d tunnels extending along the c axis. The characteristic feature is that diffuse planes are situated at positions incommensurate to the period of the framework, k=0.775c^*, 1.55c^*, and 1.775c^*. K ion density, that is the number of K ions per framework potential well, is given by n=x/2=0.775 for $K_xMg_{x/2}Ti_{8-x/2}O_{16}$ (x=1.55). The period of K ions in an arrangement with a uniform separation is c/n=c/0.775 which is related to peak positions in the diffuse scattering.

Attempts to explain the diffuse scattering or static structure factors in hollandite have been done by theoretical analyses [2-4] based on the Frenkel-Kontorova model [5], that is the system consisting of a harmonic chain of particles subject to a sinusoidal potential. Another approach by a molecular dynamics (MD) method applying classical pairwise potentials has also reproduced the principal features in static structure factors of hollandite [6]. In the present paper, the correlation between the K ions in hollandite $K_xMg_{x/2}Ti_{8-x/2}O_{16}$ (x=1.55) at elevated temperatures was examined by X-ray diffuse scattering in conjunction with the MD simulation. The evaluation of the validity of the simulation results was also done by comparing average distributions of the K ions with those from X-ray diffraction analyses up to 919 K reported by Weber and Schulz[7].

2. Experimental

Single crystals were grown by the slow-cooling method from a K_2CO_3-MoO_3 flux-melt. A single crystals of $K_xMg_{x/2}Ti_{8-x/2}O_{16}$ (x=1.55) enclosed into a quartz capillary was mounted on an automated four-circle diffractometer (Rigaku AFC 7R). The crystal was heated by a small cylinder-shaped furnace with a Pt coil set on the diffractometer as shown in Fig. 1. The inside of the furnace was separated from the outside by Al films attached on both sides of the furnace. Incident and diffracted X-ray beams penetrated Al films. The temperature of the specimen, monitored and controlled

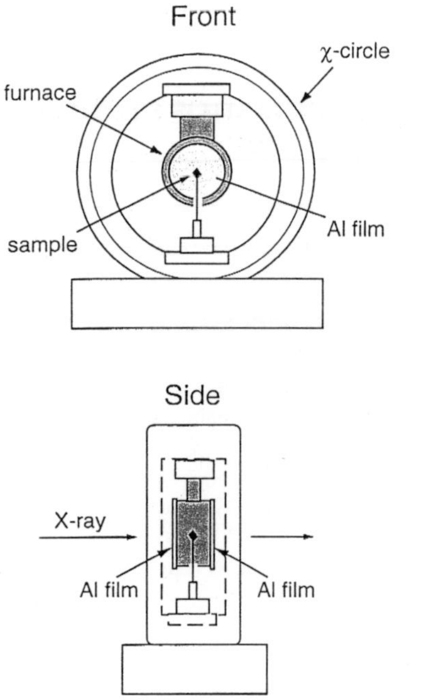

Fig. 1 Schematic representation of an equipment for heating a single crystal in the X-ray diffraction measurement.

by a pair of thermocouples placed a few millimeters away, was changed between 295 K and 673 K. The X-ray scattering profile was measured between (0, 0, 0.5) and (0, 0, 1.9) along $\boldsymbol{c}^*$. Three dimensional ordering was ignored in this incommensurate hollandite, although interchain coupling was indicated in a commensurate phase $K_{1.5}A_{1.5}Ti_{6.5}O_{16}$ with significant intensity variations in the $\boldsymbol{a}^*\boldsymbol{b}^*$ plane at room temperature and 26 K [8]. Scattering peaks from Al films were checked by comparing the profile at room temperature with that measured without Al films.

$K_xMg_{x/2}Ti_{8-x/2}O_{16}$ (x=1.55) has a body-centered tetragonal unit cell and the lattice parameters a=b=10.1496(8), c=2.9715(2) Å experimentally determined at 298 K increase to a=b=10.197(2), c=2.992(1) Å at 919 K [7]. The MD calculation was performed for a linear array of the 40 unit cells along the $\boldsymbol{c}$ axis containing 62 K, 31 Mg, 289 Ti, and 640 oxygen ions, which gives a chemical composition $K_{62/40}Mg_{31/40}Ti_{289/40}O_{640/40}$ or $K_xMg_{x/2}Ti_{8-x/2}O_{16}$ (x=1.55). The interatomic potential has the form,

$$V_{ij}(r_{ij})=z_iz_je^2/r_{ij}+f_o(b_i+b_j)\exp[(a_i+a_j-r_{ij})/(b_i+b_j)] -c_ic_j/r_{ij}^6, \quad (1)$$

where z_i is the effective charge of the ith ion, e is the unit charge, r_{ij} is the distance between the ith and jth ions, f_o is the constant (1 kcal·mol^{-1}·Å^{-1}) and a_i, b_i and c_i are concerning to size, softness and polarizability of the ion. The parameters used are listed in Table 1, which were empirically determined. The details are given in the literature [6]. The calculation of 10000 steps after an initial aging of 10000 steps with an integration time of 1.0 × 10^{-15} sec was performed using the program *MXDTRICL* [9]. The *NPT*-ensemble was adopted using the methods proposed by Andersen [10] and Nose [11]. Pressure was set to 10^5 Pa and temperature is changed between 200 K and 800 K with an interval of 50 K. Deviations of cell parameters were less than 2 % over the temperature range examined.

Table 1. Parameters used for the calculation.

	z_i	a_i(Å)	b_i(Å)	c_i(kcal$^{1/2}$mol$^{-1/2}$Å^3)
K	1.00	1.65	0.110	-
Ti	2.23	1.05	0.090	-
Mg	1.67	0.88	0.065	-
O	-1.185	2.01	0.176	70.4

3. Results and Discussion

X-ray scattering profile primarily gives three peaks at room temperature as previously reported [1]. At high temperatures, however, the estimation was impossible for one of the peaks around k=1.55c* on which one of the scattering peaks from Al films superposed as shown in Fig. 2. Temperature dependence of the two peaks is given in Fig.3. Both peaks are clearly observed even at 673 K, although intensities are decreasing with increasing temperature.

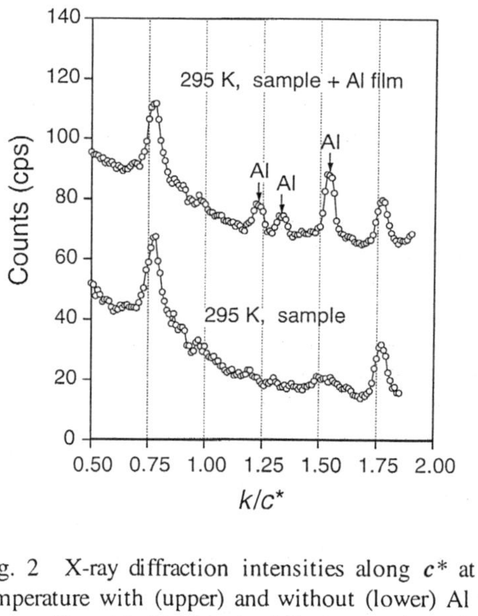

Fig. 2 X-ray diffraction intensities along c* at room temperature with (upper) and without (lower) Al films. The profile with Al films is shifted upward by 50 cps.

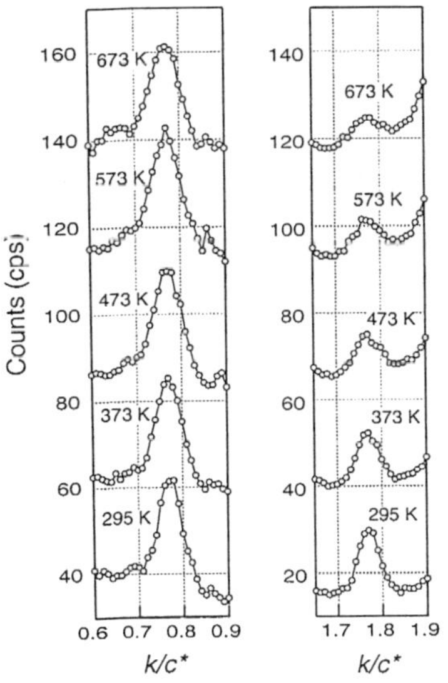

Fig. 3 Temperature dependence of the diffuse X-ray scattering profile of $K_xMg_{x/2}Ti_{8-x/2}O_{16}$ (x=1.55) along $\boldsymbol{c}^*$.

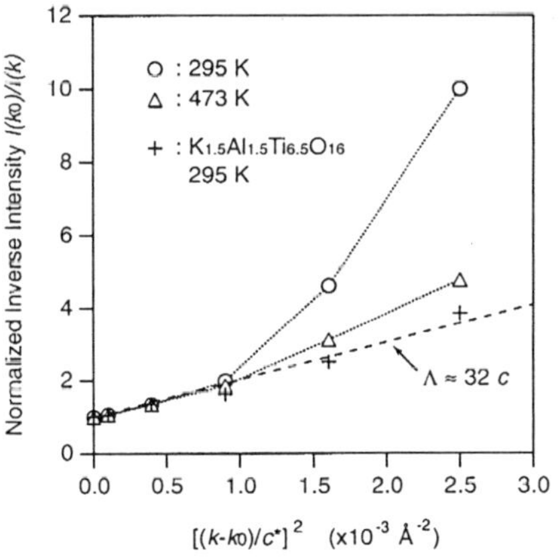

Fig. 4 Inverse intensity around $\boldsymbol{k}_0$=1.77$\boldsymbol{c}^*$ in $K_xMg_{x/2}Ti_{8-x/2}O_{16}$ (x=1.55) as a function of $(\boldsymbol{k}-\boldsymbol{k}_0)^2$. The data for $K_{1.5}A_{1.5}Ti_{6.5}O_{16}$ is also given for comparison.

Terauchi *et al.* found that the intensity profile $I(\boldsymbol{k})$ for another hollandite $K_{1.5}A_{1.5}Ti_{6.5}O_{16}$ is given by

$$I(\boldsymbol{k})=\Lambda/[(2\pi/\Lambda)^2+(\boldsymbol{k}-\boldsymbol{k}_0)^2], \quad (2)$$

where $\boldsymbol{k}_0$ is a wave vector giving the peak top and Λ is the correlation length [8]. This is, however, not the case for the present hollandite $K_xMg_{x/2}Ti_{8-x/2}O_{16}$ (x=1.55).The inversed intensity around $\boldsymbol{k}_0$=1.77$\boldsymbol{c}^*$ plotted as a function of $(\boldsymbol{k}-\boldsymbol{k}_0)^2$ (Fig. 4) remarkably deviates from a straight line, and the estimation of the correlation length, which is to be given from a slope of the linear plot, is impossible. The plot becomes closer to the linear one with increasing temperature. Reliable evaluation seemed, however, impossible at 573K and higher because of the difficulty in estimating the background. It should be noted that the data for a single crystal of $K_{1.5}A_{1.5}Ti_{6.5}O_{16}$ grown by us roughly approximates to the linear even at room temperature as shown in Fig. 4, where scattering intensities were measured in the same condition to that for $K_xMg_{x/2}Ti_{8-x/2}O_{16}$ (x=1.55). The correlation length for $K_{1.5}A_{1.5}Ti_{6.5}O_{16}$ is estimated as about 32*c*, which is very close to that obtained by Terauchi *et al.*, 33.5*c*. The reason is not clear why the two hollandites $K_{1.5}A_{1.5}Ti_{6.5}O_{16}$ and $K_xMg_{x/2}Ti_{8-x/2}O_{16}$ (x=1.55) give different types of the intensity profile, the former is the Lorentzian as given by Eq. 2 but the latter is not.

Temperature dependence of the K ion distribution along the $\boldsymbol{c}$ axis was obtained from trajectories of all K ions in MD calculations. The number of trajectories reduced in the cavity around (0, 0, 1/2) was counted at each *z* level. Three peaks were observed at *z*=1/2 and 1/2±Δz at 400 K and below, which is due to the prominent off-centering of a certain amount of K ions from the minima of potential wells. The profile, however, became monotonous with a broad peak at *z*=1/2 at 500 K and above. These are in good agreement with experimental results reported [7].

In static structure factors (Fig. 5) obtained from density correlation functions as done in the previous paper [6], the intensity of three peaks are decreasing with increasing temperature. The peaks are, however, retained even at 800 K. X-ray scattering experiments give more broad peaks (Fig. 3) than those from the MD calculations, implying that the fluctuation of K density is not enough in the calculation for the definite cell with 40 cavities along the tunnel axis. The larger MD cell, which allows more variation in the local K density, might be necessary to embody the profile quantitatively. Thus, we limit our discussion about the static structure factor to the qualitative one at this stage.

Peak positions in static structure factors are given by $\boldsymbol{k}=l_1\boldsymbol{c}^*+l_2n\boldsymbol{c}^*$ (l_1, l_2: integer) with two basis vectors $n\boldsymbol{c}^*$, corresponds to the unit separation of K ions, and $\boldsymbol{c}^*$, the period of the host framework. Indices are (l_1, l_2)=(0, 1) for the peak at $\boldsymbol{k}$=0.775$\boldsymbol{c}^*$, (0,2) for $\boldsymbol{k}$=1.55$\boldsymbol{c}^*$, and (1,1) for $\boldsymbol{k}$=1.775$\boldsymbol{c}^*$. The relation between the peak position and the K concentration was supported by the MD calculations with various K ion concentrations [12]. This interpretation derives from the assumption that K ions tend to arrange with the unit separation, *c*/*n*, and are perturbed by the framework potential with the period *c*. The peak at $\boldsymbol{k}$=0.775$\boldsymbol{c}^*$ correspond to the basic arrangement with a unit separation *c*/*n*, while the peak at $\boldsymbol{k}$=1.775$\boldsymbol{c}^*$ is due to the modulation of the former. The latter is the first-order satellite peak of the former in terms of crystallography. Such an incommensurate

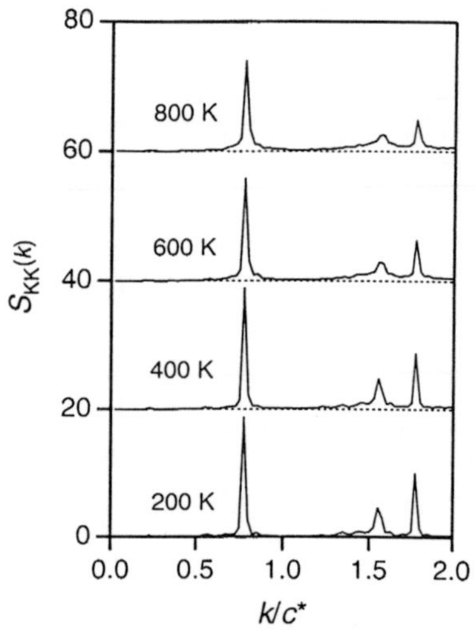

Fig. 5 Static structure factors from molecular dynamics simulations for K-hollandite $K_xMg_{x/2}Ti_{8-x/2}O_{16}$ (x=1.55) between 200 K and 800 K.

character in $K_xMg_{x/2}Ti_{8-x/2}O_{16}$ (x=1.55) seems to contributes to the conductivity along the tunnel with a low activation energy of 0.058 eV in a microwave region [13]. Dielectric function calculated by the same manner as in the previous paper [6] showed no additional modes compared with that at room temperature.

In summary, incommensurate peaks in diffuse scattering were observed even at 673 K, implying that K ions in the hollandite tunnel have tendency to arrange with a uniform interval to reduce repulsive interactions between themselves, concurrently perturbed by the host structure up to at lest 673 K. Basic characters of K-hollandite were successfully reproduced by molecular dynamics simulations with classical pairwise potentials giving results in qualitative agreement with experiments at room temperature and elevated temperatures.

References

[1] H. U. Beyeler, Phys. Rev. Lett. **37**, 1557 (1976).
[2] T.Geisel, Solid State Commun. **32**, 739 (1979).
[3] T. Ishii, J. Phys. Soc. Jpn. **52**, 4066 (1983).
[4] K. Takahashi, I. Mannari, and T. Ishii, Solid State Ionics **74**, 11 (1994).
[5] J. Frenkel and T. Kontorova, J Phys. (Moscow) **1**, 137 (1939).
[6] Y. Michiue and M. Watanabe, Phys. Rev. B **59**, 11298 (1999).
[7] H. P. Weber and H. Schulz, J. Chem. Phys. **85**, 475 (1986).
[8] H. Terauchi, T. Futamura, T. Ishii, and Y. Fujiki, J. Phys. Soc. Jpn. **53**, 2311 (1984).
[9] K. Kawamura, computer code *MXDTRICL*.
[10] H. C. Andersen, J. Chem. Phys. **72**, 2384 (1980).
[11] S. Nose, J. Chem. Phys. **81**, 511 (1984).
[12] Y. Michiue and M. Watanabe, J. Phys. Soc. Jpn. **70**, 1986 (2001).
[13] S.Yoshikado, T. Ohachi, I. Taniguchi, Y. Onoda, M. Watanabe, and Y. Fujiki, Solid State Ionics **7**, 335 (1982).

FREQUENCY DEPENDENCE OF SPIN-LATTICE RELAXATION OF ^{27}AL IN ONE-DIMENSIONAL IONIC CONDUCTOR, PRIDERITES

Y. ONODA
7-1 Onogawa, Tsukuba 305-0053, Ibaraki, Japan,
E-mail: onodaym@mail1.accsnet.ne.jp

Y. FUJIKI
410-22 Shimohirooka, Tsukuba 305-0042, Ibaraki, Japan.

Y. MICHIUE and M. TANSHO
Advanced Materials Laboratory, National Institute for Material Science,
1-1 Namiki, Tsukuba 305-0044, Ibaraki, Japan.

S. OOKI, K. HASHI, A. GOTOH and T. SHIMIZU
Tsukuba Magnet laboratory, National Institute for Material Science,
3-13 Sakura, Tsukuba 305-0003, Ibaraki, Japan.

S. YOSHIKADO and T. OHACHI
Department of Electronics, Doshisha University, 1-3 Tataramiyadai, Kyotanabe 610-0321, Japan.

Temperature and frequency dependences of T_1^* of ^{27}Al in K-Al- and Rb-Al-priderite were measured at three frequencies of 20.8, 104.2 and 130.3 MHz in the temperature range from 20 K to 909 K. The temperature dependence of K-Al-priderite is divided into three regions: region 1 below 30 K, region 2 from 30 K to 170 K where T_1^* is dominated by the ionic motion over the intrinsic barriers, and region 3 above 170 K where T_1^* is dominated by the ionic motion over the impurity barriers. The temperature and frequency dependence of T_1^* in the region 2 is well described by a trial function of $J(\omega, \tau)$ which has the $T_1^* \propto \omega^{1.0}$ dependence in the low temperature limit. But in the case of Rb-Al-priderite, the frequency dependence measured in a temperature range which corresponds to the region 2 of K-Al-priderite showed an intermediate dependence between $\omega^{1.5}$ and ω^2 dependence. These frequency dependences are compared with Ishii's calculation result of $S(\omega)$ of hopping ions in one-dimensional random lattice based on his relaxation mode theory. $E_{\rm NMR}$ obtained from the slope of the straight line in the log(T_1^*) .vs. $1/T$ plot in the region 2 of both samples showed a tendency that $E_{\rm NMR}$measured at low frequency has a little smaller value. It means that $E_{\rm NMR}$ is dependent on the frequency.

1. Introduction

One-dimensional(1d) ionic conductor with the hollandite-type structure(Fig.1), priderite, is known to show anomalous frequency and temperature dependence in the spin- lattice relaxation time T_1.[1,2] The temperature dependence of T_1^* of ^{27}Al in the framework of the K-Al- priderite(KATO) measured at 20.8 MHz in the temperature range from 30 K to 500 K was divided into two regions: low temperature (LT) region where T_1^* is dominated by the ionic motion over the intrinsic barriers, and high temperature (HT) region where T_1^* is dominated by the motion over the impurity barriers. In the LT region, T_1^* shows a straight line over two and half decades of T_1^* value in the log(T_1^*) vs. $1/T$ plot, and the slope gave a small activation energy, $E_{\rm NMR}$= 0.029 eV. The temperature dependence above 100 K shows an anomalously wide $T_1^*{}_{\rm min}$ region[1] and it was attributed to two widely spread barrier height distribution functions for the intrinsic and the random barrier.[3]

Conductivities measured at the microwave range showed a very weak frequency dependence and its log (σT) .vs. $1/T$ plot show nearly straight lines.[4] But, the conductivity measured at the radio frequency range showed comparatively a large frequency dependence described by $\sigma \propto \omega^\nu, \nu(T) = (1 - T/T_m)/(1 + T/T_m)$.[5,6] These anormalous behaviors are attributed to the random barriers in the train of periodic potentials. The random barriers divide the train into segments and anomalous behavior is caused by the ionic hopping from a segment to a neighboring segment.

We measured the frequency dependence of T_1 of ^{27}Al in KATO in the frequency range from 10.5 MHz

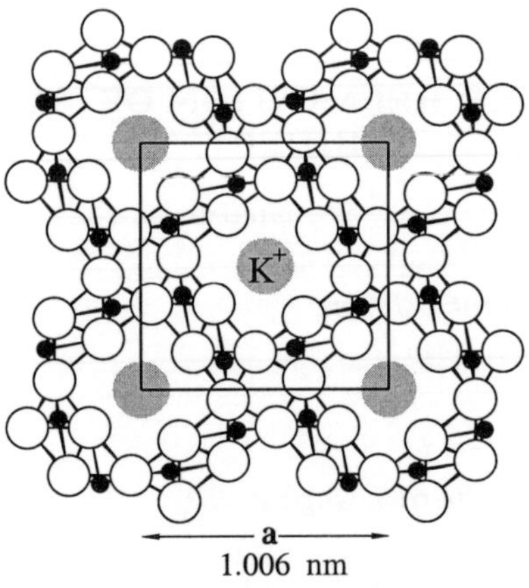

Fig.1 Projection of K-Al-priderite along c-axis. Al^{3+} ions substitute randomly for the Ti^{4+} sites denoted by small black circles.

to 55 MHz and at the midpoint temperature(=45 K) of the straight line in the $\log(T_1^*)$ .vs. $1/T$ plot, and reported an anomalous dependence approximately proportional to $\omega^{3/2}$, not proportional to BPP-type ω^2 dependence.[7] That of Rb-Al-priderite (RATO) measured in the frequency range from 10.5 to 20.8 MHz and at 125 K gave similar dependence(Fig.2A). In the materials, the recovery curve of the nuclear magnetization $M(t)$ of ^{27}Al is essentially non-exponential because of the randomness of the electric field gradient(efg) tensor(Fig.2B). Therefore, we determined T_1 in the measurement assuming that the recovery is exponential at a region where $\log((M^\infty - M(t))/M^\infty)$ is roughly from 0.2 to 0.05 as shown in Fig.2B, and T_1 thus determined was inevitable to have a considerably large error.

Recently we remeasured the temperature dependences of T_1^* of ^{27}Al in KATO and RATO at 104.2 MHz. The result showed similar temperature dependences with that measured at 20.8 MHz. Here, T_1^* is defined by the time when $(M^\infty - M(t))/M^\infty$ becomes $1/e$ and is a convenient measure of the relaxation time when the recovery of $M(t)$ is non-exponential. The weak point of T_1^* is that it often includes comparatively large systematic error. Nevertheless, its temperature dependence is thought to give a correct picture of the temperature dependence of the ionic motions. In the case of KATO, we measured T_1 which were determined as noted above, at 45 K and at 104.2 and 130.4 MHz. However, the measurement gave a negative data for the $\omega^{1.5}$ dependence.

Theoretical approach to the non-BPP behavior of priderites have been developed by Ishii based on his relaxation mode theory.[8,9] In his recent calculation in 1d random lattice with a flat barrier height distribution, the temperature dependence of the incoherent dynamical structure factor $S(q,\omega)$ and the spin-lattice relaxation rate $S(\omega)(\propto 1/T_1)$ is divided into four stages. The stage I is observed above the temperature $T(T_{1\,\min})$, and the spin-lattice relaxation is composed by the diffusive mode only. The stage II is a critical region where both of the diffusive mode and the non-diffusive mode of ionic motion contributes to $S(\omega)$. The region appears at low temperatures below $T(T_{1\,\min})$ and the $\log(S(\omega))$ vs. 1/T plot shows a straight line over two decades. The frequency dependence in this stage is estimated to be $S(\omega) \propto \omega^{s-2} (s \approx 0.6)$. $S(\omega)$ in the stage III is contributed only by the non-diffusive mode and the frequency dependence $S(\omega) \propto \omega^{s'-2} (s' \approx 1.0)$ is predicted. The temperature dependence in this stage is much weaker than that of the stage II and does not show astraight line region.

The purpose of this study is to measure the fre-

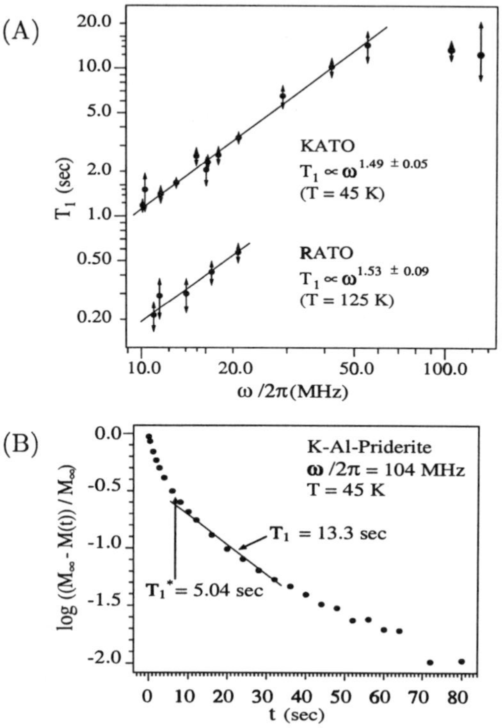

Fig.2 Frequency dependence of T_1 of ^{27}Al (Fig.2A)[2] T_1s measured at 104 and 130 MHz are new data. The measured temperatures(45 K and 125 K) are the midpoint temperatures of the straight lines. Fig.2B shows the recovery curve of the nuclear magnetization $M(t)$ of ^{27}Al in KATO measured at 104.2 MHz.

quency dependence of $T_1^*(T_1)$ of KATO and RATO at more wide frequency range and get more reliable data in order to verify the theoretical calculation by Ishii. His theory only offers us an accessible approach to explain the temperature and frequency dependences of our NMR data qualitatively and quantitatively.

2. Experimental

A flux method was used to synthesize the priderites, $A_xAl_xTi_{8-x}O_{16}$(A: K^+, Rb^+, x $\approx$ 1.5). [10] As grown needle crystals with average size of 1 mm length and 0.1 mm width were smashed into powders to avoid orientation effect, and the powders were sealed in evacuated silica tubes for NMR measurements. Temperature dependence of T_1^* was measured at two frequencies 104.2 and 130.4 MHz in the temperature range from 20 K to 909 K and the data were compared with the old temperature dependence data measured at 20.8 MHz using Bruker SXP100 spectrometer.[1] Bruker MSL400 spectrometers were used for the measurement at 104.2 MHz and Thamway's PROT1000 spectrometer was used for the measurement at 130.4 MHz.

The spin-lattice relaxation of ^{27}Al above 30 K was confirmed to be caused by the quadrupolar interaction, not by other interactions such as the dipolar coupling with paramagnetic impurities.[1]

3. Results and Discussion

3.1 Temperature Dependence

Fig.3 shows the temperature dependence of T_1^* measured at 20.8 MHz and 104.2 MHz. The temperature dependence is divided into three regions. The explanation of the temperature dependence of the region 2 and 3 are as follows. Ionic motion of K^+ ion over the intrinsic barriers begins to be observed at 30 K. As the motion becomes rapid with the increase of the temperature, T_1^* decreases, and above $T(T_{1\,min})$ the motion becomes too rapid and T_1^* begins to increase. But another slow fluctuation begins to contribute to T_1^* in the region 3. It is caused by the ion hopping from a segment to a neighboring segment over the impurity barriers. It is like a charge density fluctuation between segments.

There are two possible explanations for the temperature dependence of region 1. One is that it is caused by the coupling with the lattice vibration. In Fig.4, which compares the temperature dependences of the two samples, the dashed curve shows the $T_1 \propto T^{-2}$ dependence, and T_1^* below 30 K looks well described by the dependence which is derived from the Raman term of the phonon coupling. A serious problem of the explanation is that the dependence appears above the Debye temperature and the temperature of many materials is a few hundreds K much higher than 30 K. An assumption may clear the problem. If there is a soft phonon along the c-axis, then the critical temperature may decrease considerably as much as the dependence is observable below 30 K.

The other explanation is that the region 1 corresponds to the stage III of Ishii's result. If the straight line region shown in Fig.3 corresponds to the stage

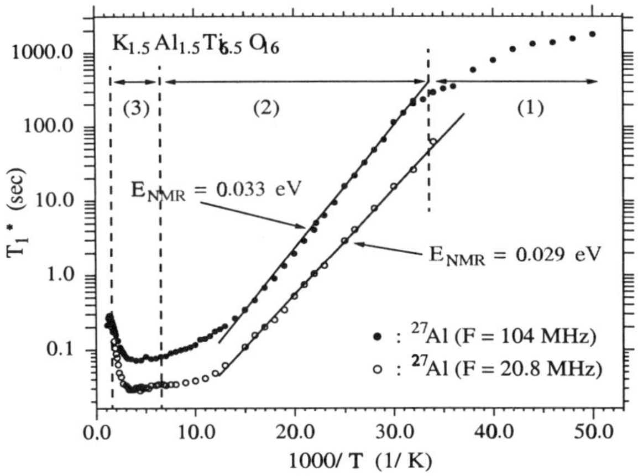

Fig. 3 Temperature dependence of T_1^* measured at 20.8 MHz(○) and 104.2 MHz(•). The properties of the three regions are described in the text.

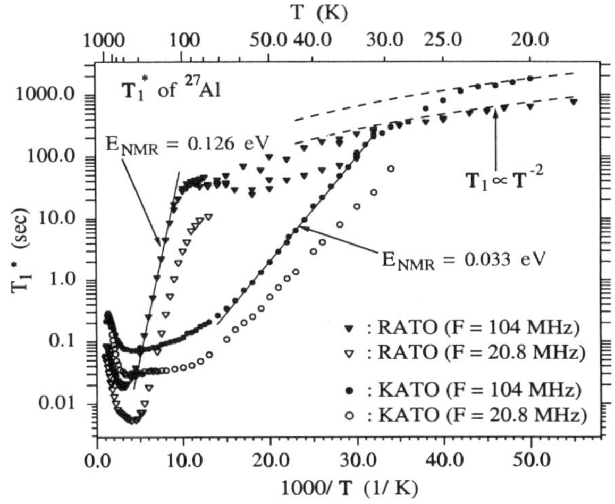

Fig. 4 Comparison of the temperature dependence of T_1^* of ^{27}Al between KATO(○: 20.8 MHz, •: 104.2 MHz) and RATO(▿: 20.8 MHz, ▾: 104.2 MHz).

II in which a straight line over two decades appears, then temperature dependence of $S(\omega)$ weakly dependent on temperature must be in the low temperature side of the region 2.

If the phonon coupling only cause the relaxation, T_1^* does not depend on the frequency, and if the local ionic motion is dominant, T_1^* will shows ω^1 dependence according to the Ishii's result. But, we have not measured the frequency dependence of T_1^* in the region 1. So, we can not judge which process is responsible for T_1^* in the region 1.

3.2 Curve fitting

We tried to explain the temperature dependence of T_1^* with so wide $T(T_{1\,\mathrm{min}}^*)$ considering the two contributions of the ionic motions over the periodic potentials and over the random barriers as was done for the explanation of the non-BPP behavior of Na-β-alumina.[11] Then, the relaxation rate is given by

$$1/T_1 = (1/T_1)_{\mathrm{intrinsic}} + (1/T_1)_{\mathrm{random}}. \qquad (1)$$

We assumed that each term has the same form as shown next,

$$(1/T_1) = C \int_{\Delta_1}^{\Delta_2} G(\Delta)\, J(\omega, \tau)\, d\,\Delta. \qquad (2)$$

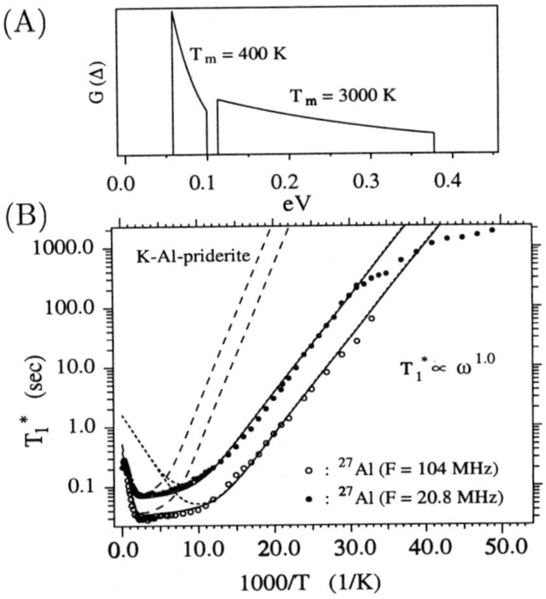

Fig.5 A curve fitting trial of the temperature dependence of T_1^* of ^{27}Al in KATO measured at 20.8 and 104.2 MHz (Fig.5B) and the barrier distribution function $G(\Delta)$ for the intrinsic and random barriers(Fig.5A). In the fitting, we used a trial function which gives $T_1^* \propto \omega^{1.0}$ dependence in the LT limit.

where $J(\omega, \tau)$ is the power spectrum of the fluctuation and it corresponds to $S(\omega)$. $G(\Delta)$ is the barrier height distribution function with the form given by $G(\Delta) \propto \exp(-\Delta/kT_m)$, which was used for the explanation of the anomalous frequency dependence of the conductivity ($\sigma(\omega) \propto \omega^{-\nu}$) in the random barrier model.[3] Δ_1 and Δ_2 are cut-off values.

As for $J(\omega, \tau)$, we tried four functional forms as shown next,

$$J(\omega, \tau) \propto \frac{\tau}{1+\omega^2\tau^2} \propto \omega^{-2}\tau^{-1}, \qquad (3)$$

$$J(\omega, \tau) \propto \frac{(\tau/\omega)^{1/2}}{1+\omega\tau} \propto \omega^{-1.5}\tau^{-1/2}, \quad (\omega\tau \gg 1) \qquad (4)$$

$$J(\omega, \tau) \propto \frac{\tau^{1/2}}{1+\omega\tau} \propto \omega^{-1}\tau^{-1/2}, \qquad (5)$$

and

$$J(\omega, \tau_0) \approx \sum_{i=1}^{n} g_i \cdot \frac{\tau_i}{1+\omega^2\tau_i^2}, \quad \tau_i = c_i\,\tau_0. \qquad (6)$$

In the equations, τ is the correlation time given by $\tau = \tau_0 \exp(E/kT)$. It was meaningless to include $W2$ process in $J(\omega, \tau)$, so the term is omitted in the equations. The functional forms of eqs. 3, 4 and 5 have symmetrical forms in the $\log(T_1^*)$ vs. $1/T$ plot. But except BPP form(eq.3), eqs 4 and 5 have no physical background, since we have no information of the τ dependence. These were introduced as the functions which give $\omega^{3/2}$ and ω^1 dependence in the LT limit. Eq.6 represents the Fourier transform of the stretched exponential function[12] but this function has been known to be incompetent for the explanation of our data even if the value of $\beta = 0.5$ is used.

Fig.5B shows a fitting result of the temperature dependence of T_1^* of KATO using eq. 5 for the intrinsic and the random terms. The dotted and dashed curves in Fig.5B come from the first and the second term in the eq.1, respectively. Barrier height distribution functions are plotted in Fig.5A. In the fitting we assumed that the two distribution does not overlap each other, but we think that it is possible for the distributions to overlap. Δ_1 is about double of $E_{\mathrm{NMR}}(\approx 0.03$ eV), since τ dependence of T_1^* in the LT and HT limit is $\tau^{1/2}$ and $\tau^{-1/2}$, respectively.

From the fitting results, we can say following facts: 1. The temperature dependence is well fitted if we use two distribution functions. 2. The distribution of the intrinsic barrier is comparatively small, that is, the characteristic temperature T_m is small. 3. The frequency dependence of T_1^* in the region 2 is well described by eq. 5. 4. But, T_1^* of the region 3 looks like to depends on the frequency more strongly.

3.3 Frequency Dependence

Fig. 6 shows a fitting result of the temperature dependence of T_1^* measured at three frequencies. For the simplification, s single distribution function is used in the fitting because we are interested in the temperature and the frequency dependence in the LT region below 100 K. The data measured at 130.4 MHz does not cover wide temperature range and the overall reliability of the data is not good. In the fitting, parameters were chosen so that the data measured at 104.2 MHz has the best fitting since the data is most reliable. The three temperature dependences in the region 2 look like well described by eq.5. It indicates that the frequency dependence of T_1^* is proportional to $\omega^s(s \approx 1.0)$. The dependence is almost the same value with that of the stage III in Ishii's calculation where the spin-lattice relaxation is caused by the fluctuation by the local ion hopping. But, the temperature dependence in this stage does not show straight line in the $\log(T_1)$ vs. 1/T plot as noted in the introduction. If we assume that the region 2 corresponds to the stage II since the temperature dependence has a straight line range in this stage, the frequency dependence must be $\omega^s(s \approx 1.4)$.

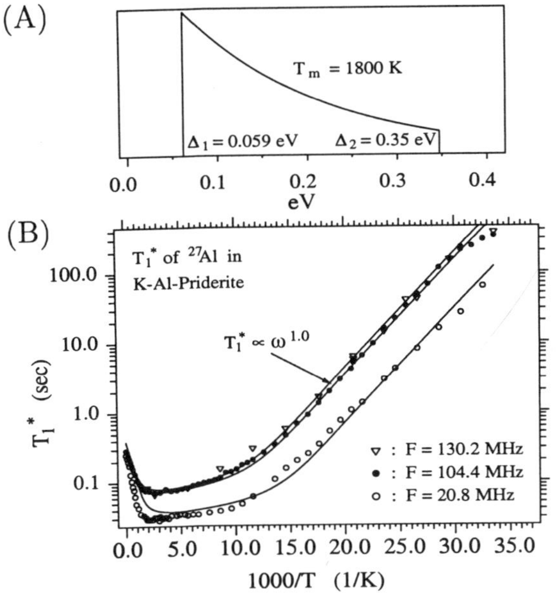

Fig.6 A curve fitting result of the temperature dependence of T_1^* measured at three frequencies(B) and the barrier height distribution(B). A single distribution function is considered for the simplicity. A function $\tau^{1/2}/(1+\omega\tau)$, which has the ω^{-1} dependence in the LT limit, was used as the trial function for $J(\omega,\tau)$. The result indicates T_1^* is proportional to $\omega^s(s \approx 1.0)$.

Fig.7 shows a result when we used eq. 4 as the trial function. The calculation curve at 20.8 MHz differs much from the experimental data. The curve at 130.4 MHz looks like also to differ from the data. The disagreement corresponds to the data shown in Fig.2A. The result looks like to deny the $\omega^s(s \approx 1.4)$ dependence if we rely on our NMR data measured by T_1^*. As a matter of course, the calculation curves based on the BPP- type function(eq.3) disagrees much more.

In the case of RATO, the temperature and the frequency dependences of T_1^* shows different frequency dependence from KATO as shown in Fig.8. Why the frequency dependence is so different between KATO and RATO?

Ishii's theory gives a reasonable explanation for the problem,[10] if we stand on the assumption that the region 2 corresponds to the stage II though the frequency dependence of KATO is different from the

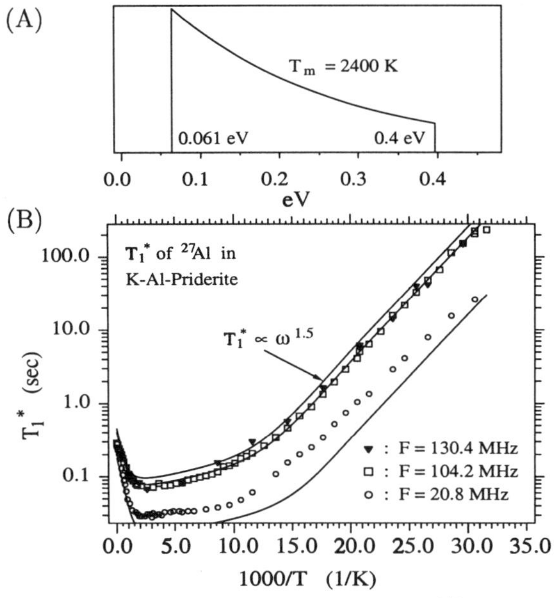

Fig. 7 A fitting result when a function $(\tau/\omega)^{1/2}/(1+\omega\tau)$ was used for $J(\omega,\tau)$. The result indicates that the frequency dependence in the region 2 is weaker than the $\omega^{3/2}$ dependence.

Ishii's prediction. [6] In the stage II, the frequency and the temperature dependence of $S(\omega)$ is contributed by two modes, diffusive and nondiffusive mode. If the randomness of the system is smaller, the contribution by the diffusive mode becomes dominant, and the frequency dependence of $S(\omega)$ approaches ω^{-2}. Rb^+ ion has much larger E_{NMR} value than K^+ ion in as shown in Fig.4. It means that the ratio of the barrier height of the periodic potential to the Coulomb interaction is much larger than that of K^+ and it means that the randomness in RATO is smaller than KATO.

3.4 Is E_{NMR} dependent on the ω?

E_{NMR} of K^+ ion measured at 20.8 MHz (E_{NMR}=0.029 eV) is different from the value mea- sured at 104.2 MHz(E_{NMR}=0.033 eV) as shown in Fig.3. In the case of Rb^+ ion in RATO, E_{NMR} measured at 20.8 MHz(= 0.12 eV) is a little smaller than the value measured at 104.2 MHz (=0.126 eV) as shown in Fig.8. The tendency observed in both samples suggests that E_{NMR} is dependent on the measuring frequency, or as shown in Figs. 6, 7 and 8, the frequency dependence is dependent on temperature. As the temperature decreases, the frequency dependence becomes larger. It may suggest that the frequency dependence finally approaches to the ω^2 dependence.

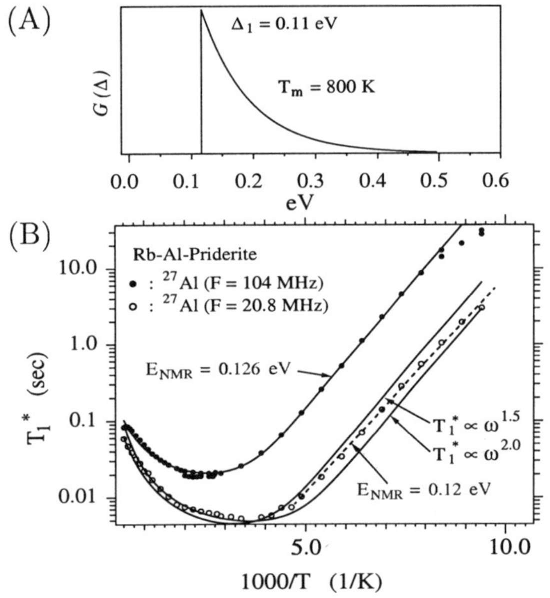

Fig. 8 A fitting result of the temperature and frequency dependence of T_1^* of RATO(B). A single barrier height distribution is used also for the simplicity. The frequency dependence is $\omega^s (s \approx 1.7)$.

Conclusion

The temperature dependence T_1^* of ^{27}Al in KATO is divided into three regions. The Curve fitting analysis of the temperature dependence above 30 K indicates that at least two barrier height distribution functions

must be introduced to obtain a good fitting. The frequency dependence of T_1^* in KATO and RATO showed strong non-BPP behavior. The frequency dependence in the region 2 is well described by the trial function which gives ω^1 dependence. The dependence is different from the ω^s, $(s \approx 1.7)$ dependence in the stage II estimated by Ishii. Rather, it resembles that of the stage III. However, as noted in the introduction, T_1^* often has comparatively large systematic errors, and we must be careful to derive a correspondence between the region 2 and the stage III. The frequency dependence of RATO shows a different dependence of $\omega^s (s \approx 1.7)$ and the difference may be explained by the difference in the contribution ratio of the diffusive mode to the non-diffusive mode in both samples if the region 2 corresponds to the stage II. E_{NMR} showed a tendency that it may depends on the measuring frequency.

References

1. Y.Onoda *et al.*, *Solid State Ionics*, **9 & 10** 1311 (1983).
2. Y.Onoda *et al.*, *Solid State Ionics*, **17** 127 (1985).
3. Y.Onoda *et al.*, *Solid State Ionics*, **18 & 19** 878 (1986).
4. S.Yoshikado *et al.*, *Solid State Ionics*, **7** 335 (1982).
5. J.Bernasconi *et al.*, *Phys.Rev.Let.*, **42** 819 (1979).
6. S.Yoshikado *et al.*, *Solid State Ionics*, **9 & 10** 1305 (1983).
7. A.Abragam, "The Principles of Nuclear magnetism", 123 (ClarendonPress, Oxford, 1961).
8. T.Ishii and T.Abe, *Solid State Ionics*, **154-155** 203 (2002).
9. T.Ishii, *J. Phys. Soc. Jpn.*, **72** 335 (2003).
10. Y.Fujiki *et al.*, *Solid State Ionics*, **25** 131 (1987).
11. R.E.Walstedt *et al.*, *Phys.Rev.* B **15** 3442 (1977).
12. C.T.Moynian *et al.*, *Phys. and Chem. of Glass* **14** 122 (1973).

ION CONDUCTION IN HOLLANDITE-TYPE ONE-DIMENSIONAL SUPERIONIC CONDUCTOR (K,Cs)-PRIDERITES

Shinzo Yoshikado[1], Yuichi Michiue[2], Yoshito Onoda[2] and Mamoru Watanabe[2]

[1]Department of Electronics, Doshisha University, Kyotanabe 610-0321, Japan
Fax:+81774-65-6801, Email:syoshika@mail.doshisha.ac.jp
[2]Advanced Materials Laboratory, National Institute for Materials Science
1-1 Namiki,Tsukuba, Ibaraki, 305-0044, Japan

The conduction of mobile alkali ions in single-crystal one-dimensional (1D) superionic conductor (K-Cs)-priderites, $K_xCs_yMg_{(x+y)/2}Ti_{8-(x+y)/2}O_{16}$, was studied by means of ac complex electrical conductivity measurement from 100 Hz to 9.8 GHz. The ion conduction behavior can be described by the moving-box model [1]. The motion of K^+ ions was different from that in K-priderite, though the average distance L between the two succeeding Cs^+ ions was large. This result shows that the correlation length λ_c for K^+ ions is longer than L.

Keywords: one-dimensional ion conduction, hollandite type, (K,Cs)-priderite, impedance spectroscopy, frequency dependence

1. Introduction

Hollandite-type superionic conductors K-Cs-priderites, $K_xCs_yMg_{(x+y)/2}Ti_{8-(x+y)/2}O_{16}$ (KCMTO, x~1.5), are ionic conductors having a one-dimensional (1D) conduction path along which K^+ and Cs^+ ions move. By adding small amounts of Cs atoms to the K-priderite, $K_{1.6}Mg_{0.8}Ti_{7.2}O_{16}$ (KMTO), it is expected that K^+ ions with a smaller ionic radius than that of Cs^+ ions are enclosed by two Cs^+ ions and move having a correlation length equal to the distance between two succeeding Cs^+ ions [1-3]. The purpose of this study is to measure the complex electrical conductivity in the wide-frequency region, determine the parameter values concerning the ion conduction using an equivalent circuit method, and investigate the motion of both K^+ ion clusters and Cs^+ ions.

2. Experiments

Single crystals of KCMTO were provided by Dr. Beyeler, of the Brown Boveri Research Center. The details of the preparation of KCMTO are described in ref.[1]. The crystals are transparent and have a needle shape in the direction of the 1D tunnel axis (a *c*-axis). The crystals having Cs mol ratio of approximately 0.0133 and 0.05 are labeled KCMTO(1) and KCMTO(2), respectively. Content of Cs was determined using X-ray fluorescence analysis [1]. The single crystal was processed into the appropriate shape with two parallel end faces (perpendicular to the *c*-axis) to measure the ac complex electrical conductivity. Vacuum-evaporated gold blocking electrodes were applied to the two end faces. The commercial impedance meters were used for measurements in the frequency region of 100 Hz to 1 GHz. The magnitude of the ac signal was adjusted such that the electrical response of the sample was linear [3]. All measurements were carried out in the temperature range of 153 to 293 K in a dry nitrogen atmosphere. At

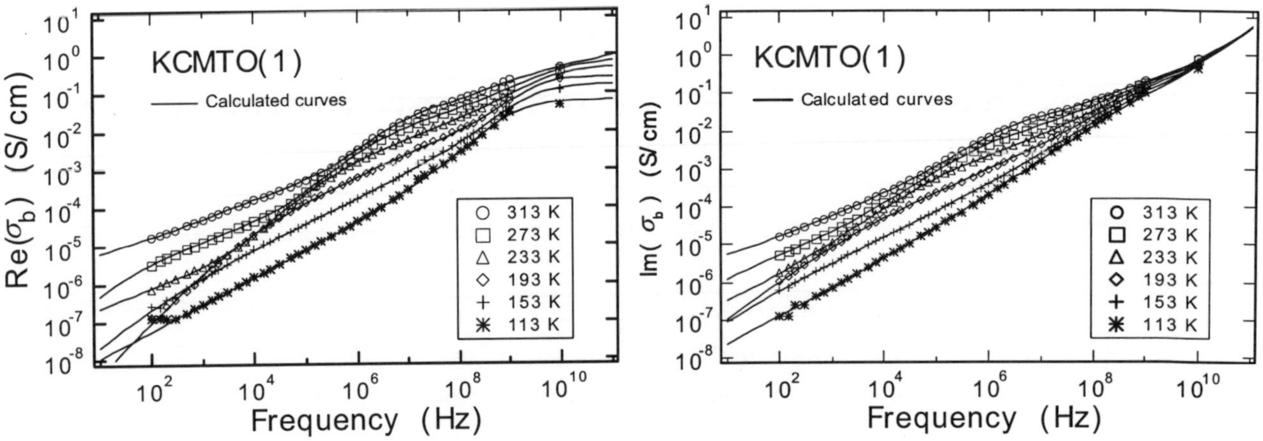

(a) (b)

Fig. 1 Frequency dependence of (a) the real and (b) the imaginary parts of the complex electrical conductivity for (K,C_S)-priderite, KCMTO(1).

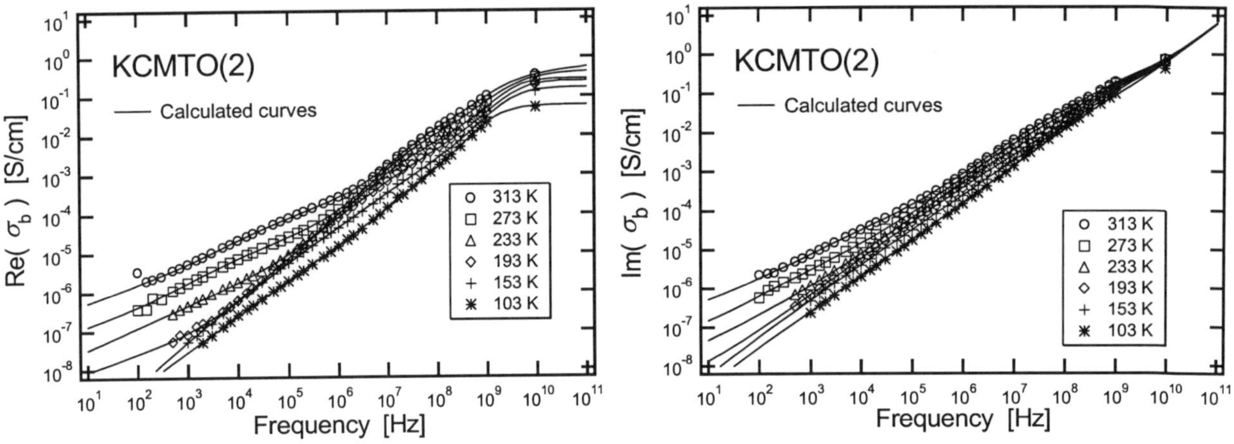

(a) (b)

Fig. 2 Frequency dependence of (a) the real and (b) the imaginary parts of the complex electrical conductivity for (K,C_S)-priderite, KCMTO(2).

9.375 GHz, each single crystal was fitted symmetrically to the E-plane of a rectangular waveguide by silver paste and the complex conductivity was determined by the reflection method [4].

3. Discussion

The frequency dependences of the real and imaginary parts of the bulk complex conductivity σ_b of KCMTO(1) and KCMTO(2) at several temperatures are shown in Figs.1 and 2, respectively. Bends are observed in the low-frequency region at high temperatures. For reference, the frequency dependences of the real and imaginary parts of σ_b of KMTO and $Cs_{1.3}Mg_{0.65}Ti_{7.35}O_{16}$(CMTO) single crystals are shown in Figs.3(a) and (b), respectively. The magnitude of σ_b of KCMTO(1) with a smaller Cs content is larger than that of KCMTO(2). The frequency dependence of both samples in the

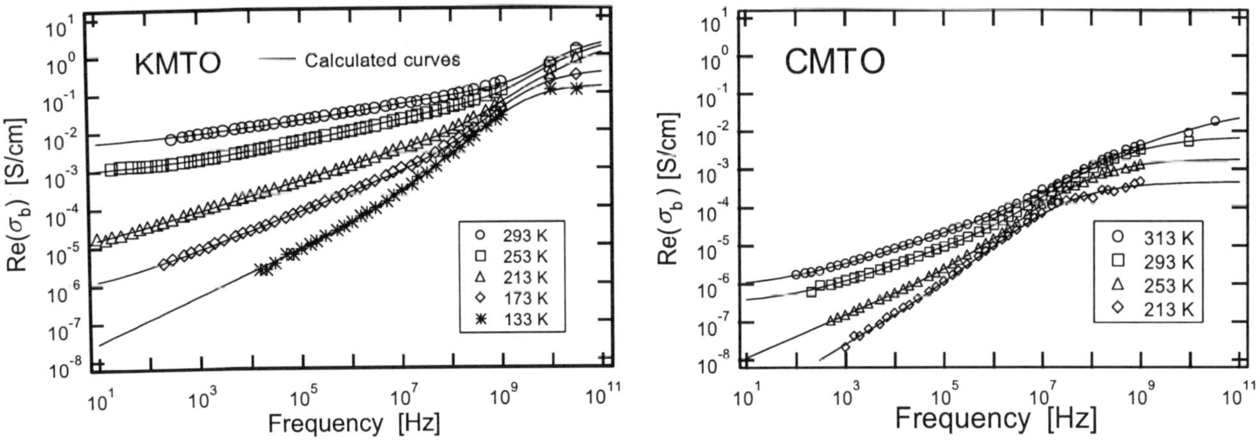

(a) (b)

Fig. 3: Frequency dependence of the real part of the complex electrical conductivity for (a) KMTO and (b) CMTO.

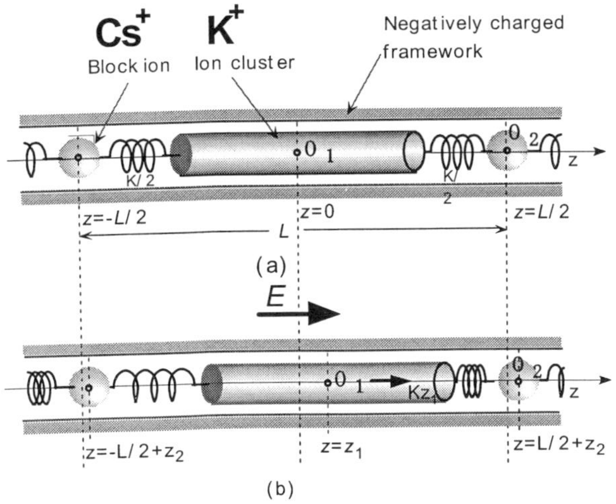

Fig.4 Ion conduction model for KCMTO.

low-frequency region at high temperatures is mutually analogous to that of CMTO, because the mobility of K^+ ions becomes larger and then the conductivity is determined by the motion of Cs^+ ions. On the other hand, the frequency dependence for both samples in the high-frequency region is analogous to that of KMTO, because the motion of each ion is localized and the conductivity is determined by the motion of K^+ ions whose concentration is much larger than that of Cs^+ ions [1-3]. The ion conduction model suitable for measured frequency dependence is shown in Fig.4 [1,2]. The model is known as the moving box model [1]. A K^+ ion cluster polarizes under an electric field in a 1D box having the average length L terminated by two Cs^+ ions with a much smaller mobility than that of a K^+ ion in the low-frequency region. On the other hand, the motion of each K^+ ion contributes to the conductivity, because the mean-square displacement of a K^+ ion decreases in the high-frequency region. Assuming that the interaction between a Cs^+ ion and a K^+ ion cluster is simply described by the force constant f, the motion of a K^+ ion cluster under the applied electric field parallel to the 1D conduction path is given by

$$\gamma \frac{dz_1(t)}{dt} = -f\left[z_1(t) - z_2(t)\right] + q_1 E(t). \qquad (1)$$

Here, $z_1(t)$ and $z_2(t)$ are the displacements of the centers of a K^+ ion cluster and a Cs^+ ion from $z=0$ and $z=\pm L/2$, respectively. q_1 is the electric charge of a K^+ ion cluster. γ is the damping factor for the motion of a K^+ ion cluster. The motion of a Cs^+ ion may also be given by

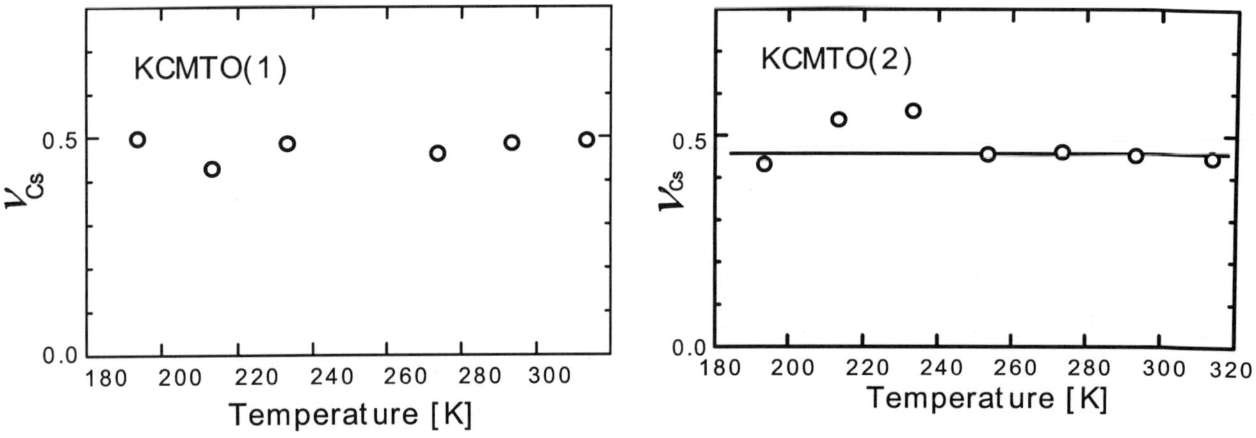

(a) (b)

Fig. 5 Temperature dependence of exponents ν_{Cs} for the ion conduction of Cs^+ ions for (a) KCMTO(1) and (b) KCMTO(2).

$$\Gamma\frac{dz_2(t)}{dt} = -f[z_2(t)-z_1(t)]+q_2E(t), \tag{2}$$

where q_2 is the electric charge of a Cs^+ ion. Γ is the damping factor for the motion of a Cs^+ ion. γ and Γ are described as

$$\gamma = \frac{q_1^2 n_B}{\sigma_{\mathrm{K}}}, \tag{3}$$

$$\Gamma = \frac{q_2^2 n_B}{\sigma_{Cs}}, \tag{4}$$

respectively. n_{B} is the density of the number of moving boxes. σ_{K} and σ_{Cs} are the complex conductivities of K^+ and Cs^+ ions of KMTO and CMTO, respectively, and are qualitatively given by

$$\sigma_{\mathrm{K}} = \frac{\left[i\omega\varepsilon_{\mathrm{PK}}\varepsilon_0 + C_{1\mathrm{K}}(i\omega)^{\nu_{\mathrm{K}}}\right]\left[C_{2\mathrm{K}}(i\omega)^{\nu_{\mathrm{K}}}\right]}{i\omega\varepsilon_{\mathrm{PK}}\varepsilon_0 + C_{1\mathrm{K}}(i\omega)^{\nu_{\mathrm{K}}} + C_{2\mathrm{K}}(i\omega)^{\nu_{\mathrm{K}}}}, \tag{5}$$

$$\sigma_{\mathrm{Cs}} = \frac{\left[i\omega\varepsilon_{\mathrm{PCs}}\varepsilon_0 + C_{1\mathrm{Cs}}(i\omega)^{\nu_{\mathrm{Cs}}}\right]\left[C_{2\mathrm{Cs}}(i\omega)^{\nu_{\mathrm{Cs}}}\right]}{i\omega\varepsilon_{\mathrm{PCs}}\varepsilon_0 + C_{1\mathrm{Cs}}(i\omega)^{\nu_{\mathrm{Cs}}} + C_{2\mathrm{Cs}}(i\omega)^{\nu_{\mathrm{Cs}}}} \tag{6}$$

[3,5,6]. The force constant f is given by

$$f = \frac{q_1^2 n_B}{\varepsilon_{\mathrm{K}}\varepsilon_0} = \frac{q_2^2 n_B}{\varepsilon_{\mathrm{Cs}}\varepsilon_0}. \tag{7}$$

Here ε_{K} and $\varepsilon_{\mathrm{Cs}}$ are the permittivities due to the polarization of K^+ ion clusters in the box and Cs^+ ions, respectively. ε_0 is the permittivity in free space. The total polarization in unit volume is given by

$$P(t) = n_B q_1 z_1(t) + n_B q_2 z_2(t). \tag{8}$$

The polarization current density is given by

$$J_P(t) = \frac{dP(t)}{dt} = \sigma(\omega)E(t), \tag{9}$$

where $\sigma(\omega)$ is the total complex conductivity and $E(t)$ is the strength of the electric field. Assuming that all mobile ions move linearly under an ac electric field, z_1 and z_2 are solved using eqs.(1) and (2). From eqs.(3)-(7), (8), and (9), the total complex conductivity considering both K^+ and Cs^+ ions is given by

$$\sigma(\omega) = \frac{i\omega\varepsilon_{\mathrm{Cs}}\varepsilon_0(\sigma_{\mathrm{K}}+\sigma_{\mathrm{Cs}}) + \sigma_{\mathrm{K}}\sigma_{\mathrm{Cs}}(1+m)^2}{i\omega\varepsilon_{\mathrm{Cs}}\varepsilon_0 + m^2\sigma_{\mathrm{K}} + \sigma_{\mathrm{Cs}}}. \tag{10}$$

m is the ratio of the concentration of K^+ to Cs^+ ions and is q_2/q_1. Values of m for KCMTO(1) and KCMTO(2) are approximately 0.0133 and 0.05, respectively. The bulk complex conductivity σ_b is given by

$$\sigma_b(\omega) = i\omega\varepsilon_{\mathrm{fw}}\varepsilon_0 + \sigma(\omega). \tag{11}$$

Here $\varepsilon_{\mathrm{fw}}$ is the permittivity of the framework of tunnels. The solid lines in Figs.1 and 2 are best fitting values calculated by the least-squares method using eq.(11). These agree with the measured values. Figures 5(a) and (b) show the temperature dependence of the frequency index ν_{Cs} of σ_{Cs} for KCMTO(1) and KCMTO(2), respectively. Although ν_{Cs} for CMTO decreases with

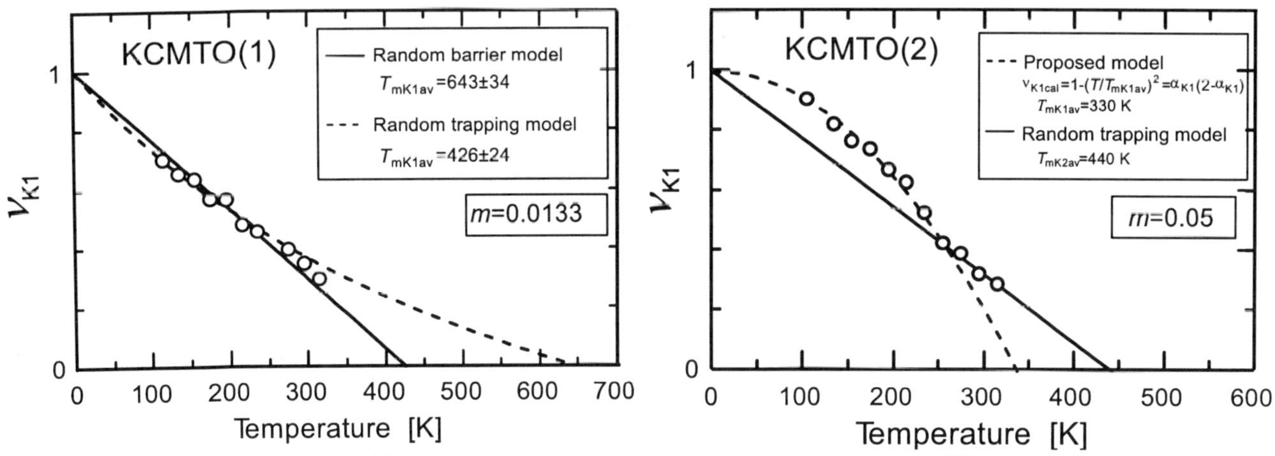

(a) (b)

Fig. 6 Temperature dependence of exponents ν_K for the ion conduction of K^+ ions for (a) KCMTO(1) and (b) KCMTO(2).

increasing temperature obeying the random barrier model as [6]

$$\nu_{Cs} = \frac{1 - T/T_m}{1 + T/T_m}. \tag{12}$$

The specific temperature T_m in eq.(12) is called the mobility transition temperature and shows the degree of the distribution of potential energy for Cs^+ ion conduction, and is approximately 880 K for CMTO. ν_{Cs} for both KCMTO(1) and KCMTO(2) are independent of the temperature and are approximately 0.5 [3,5,6]. This is due to the fact that the Cs^+ ion system is considered as solitary system, because the mobility of Cs^+ ions is much smaller than that of K^+ ions and the distance L between two succeeding Cs^+ ions is large.

The temperature dependence of the frequency index ν_K for σ_K in eq.(5) is shown in Fig.6. A clear difference between KCMTO(1) and KCMTO(2) was observed. The temperature dependence of ν_K below approximately 250 K for KCMTO(1) is analogous to that (dashed line) obeying eq.(12), which is similar to that for KMTO or CMTO. T_m for KMTO is between 470 and 490 K. On the other hand, T_m for KCMTO(1) is approximately 640 K and is higher than that for KMTO. It is speculated that the distribution of the potential energies for K^+ ion conduction become wide, because Cs ions modulate the size of bottlenecks formed by four oxygen ions in the 1D conduction path due to the larger ionic radius of Cs^+ ions. Above 250 K, the temperature dependence of ν_K deviates from eq.(13) and obeys the equation derived from the random trapping model given by [7]

$$\nu_K = 1 - T/T_m. \tag{13}$$

This model is applicable the conduction system in which not the height but the depth of the potential energy has the distribution.

The temperature dependence of ν_K for KCMTO(2) also deviates from eq.(12) and shows a different temperature dependence at approximately 250 K. Below 250 K, ν_K obeys the equation given by

$$\nu_K = 1 - (T/T_m)^2, \tag{14}$$

and above 250 K, obeys eq.(13). The specific temperatures T_m for KCMTO(1) and KCMTO(2) are approximately 426 and 440 K, respectively, and both are

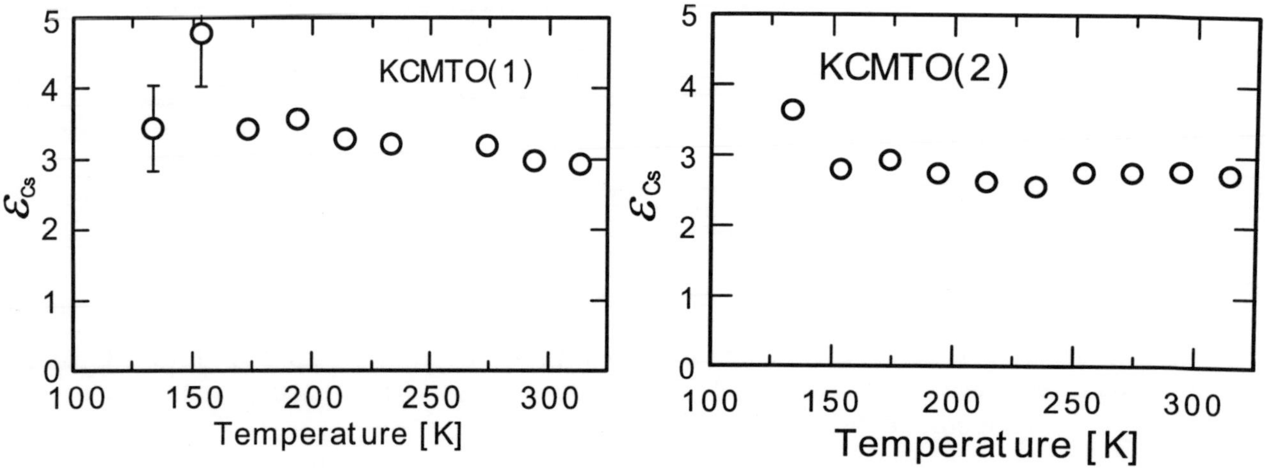

(a) (b)

Fig. 7 Temperature dependence of exponents ε_{Cs} for the ion conduction of Cs^+ ions for (a) KCMTO(1) and (b) KCMTO(2).

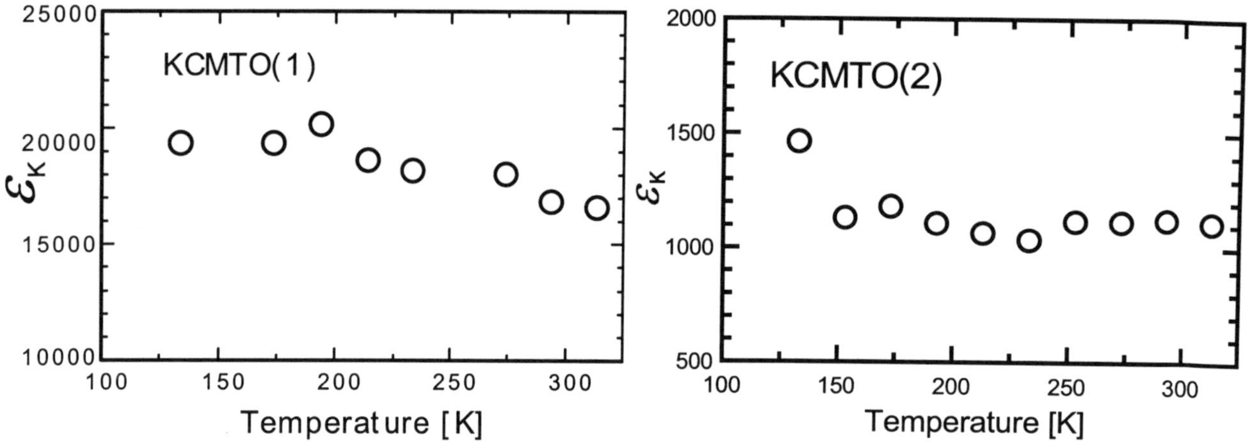

(a) (b)

Fig. 8 Temperature dependence of exponents ε_K for the ion conduction of K^+ ions for (a) KCMTO(1) and (b) KCMTO(2).

agreement. It is speculated that the deviation of ν_K from eq.(12) becomes large with increasing the Cs^+ content, because the motions of the K^+ ion cluster are limited by the motion of Cs^+ ions and the correlation length λ_c for the motion of K^+ ions is determined by the distance L between two succeeding Cs^+ ions. λ_c expresses the range which the relaxation of ion movement attains to. Because λ_c becomes short with increasing temperature, λ_c is smaller than L above approximately 250 K and the motion of K^+ ions is affected by the trap-type potentials. It is speculated that λ_c is above 20 sites (approximately 6 nm).

The temperature dependency of the relative permittivity ε_{Cs} due to the polarization of Cs^+ ions in eq.(7) or (10) is shown in Fig.7. Both values of ε_{Cs} of KCMTO(1) and KCMTO(2) are almost the same and

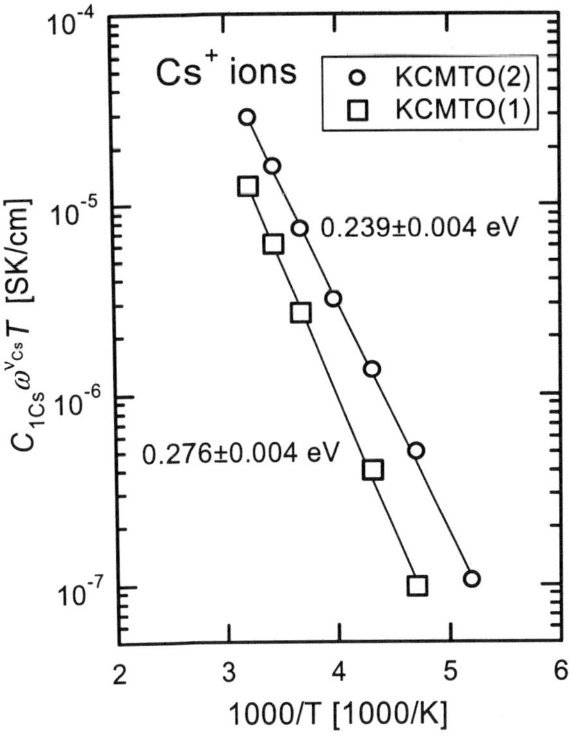

Fig.9 Temperature dependences of $C_{1Cs}\,\omega^{\nu_{Cs}}T$ of KCMTO(1) and KCMTO(2) at the frequency of 10^4Hz.

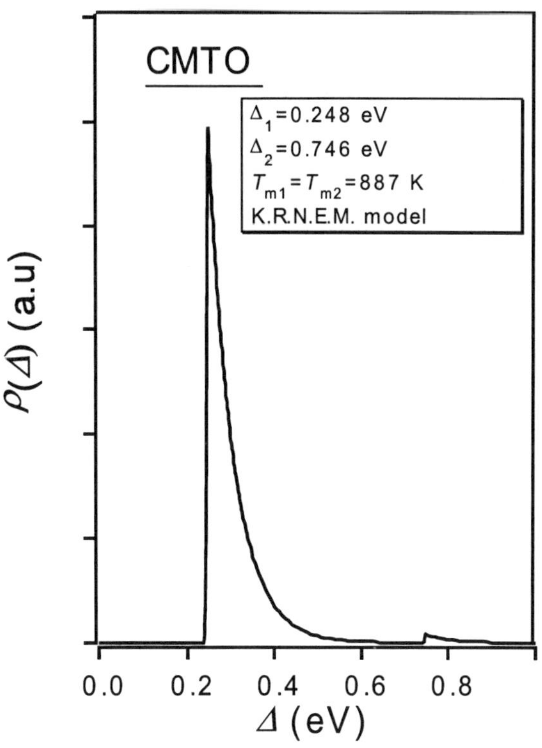

Fig.10 Distribution of the intrinsic and the random barriers for CMTO obtained using Kirkpatrick's resistor network effective medium model.

are independent of temperature. Therefore, it is found that the force constant f of KCMTO (2) is approximately four times as large as that of KCMTO (1). Moreover, the temperature dependence of the relative permittivity ε_K due to the polarization of K^+ ion clusters calculated from eq.(7) is shown in Fig.8.

The temperature dependence of $C_{1CS}\omega^{\nu_{Cs}}T$ at the frequency of 10^4Hz is shown in Fig.9. Both values of ν_{Cs} for both KCMTO(1) and KCMTO(2) were independent of temperature as shown in Fig.5 and then the temperature dependence of $C_{1CS}\omega^{\nu_{Cs}}T$ was the Arrhenius type. Therefore, the inclination in the Arrhenius plot did not depend on frequency, and the calculated activation energies were 0.28eV and 0.24eV for KCMTO (1) and KCMTO(2), respectively, and show the almost the same value. Moreover, these values were almost equal to the value of 0.25 eV, which was the lower limit of the potential barrier for CMTO calculated using Kirkpatrick's resistor network effective medium model as shown in Fig.10 [7,8]. Δ shows the barrier height of the random barriers. $\rho(\Delta)$ is the distribution function of Δ. This result is due to the fact that both the lattice constants of KCMTO (1) and KCMTO (2) are almost equal to that of CMTO.

References

[1] H. U. Beyeler and S. Sträsler, Phys. Rev. **B24** (1981) 2121.

[2] S. Yoshikado, T. Ohachi, I. Taniguchi, Y. Onoda, M. Watanabe, and Y. Fujiki, Solid State Ionics **18&19** (1986) 507.

[3] S. Yoshikado, I. Taniguchi, M. Watanabe, Y. Onoda, and Y. Fujiki, Solid State Ionics **79** (1995) 34.

[4] S. Yoshikado and I. Taniguchi, IEEE Trans. Microwave Theory Tech. **MTT 37** (1989) 984.

[5] H. U. Beyeler, J. Bernasconi, and S. Sträsler, Fast Ion Transport in Solid, edited byVashishta, P., Mundy, J. N., and Shenoy, G. K. (North-Holland, New York, 1979) 503.

[6] J. Bernasconi, H. U. Beyeler, and S. Sträsler, Phys. Rev. Lett. **42** (1979) 819.

[7] S. Alexander, J. Bernasconi, W. R. Schneider, and R. Orbach, Rev. Mod. Phys. **53** (1981) 177.

[8] S. Kirkpatrick, Rev. Mod. Phys. **69** (1983) 1031.

Author Index

S. Adams
Universität Goettingen
GZG, Abt. Kristallographie

W. Dieterich
Fachbereich Physik
Universität Konstanz

J. R. Dygas
Faculty of Physics
Warsaw University of Technology

J. C. Dyre
Department of Mathematics and Physics (IMFUFA)
Roskilde University

J. B. Metson
Department of Chemistry and MacDiarmid Institute of Material Science and Nanotechnology
The University of Auckland

S. Selvasekarapandian
Solid State and Radiation Physics Laboratory
Department of Physics
Bharathiar University

S. Sharma
Institute for Theoretical Physics
Karl-Franzens-Universität Graz

D. L. Sidebottom
Department of Physics
Creighton University

M. Aniya
Department of Physics
Faculty of Science
Kumamoto University

A. Fujii
Shockwave and Condensed Matter Research Center
Kumamoto University

F. Fujishiro
Department of Physics
College of Humanities and Science
Nihon University

T. Hattori
Institute of Multidisciplinary Research for Advanced Materials (IMRAM)
Tohoku University

K. Ibuki
Department of Molecular Science and Technology,
Faculty of Engineering
Doshisha University

N. Inoue
Department of Physics, Faculty of Science
Ehime University

T. Ishii
Faculty of Engineering
Okayama University

Y. Ito
Institute for Chemical Research
Kyoto University

K. Kamada
Department of Applied Chemistry and Biochemistry
Faculty of Engineering
Kumamoto University

T. Kanashiro
Department of Physics
Faculty of Engineering
Tokushima University

I. Katayama
Institute of Particle and Nuclear Studies
High Energy Accelerator Research Organization (KEK)

J. Kawamura
Institute of Multidisciplinary Research for Advanced Materials (IMRAM)
Tohoku University

M. Kobayashi
Department of Physics
Niigata University

Y. Kowada
Hyogo University of Teacher Education

H. Koyama
Institute of Multidisciplinary Research for Advanced Materials (IMRAM)
Tohoku University

Y. Matsuo
Department of Applied Physics
Faculty of Science
Tokyo University of Science

Y. Michihiro
Department of Physics
Faculty of Engineering
Tokushima University

I Y. Michiue
Advanced Materials Laboratory
National Institute for Materials Science

S. Mochizuki
Department of Physics, College of Humanities and Science
Nihon University

K. Nakamura
Department of Physics
Faculty of Engineering
Tokushima University

S. Ono
Graduate School of Science and Technology
Niigata University

M. Onoda
Advanced Materials Laboratory
National Institute for Materials Science

Y. Onoda

T. Sakuma
Department of Physics
Faculty of Science
Ibaraki University

F. Shimojo
Department of Physics
Faculty of Science
Kumamoto University

K. Takahashi
Kobe Women's University
Seto Junior College

S. Takeda
Department of Physics
Faculty of Science
Kyushu University

T. Tojo
Materials and Structures Laboratory
Tokyo Institute of Technology

T. Tomoyose
Department of Physics
Ryukyu University

T. Usuki
Faculty of Science
Yamagata University

S. Yoshikado
Department of Electronics
Faculty of Engineering
Doshisha University

W. Yu
Graduate School of Science and Technology
Kumamoto University

Y. Zou
Department of Physics
Faculty of Science
Ehime University

A. Imai
Nagoya Industrial Science Research Institute

T. Ishiguro
Doshisha Research Center for Human Security
Doshisha University

O. Kamishima
Institute of Multidisciplinary Research for Advanced Materials (IMRAM)
Tohoku University

A. Ueda
Prof. Emeritus
Kyoto University